VANET

VANET: Vehicular Applications and Inter-Networking Technologies

Hannes Hartenstein
Karlsruhe Institute of Technology (KIT), Germany

Kenneth P Laberteaux
Toyota Technical Center, USA

A John Wiley and Sons, Ltd, Publication

This edition first published 2010
© 2010 John Wiley & Sons Ltd.

Registered office
John Wiley & Sons Ltd, The Atrium, Southern Gate, Chichester, West Sussex, PO19 8SQ, United Kingdom.

For details of our global editorial offices, for customer services and for information about how to apply for permission to reuse the copyright material in this book please see our website at www.wiley.com.

The right of the author to be identified as the author of this work has been asserted in accordance with the Copyright, Designs and Patents Act 1988.

All rights reserved. No part of this publication may be reproduced, stored in a retrieval system, or transmitted, in any form or by any means, electronic, mechanical, photocopying, recording or otherwise, except as permitted by the UK Copyright, Designs and Patents Act 1988, without the prior permission of the publisher.

Wiley also publishes its books in a variety of electronic formats. Some content that appears in print may not be available in electronic books.

Designations used by companies to distinguish their products are often claimed as trademarks. All brand names and product names used in this book are trade names, service marks, trademarks or registered trademarks of their respective owners. The publisher is not associated with any product or vendor mentioned in this book. This publication is designed to provide accurate and authoritative information in regard to the subject matter covered. It is sold on the understanding that the publisher is not engaged in rendering professional services. If professional advice or other expert assistance is required, the services of a competent professional should be sought.

Library of Congress Cataloging-in-Publication Data

Hartenstein, Hannes.
    VANET: vehicular applications and inter-networking technologies / Hannes Hartenstein, Kenneth Laberteaux, editors.
      p. cm.
    Includes index.
    ISBN 978-0-470-74056-9 (cloth)
1. Vehicle ad hoc networks (Computer networks) I. Hartenstein, Hannes. II. Laberteaux, Kenneth.
    TE228.37.V36 2010
    388.3'12–dc22
                                                                                            2009026531

A catalogue record for this book is available from the British Library.

ISBN 9780470740569 (H/B)

Set in 10/12pt Palatino by Sunrise Setting Ltd, Torquay, UK.
Printed and Bound in Singapore by Markono Print Media Pte Ltd.

Dedication

To my parents. HH

To Maria. KPL

Contents

Foreword	xv
About the Editors	xix
Preface	xxi
Acknowledgments	xxv
List of Contributors	xxvii

1 Introduction   1
Hannes Hartenstein and Kenneth P Laberteaux

- 1.1 Basic Principles and Challenges 1
- 1.2 Past and Ongoing VANET Activities 4
  - 1.2.1 From the beginning to the mid 1990s 5
  - 1.2.2 From the mid 1990s to the present 7
  - 1.2.3 Examples of current project results 10
- 1.3 Chapter Outlines 14
- References 17

2 Cooperative Vehicular Safety Applications   21
Derek Caveney

- 2.1 Introduction 21
  - 2.1.1 Motivation 21
  - 2.1.2 Chapter outline 22
- 2.2 Enabling Technologies 23
  - 2.2.1 Communication requirements 23
  - 2.2.2 Vehicular positioning 23
  - 2.2.3 Vehicle sensors 25
  - 2.2.4 On-board computation platforms 26
- 2.3 Cooperative System Architecture 26
- 2.4 Mapping for Safety Applications 28
  - 2.4.1 Non-parametric path prediction 30
  - 2.4.2 Parametric path prediction 31
  - 2.4.3 Stochastic path prediction 35

	2.5	VANET-enabled Active Safety Applications	37
	2.5.1	Infrastructure-to-vehicle applications	40
	2.5.2	Vehicle-to-vehicle applications	41
	2.5.3	Pedestrian-to-vehicle applications	47
	References		48

3 Information Dissemination in VANETs — 49
Christian Lochert, Björn Scheuermann and Martin Mauve

- 3.1 Introduction — 49
- 3.2 Obtaining Local Measurements — 50
- 3.3 Information Transport — 54
  - 3.3.1 Protocols for information transport — 54
  - 3.3.2 Improving network connectivity — 59
  - 3.3.3 What to transport — 61
- 3.4 Summarizing Measurements — 63
- 3.5 Geographical Data Aggregation — 66
- 3.6 Conclusion — 75
- References — 76

4 VANET Convenience and Efficiency Applications — 81
Martin Mauve and Björn Scheuermann

- 4.1 Introduction — 81
- 4.2 Limitations — 82
  - 4.2.1 Capacity — 82
  - 4.2.2 Connectivity — 83
  - 4.2.3 Competition — 87
- 4.3 Applications — 87
- 4.4 Communication Paradigms — 89
  - 4.4.1 Centralized client/server systems — 89
  - 4.4.2 Infrastructure-based peer-to-peer communication — 90
  - 4.4.3 VANET communication — 92
- 4.5 Probabilistic, Area-based Aggregation — 93
  - 4.5.1 FM sketches — 94
  - 4.5.2 Using sketches for data aggregation in VANETs — 94
  - 4.5.3 Soft-state sketches — 96
  - 4.5.4 Forming larger area aggregates — 96
  - 4.5.5 Application study — 97
- 4.6 Travel Time Aggregation — 99
  - 4.6.1 Landmark-based aggregation — 99
  - 4.6.2 Judging the quality of information — 101
  - 4.6.3 Hierarchical landmark aggregation — 101
  - 4.6.4 Evaluation — 103
- 4.7 Conclusion — 104
- References — 105

5 Vehicular Mobility Modeling for VANET — 107
Jérôme Härri

- 5.1 Introduction — 107
- 5.2 Notation Description — 112
- 5.3 Random Models — 113
- 5.4 Flow Models — 115
  - 5.4.1 Microscopic flow models — 116
  - 5.4.2 Macroscopic flow models — 121
  - 5.4.3 Mesoscopic flow models — 123
  - 5.4.4 Lane changing models — 124
  - 5.4.5 Intersection management — 128
  - 5.4.6 Impact of flow models on vehicular mobility — 129
- 5.5 Traffic Models — 131
  - 5.5.1 Trip planning — 132
  - 5.5.2 Path planning — 133
  - 5.5.3 Influence of time — 134
  - 5.5.4 Impact of traffic models on vehicular mobility — 134
- 5.6 Behavioral Models — 135
- 5.7 Trace or Survey-based Models — 137
- 5.8 Integration with Network Simulators — 139
  - 5.8.1 Network simulators — 139
  - 5.8.2 Isolated mobility models — 141
  - 5.8.3 Embedded mobility models — 141
  - 5.8.4 Federated mobility models — 143
  - 5.8.5 Application-centric versus network-centric simulations — 145
  - 5.8.6 Discussion — 146
- 5.9 A Design Framework for Realistic Vehicular Mobility Models — 147
  - 5.9.1 Motion constraints — 147
  - 5.9.2 Traffic generator — 148
  - 5.9.3 Application-based level of realism — 149
- 5.10 Discussion and Outlook — 150
- 5.11 Conclusion — 151
- References — 152

6 Physical Layer Considerations for Vehicular Communications — 157
Ian Tan and Ahmad Bahai

- 6.1 Standards Overview — 158
  - 6.1.1 A brief history — 158
  - 6.1.2 Technical alterations and operation — 159
- 6.2 Previous Work — 165
- 6.3 Wireless Propagation Theory — 166
  - 6.3.1 Deterministic multipath models — 166
  - 6.3.2 Statistical multipath models — 171
  - 6.3.3 Path loss modeling — 173

6.4	Channel Metrics		174
	6.4.1	Delay spread	174
	6.4.2	Coherence bandwidth	175
	6.4.3	Doppler spread	179
	6.4.4	Coherence time	181
	6.4.5	Impact on OFDM systems	182
6.5	Measurement Theory		184
6.6	Empirical Channel Characterization at 5.9 GHz		188
	6.6.1	Highway environments	188
	6.6.2	Urban environments	193
	6.6.3	Rural LOS environments	199
	6.6.4	Results summary	200
	6.6.5	Analysis	201
6.7	Future Directions		206
6.8	Conclusion		207
6.9	Appendix: Deterministic Multipath Channel Derivations		208
	6.9.1	Complex baseband channel representation – continuous time	208
	6.9.2	Complex baseband channel representation – discrete time	209
6.10	Appendix: LTV Channel Response		210
6.11	Appendix: Measurement Theory Details		212
	6.11.1	PN sequence bits	212
	6.11.2	Generation of LTI channel estimates	212
	6.11.3	Generation of Ricean K-factor estimates	214
References			215

7 MAC Layer and Scalability Aspects of Vehicular Communication Networks 219

Jens Mittag, Felix Schmidt-Eisenlohr, Moritz Killat, Marc Torrent-Moreno and Hannes Hartenstein

7.1	Introduction: Challenges and Requirements		219
7.2	A Survey on Proposed MAC Approaches for VANETs		221
	7.2.1	Time-division multiple access based approaches	222
	7.2.2	Space-division multiple access based approaches	223
	7.2.3	Code-division multiple access based approaches	224
7.3	Communication Based on IEEE 802.11p		225
	7.3.1	The IEEE 802.11 standard	225
	7.3.2	IEEE 802.11p: towards wireless access in vehicular environments	228
	7.3.3	Modeling and simulation of IEEE 802.11p-based networks	232
7.4	Performance Evaluation and Modeling		237
	7.4.1	Performance results of IEEE 802.11p-based active safety communications	238
	7.4.2	Computational costs of simulation	241
	7.4.3	Analytical models for performance of IEEE 802.11 networks	241

		7.4.4	An empirical model for performance of IEEE 802.11p networks 245

        7.4.5   Conclusion 255
  7.5 Aspects of Congestion Control 255
        7.5.1   The need for congestion control 256
        7.5.2   Congestion control by means of transmit power control 258
        7.5.3   Congestion control by means of rate control 265
  7.6 Open Issues and Outlook 267
  References 269

8 Efficient Application Level Message Coding and Composition   273
Craig L Robinson

  8.1 Introduction to the Application Environment 274
        8.1.1   Safety applications and data requirements 274
        8.1.2   Desirable architectural features 276
        8.1.3   Broadcast characteristics 277
  8.2 Message Dispatcher 278
        8.2.1   Data element dictionary 279
        8.2.2   Message construction 280
        8.2.3   What and when to send 280
  8.3 Example Applications 281
        8.3.1   Emergency brake warning 281
        8.3.2   Intersection violation warning 281
        8.3.3   Message composition 284
        8.3.4   Implementation 285
        8.3.5   Analysis 285
  8.4 Data Sets 286
  8.5 Predictive Coding 287
        8.5.1   Linear predictive coding 288
        8.5.2   System model 289
        8.5.3   Tolerable error 289
        8.5.4   Predictive coding transmission policies 290
        8.5.5   Predictive coding results 291
  8.6 Architecture Analysis 294
  8.7 Conclusion 296
  References 296

9 Data Security in Vehicular Communication Networks   299
André Weimerskirch, Jason J Haas, Yih-Chun Hu and Kenneth P Laberteaux

  9.1 Introduction 299
        9.1.1   Outline 301
        9.1.2   State of the art 301
  9.2 Challenges of Data Security in Vehicular Networks 302
  9.3 Network, Applications, and Adversarial Model 304
        9.3.1   Network model 304
        9.3.2   Applications model 305
        9.3.3   Attacker model 309

9.4 Security Infrastructure 313
    9.4.1 Cryptography services 313
    9.4.2 Key management 316
9.5 Cryptographic Protocols 325
    9.5.1 Certificate verification 325
    9.5.2 Encryption 326
    9.5.3 Key agreement 327
    9.5.4 Authentication 329
    9.5.5 Secure positioning 334
    9.5.6 Identification of misbehaving nodes 335
    9.5.7 Summary 337
9.6 Privacy Protection Mechanisms 337
    9.6.1 Properties 340
    9.6.2 Key assignment 344
    9.6.3 Tracking vehicles 354
    9.6.4 Evaluation 355
9.7 Implementation Aspects 356
    9.7.1 Cryptographic schemes and key length 356
    9.7.2 Physical security 357
    9.7.3 Organizational aspects 359
    9.7.4 Update of software and renewal of certificates 359
9.8 Outlook and Conclusions 360
References 360

10 Standards and Regulations   365
John B Kenney

10.1 Introduction 365
10.2 Layered Architecture for VANETs 366
    10.2.1 General concepts and definitions 366
    10.2.2 A protocol stack for DSRC 367
10.3 DSRC Regulations 369
    10.3.1 DSRC in the United States 370
    10.3.2 DSRC in Europe 374
10.4 DSRC Physical Layer Standard 376
    10.4.1 OFDM physical medium dependent (PMD) function 377
    10.4.2 OFDM physical layer convergence procedure (PLCP) function . 381
10.5 DSRC Data Link Layer Standard (MAC and LLC) 383
    10.5.1 Medium access control (MAC) sublayer 383
    10.5.2 Logical link control (MAC) sublayer 392
10.6 DSRC Middle Layers 393
    10.6.1 MAC extension for multi-channel operation: IEEE 1609.4 395
    10.6.2 Network services for DSRC: network and transport layers, IEEE 1609.3 399
    10.6.3 WSA length summary 407
    10.6.4 Middle layer security: IEEE 1609.2 409

10.7 DSRC Message Sublayer 414
   10.7.1 SAE J2735 DSRC message sets 414
   10.7.2 Case study: The basic safety message 417
   10.7.3 Case study: The probe vehicle data message 422
   10.7.4 Case study: The roadside alert message 422
10.8 Summary 422
10.9 Abbreviations and Acronyms 425
References 428

Index   **431**

Foreword

In August 1997 the National Automated Highway System Consortium (NAHSC) demonstrated several highway automation technologies on interstate I-15 in San Diego. Several thousand people got rides in automated cars and buses featuring vision-based lane-keeping and car-following. The highlight of the event was a fully automated ('hands-off, feet-off, brains-off') highway system (AHS). The goal of the AHS demonstration was a proof-of-concept of an AHS architecture that enhanced highway *capacity* and *safety*. Capacity increase was achieved by organizing the movement of vehicles in closely spaced platoons. The AHS demonstration involved a seven-car vehicle, with vehicles spaced 22ft apart, driving at 60 mph. Taking inter-platoon separation into account, this gives an average inter-vehicle distance (bumper-to-bumper) of 60 feet. At a speed of 60 mph, this amounts to a maximum flow or capacity of 5,280 vehicles/hour, compared with a capacity of 2,000 vehicles/hour in today's highways. Each vehicle had electronic actuators – steering, braking and throttle – that were controlled by the vehicle's own computer.

Safety was increased because the computer was connected to sensors that provided (1) measurements about the vehicle itself (speed, acceleration, tire slip), (2) the vehicle's location within the lane, (3) the relative speed and distance between the vehicle and the vehicle in front. Most importantly, an inter-vehicle communication system formed a local area network to exchange information with other vehicles in the neighborhood, as well as to permit a protocol among neighboring vehicles to support *cooperative* maneuvers such as lane-changing, joining a platoon, and sudden braking. Computer-controlled driving eliminated driver misjudgement, which is a major cause of accidents today. At the same time, a suite of safety control laws ensured fail-safe driving despite sensor, communication and computer faults. Although not part of its original goal, the AHS experiment also showed that it could significantly reduce fuel consumption by greatly reducing driver-induced acceleration and deceleration surges during congestion.

The AHS experiment met its goals, but there was no direct, practicable path towards its wide-scale deployment. The main obstacle is that the AHS concept required all vehicles to be automated: mixing manually driven and automated vehicles reduces the capacity increase and creates safety issues that were not addressed in depth in the NAHSC project. However, the project inspired a series of follow-on USDOT research programs to improve mobility and safety through better sensing, communication, and cooperative control.

The first follow-on program was dubbed Vehicle-Infrastructure Integration or VII, later renamed SafeTrip-21, and now called IntelliDrive. The different names signal subtle shifts in assumptions and objectives.

A major push behind VII was the Federal Communications Commission (FCC) ruling dedicating a 75 MHz spectrum in the 5.9 GHz band for the exclusive use of automotive applications. The spectrum became known as Direct Short-Range Communications or DSRC. The permissible power levels give DSRC signals a range of 1 km with data rates of 6 to 27 Mbps. A community of researchers set about developing DSRC standards, including the PHY and MAC layers, a communications architecture, and mobility and safety applications. The architecture envisaged ad-hoc communications among on-board units (OBUs) in vehicles and roadside units (RSU). The RSUs would function as data repositories and as repeaters. Mobility applications dealt with providing traveler information, where safety applications used DSRC to alert drivers about potentially conflicting situations based on information obtained from neighboring vehicles and the roadside.

As the DSRC community was making slow progress in standardization and applications, the tremendous advances in consumer electronics led to a range of wireless-based hand-held devices (GPS-equipped cellphones and PDAs), with location-based services, including navigation aids and traffic information. SafeTrip-21 recognized these changes by expanding the focus from DSRC to include these alternative channels of information. However, the expanded focus also led to a dilution of effort. In its reincarnation as IntelliDrive, the focus has narrowed again to safety concerns, with DSRC as the main communication medium.

This US-centric account should not obscure parallel and coordinated developments in Europe and Japan, where there are active programs based on DSRC. (The only difference is that Japanese DSRC is in the 5.8 GHz band.)

The international academic and industry research community meanwhile has organized itself under the banner of VANET–Vehicle Ad-hoc Networks, holding annual workshops that bring together researchers from communication networks, computer science, electrical engineering, automotive engineering, and transportation. The VANET workshops have served to catalyze and consolidate a large body of research. VANET is now a well-recognized field of research, concerned with all technical aspects of vehicular ad-hoc networks, from radio propagation to network design, and performance to applications. The field concerns itself as well with issues of security and privacy, data reliability and aggregation.

The editors, Hannes Hartenstein and Kenneth P Laberteaux, respectively representing the academic and industry research communities, are eminent scholars in the field. Both have served important roles in the creation of this research community: Ken Laberteaux, who coined the VANET term, was the driving force in launching the first VANET Workshop in 2004, and served as the general co-chair for the Workshop's first two years. Hannes Hartenstein was an active contributor to the European Fleetnet Project (which preceded the first VANET Workshop) and served as VANET Workshop general co-chair in 2005 and technical program co-chair in 2006. In addition to this book, both editors continue to actively publish in the field.

VANET has become an exciting field of research, with a large body of knowledge and many open problems. But for the newcomer, whether graduate student or professional, the VANET literature is too vast to master in a reasonable amount of time. The publication of this volume comes at an opportune moment. The editors have created a volume that covers the most important contributions to VANET over the past decade. The book will serve well both as a classroom text that can be used in a graduate course for electrical engineering and computer science, and as a reference text for the practicing professional.

Pravin Varaiya
Berkeley

About the Editors

Hannes Hartenstein is a professor for decentralized systems and network services at the Karlsruhe Institute of Technology (KIT), Germany, which is formed by the University of Karlsruhe and the Research Center Karlsruhe, and a director of the KIT Steinbuch Centre for Computing. Prior to joining the University of Karlsruhe, he was a senior research staff member with NEC Europe. He was NEC's project leader (2001–03) for the 'FleetNet – Internet on the Road' project partly funded by the German Ministry of Education and Research (BMBF), and involved in the 'NOW: Network on Wheels' project (2004–08), also funded by BMBF. He is currently actively participating in the EU FP7 project PRE-DRIVE-C2X. He was General Co-Chair of the ACM International Workshop on Vehicular Ad-Hoc Networks (VANET) in 2005, technical co-chair of ACM VANET in 2006, technical co-chair of the IEEE Symposium on Wireless Vehicular Communications (WiVeC) 2007, and technical co-chair of the IFIP/IEEE Conference on Wireless On-Demand Network Systems and Services (WONS) in 2008. He is a member of the scientific directorate of the Center for Informatics, Schloss Dagstuhl. His research interests include mobile networks, virtual networks, and IT management. He holds a diploma in mathematics and a doctoral degree in computer science, both from Albert-Ludwigs-Universität, Freiburg, Germany.

Kenneth P Laberteaux is a senior principal research engineer for the Toyota Technical Center in Ann Arbor, MI. His research focus is information-rich vehicular safety systems, focusing on architecture, security, and protocol design for vehicle-to-vehicle and vehicle-to-roadside wireless communication. He was a founder and two-year (2004–05) general co-chair of the highly selective, international Vehicular Ad-hoc Networks (VANET) workshop. He serves as the architect and technical lead for communications research within a multi-year, multi-million dollar Vehicle Safety Communications-Applications collaboration project between the US government and several automotive companies. He completed his MSc (1996) and PhD (2000) degrees in electrical engineering at the University of Notre Dame, focusing on adaptive control for communications. In 1992, he received his BSE (summa cum laude) in electrical engineering from the University of Michigan, Ann Arbor.

Preface

The field of vehicular applications and inter-networking technologies (VANET) with its radio-based direct vehicle-to-vehicle and vehicle-to-infrastructure communication strives to harness the power of ubiquitous communication for the sake of traffic safety and transport efficiency. This book addresses the applications and technical aspects of VANET that can be established by short- and medium-range communication primarily based on wireless local area network technology. The distinctive set of candidate applications (e.g., collision warning and local traffic information for drivers), resources (e.g., licensed spectrum, rechargeable power source), and the environment (e.g., vehicular traffic flow patterns, privacy concerns) make VANET a unique area of wireless communication.

With about ten years of intense research activity and progress in the field of VANET, deep insights were gained into how to design VANET, and a large number of communication methods and protocols were proposed. In this book, leading experts in this field survey and evaluate the state of the art in vehicular inter-networking. Since VANET are still an actively evolving field, the chapters of this book include latest research results and point to open and future issues. Thus, the book is intended to serve as a consolidated reference to the current state of research and should enable the reader to assess the potential and the future options of VANET deployments. There are plenty of exciting research challenges yet to be solved and, furthermore, there are various deployment challenges to be addressed as innovation heavily depends on acceptance of technology.

Here is a brief synopsis of each chapter:

Chapter 1 provides an introduction to the basic principles and challenges of VANET, presents a short history of VANET activities, and outlines the contributions of the various chapters.

Chapter 2 discusses foreseen safety applications and indicates the corresponding requirements of the communication system.

Chapter 3 describes methods for information dissemination and aggregation within VANET required to support efficiency and convenience applications.

Chapter 4 builds on Chapter 3 and focuses on 'non-safety' applications, thus, on applications which primarily target efficiency and convenience.

Chapter 5 contains a survey and taxonomy of vehicular mobility models. It emphasizes the application of those models to the field of VANET.

Chapter 6 details the physical layer aspects of the foreseen IEEE 802.11-based communication system.

Chapter 7 discusses aspects of medium access control, based on IEEE 802.11 and the foreseen 802.11p standard, and provides a treatment of congestion control for the wireless medium.

Chapter 8 details the middleware aspects of VANET as a basis for efficient and semantically understandable communication.

Chapter 9 deals with security and privacy aspects of VANET.

Chapter 10 covers the current status on standardization activities and regulation in the field of VANET.

The book is designed i) for a survey course for college engineering students ranging from third-year undergraduate to first-year graduate, ii) for providing a valuable tool to professional automotive technologists, and iii) as a concise primer for researchers attracted to this field. It is assumed that the reader has a basic understanding of mobile networks and of IEEE 802.11-based wireless LANs. This book does not cover the issue of geographical positioning nor does it deal with hardware implementation issues.

The term VANET as an acronym for vehicular ad-hoc networks was originally adopted to reflect the ad-hoc nature of these highly dynamic networks. However, because the term 'ad-hoc network' was associated widely with unicast-routing-related research, we decided to redefine the acronym VANET to deemphasize ad-hoc networking. In this book we use the term VANET for the whole field of research, but also when referring to an instance of a vehicular inter-network.

This book has its roots in a tutorial we presented jointly at the ACM International Conference on Mobile Computing and Networking and at the ACM International Symposium on Mobile Ad-Hoc Networking and Computing in Montreal, Quebec, Canada, in 2007. An 'extract' of the tutorial was published under the title 'A Tutorial Survey on Vehicular Ad-Hoc Networks' in the June 2008 issue of the IEEE Communications Magazine.

As this book goes to press, there remain several open issues, especially in the areas of standards and regulations. We endeavor to present updates as well as errata on this book via the following web page:

http://www.vanetbook.com

The past decade has produced significant VANET research and technology creation, but the next one or two decades will be crucial in determining whether VANET will soon become a reality. Several automotive companies, research institutions, and government organizations are currently engaged in significant evaluation, modification, and engineering of VANET systems. Demonstrations continue to show the basic soundness of the underlying VANET technology; although not perfected, near-term VANET technology appears to be 'good enough' to 'be useful.' Therefore, it seems likely that a 'first generation' of VANET will soon coalesce, most likely around the nearly completed IEEE 802.11p and 1609 standards. In addition, more attention will be given to other topics, such as application refinement, human–machine interfaces, market acceptance/penetration rates, business cases, and of overall system effectiveness.

We contend that the fundamental opportunities and synergies of interconnected vehicles and infrastructures will one day make VANET a reality. However, it is difficult to predict whether the first widely deployed VANET will be based on near-term technology, or on the fruit of future VANET research. Over future decades, we foresee significant progress in other technologies, such as distributed control, artificial intelligence, vehicle sensors, and energy management. These advances will enable the next generation (and the generation after that) of safety, efficiency, and convenience applications. It seems all but certain that the future will bring a sharp increase of transportation demand while transportation resources ebb. In light of these challenges, timely, accurate, and trustworthy information becomes increasingly necessary. And so our work continues, and with it, the hope that VANET will bring a better tomorrow.

Acknowledgments

We would like to express our gratitude to the following persons:

- all co-authors, whose contributions made this book possible

- the persons who shaped our thinking about VANET. We would particularly like to mention (in alphabetical order): Wieland Holfelder, Yih-Chun Hu, Jean-Pierre Hubaux, Daniel Jiang, John Kenney, PR Kumar, Martin Mauve, Sam Oyama, Paolo Santi, Raja Sengupta, Pravin Varaiya, Andre Weimerskirch

- the Wiley staff: Birgit Gruber, Tiina Ruonamaa, Anna Smart, Sarah Tilley, Brett Wells

- the Decentralized Systems and Network Services research group at Karlsruhe Institute of Technology and particularly Moritz Killat who spent a significant amount of time in skillfully preparing the final LaTeX manuscript

- last, but definitely not least, our families and friends.

Hannes Hartenstein
Kenneth P Laberteaux
Karlsruhe and Ann Arbor

List of Contributors

Ahmad Bahai is a fellow and chief technologist at National Semiconductor. He is also director of National Semiconductor Labs (NS Labs). Bahai is a consulting professor at Stanford University and an adjunct professor and member of the University of California at Berkeley's Electrical Engineering and Computer Science Industrial Advisory Board. Prior to joining National, he was chief technology officer at Algorex and a technical manager at AT&T Bell Labs Advanced Wireless Communications Labs. Bahai co-invented a multi-carrier spread spectrum theory which is being used in most modern wireless systems and standards. He is the author of a textbook on OFDM, 'Multi-carrier Digital Communications' and served as the associate editor of IEEE Communication Letters for five years. Bahai has authored more than 60 papers in journals/conferences and holds several patents in wireless, analog, and mixed-signal processing systems. Bahai received his master's degree in electrical engineering from Imperial College, University of London in 1988 and a PhD in electrical engineering from the University of California at Berkeley in 1993.

Derek Caveney is a principal research scientist with the Toyota Technical Center, Toyota Motor Engineering & Manufacturing North America, in Ann Arbor, MI. He received the BScE degree in applied mathematics from Queen's University, Kingston, ON, Canada, in 1999 and the MSc and PhD degrees in mechanical engineering from the University of California, Berkeley, in 2001 and 2004, respectively. From 2004 until 2005, he was a visiting postdoctoral scholar with the Center for Collaborative Control of Unmanned Vehicles. His interests include cooperative control of vehicles for safety and mobility.

Jérôme Härri received a MSc degree and a Dr. ès sc. degree in telecommunication from the Swiss Institute of Technology (EPFL), Lausanne, Switzerland. He is an assistant professor at the Institute of Telematics at Universität Karlsruhe (TH), Germany. Previously, he was a research assistant and PhD student at EURECOM in Sophia-Antipolis, France, working on mobility modeling and management for mobile wireless ad-hoc networks. His research interests include inter-vehicular communication, vehicular mobility modeling, and intelligent transportation systems. He is currently involved in the activities of the CAR 2 CAR Communication Consortium.

Jason Haas is currently a PhD candidate at the University of Illinois at Urbana-Champaign, studying under Yih-Chun Hu. He received his BSc degree from the

University of Wisconsin-Madison in 2003 in electrical and computer engineering and in physics. He received his MSc degree from the University of Illinois at Urbana-Champaign in 2007 in electrical and computer engineering. His research interests include network security, vehicular networking, physical security, and cyber-physical systems.

Yih-Chun Hu is an assistant professor of electrical and computer engineering at the University of Illinois at Urbana-Champaign. He got his PhD from Carnegie Mellon University in 2003 as a student of David B Johnson and was a postdoctoral researcher at the University of California, Berkeley under Doug Tygar. His research interests are in network security and wireless networks, and he has served as the technical program co-chair of ACM's VANET conference in 2007 and 2008.

John B Kenney is a communications consultant who specializes in data networks. He recently represented Toyota in the Vehicle Safety Communications consortium where he led the communications research group. He is an active contributor to standards efforts in the IEEE 802.11 Working Group, IEEE 1609 Working Group, and SAE DSRC Technical Committee. Prior to his work with Toyota he led a networking research group at the Tellabs Research Center. There he worked on router architectures and protocols to support quality of service in high performance IP and ATM networks. He is also an adjunct professor of electrical engineering at the University of Notre Dame. He holds a PhD in electrical engineering from the University of Notre Dame and an MSEE from Stanford University.

Moritz Killat studied computer science at the University of Passau and the University of Karlsruhe (TH), Germany. Since 2005 he has been with the Institute of Telematics, University of Karlsruhe (TH), and received a PhD in computer science in 2009. His research interests address the analysis and development of combined application and communication simulation for car-to-x communication.

Christian Lochert is an IT consultant working with PSI Transcom GmbH in Düsseldorf, Germany. He received his diploma degree in information systems in 2003 from the University of Mannheim, Germany and in 2008 his PhD degree from the Heinrich Heine University, Düsseldorf, Germany. His current research interests include vehicular ad-hoc networks, congestion control in mobile ad-hoc networks and real-time communication in cellular networks.

Martin Mauve received the MS and PhD degrees in computer science from the University of Mannheim, Germany, in 1997 and 2000 respectively. From 2000 to 2003, he was an assistant professor at the University of Mannheim. In 2003, he joined the Heinrich Heine University, Düsseldorf, Germany, as a full professor and head of the research group for computer networks and communication systems. His research interests include distributed multimedia systems, mobile ad-hoc networks and inter-vehicle communication.

Jens Mittag holds a diploma in computer science from the University of Karlsruhe, Germany, and is currently pursuing his PhD within the Decentralized Systems and Network Services Research Group at the Karlsruhe Institute of Technology (KIT), Germany. His research interests include mobile networks, simulation environments and, more recently, the modeling of wireless lower physical layers. Before joining the Decentralized Systems and Network Services Research Group as a student assistant in 2005, he was with the Process- and Data-Management in Engineering Research Group at the Research Center for Information Technology, Karlsruhe, Germany.

Craig L Robinson received his undergraduate education at the University of the Witwatersrand, South Africa in 2000. In 2001 he was awarded a Fulbright Scholarship through which he completed a PhD in control systems at the University of Illinois at Urbana-Champaign in 2008. His research interests in vehicular networks include communication protocols, estimation and control in the presence of packet losses, and system architecture. Craig developed early inter-vehicular communications applications at Toyota Technical Center in 2005 and is currently doing prototype development and research at Mercedes Benz Research and Development North America.

Björn Scheuermann is a junior professor at the Heinrich Heine University, Düsseldorf, Germany. He studied mathematics and computer science at the University of Mannheim, Germany, where he received a Bachelors and a Masters degree, both in 2004. He got a scholarship from the German National Merit Foundation. In 2007 he received his PhD in computer science from the Heinrich Heine University, Düsseldorf, Germany. His research interests include inter-vehicle communications and mobile networking.

Felix Schmidt-Eisenlohr has been a PhD candidate and a research assistant at the Institute of Telematics at Universität Karlsruhe (TH), Germany, since 2006. He received his diploma degree in computer sciences from Universität Karlsruhe (TH) in 2005. His research interest is on understanding and optimizing wireless communication in vehicular communication networks and on methodologies that allow realistic simulations of vehicle-to-vehicle communication.

Ian Tan received a BSc in computer engineering from the University of Illinois at Urbana-Champaign in 2005 and a MSc in electrical engineering from the University of California at Berkeley in 2008. He has worked for a number of commercial and educational entities, including ViaSat, Motorola, and the Tampere University of Technology in Tampere, Finland. He is currently employed as an associate engineer with the Radar Science and Engineering section of the Jet Propulsion Laboratory in Pasadena, CA. His research interests include vehicular wireless communications, signal processing, synthetic aperture radar, and digital hardware design.

Marc Torrent-Moreno holds an MSc in telecommunication engineering from the Polytechnic University of Catalonia and a PhD in computer science from the University of Karlsruhe in Germany. He has participated since 2001 in several research projects in the field of mobile networks, as part of different companies and universities. Of note are British Telecom, NEC Deutschland, DaimlerChrysler Research and Technology North America, University of California at Berkeley, and University of Karlsruhe. Currently, he is head of the R&D Mobility group in Barcelona Digital Centre Tecnològic, developing ICT projects in the area of smart environments, intelligent vehicles and mobile services.

André Weimerskirch is chief executive officer and president of American-based escrypt Inc. and is in charge of the international activities of escrypt. From 2004 to 2007 Weimerskirch held the position of chief technology officer of escrypt GmbH where he significantly shaped their technological strategy. Previously, Weimerskirch worked with several renowned companies in the areas of research, development, and consulting, including Accenture, Deutsche Post, Philips, and Sun. He studied business information technology as well as mathematics at Darmstadt Technical University before receiving his MSc in computer science at Worcester Polytechnic Institute, USA. He then received a PhD at Ruhr-University of Bochum in the area of applied data security. He has led several national and international projects in the areas of data security and has published numerous articles.

1

Introduction

1.1 Basic Principles and Challenges

The basic concept of VANET is straightforward: take the widely adopted and inexpensive wireless local area network (WLAN) technology that connects notebook computers to each other and the Internet, and, with a few tweaks, install it on vehicles. Of course, if it were truly that straightforward, the active VANET research community would likely never have formed and this book would never have been written. As the reader likely understands (especially if they continue reading), the vehicular environment creates unique opportunities, challenges, and requirements. This book documents the early years and the current state of the art of this exploration.

First, consider the opportunities. If vehicles can directly communicate with each other and with infrastructure, an entirely new paradigm for vehicle safety applications can be created. Even other non-safety applications can greatly enhance road and vehicle efficiency. Second, new challenges are created by high vehicle speeds and highly dynamic operating environments. Third, new requirements, necessitated by new safety-of-life applications, include new expectations for high packet delivery rates and low packet latency. Further, customer acceptance and governmental oversight bring very high expectations of privacy and security.

Even today, vehicles generate and analyze large amounts of data, although typically this data is self-contained within a single vehicle. With a VANET, the 'horizon of awareness' for the vehicle or driver drastically increases. The VANET communication can be either done directly between vehicles as 'one-hop' communication, or vehicles can retransmit messages, thereby enabling 'multihop' communication. To increase coverage or robustness of communication, relays at the roadside can be deployed. Roadside infrastructure can also be used as a gateway to the Internet and, thus, data and context information can be collected, stored and processed 'somewhere', e.g., in upcoming Cloud infrastructures.

It warrants repeating that the interest in vehicular inter-networks is strongly motivated by the wealth of applications that could be enabled. First of all, active safety applications, i.e., accident prevention applications, would benefit from this most direct form of communication. Second, by collecting traffic status data from a wider area, traffic flow could be improved, travel times could be reduced as well as emissions from the vehicles. As it was concisely stated as the tenet of the Intelligent Transportation System World Congress in 2008: save time, save lives.

The application classes 'Safety' and 'Efficiency' can be used to classify applications based on their primary purpose. However, the aspects of safety and efficiency cannot be seen as completely disjoint sets of features. Obviously, vehicle crashes can lead to traffic jams.[1] A message reporting an accident can be seen as a safety message from the perspective of near-by vehicles. The same message can be seen by further-away vehicles as an input to calculate an alternative route within a transport efficiency application. Figure 1.1 schematically illustrates the aspects of hazard warning and traffic information.

While being conceptually straightforward, design and deployment of VANET is a technically and economically challenging endeavour. As described in the following chapters, key technical challenges include the following issues:

- Inherent characteristics of the radio channel. VANET present scenarios with unfavorable characteristics for developing wireless communications, i.e., multiple reflecting objects able to degrade the strength and quality of the received signal. Additionally, owing to the mobility of the surrounding objects and/or the sender and receiver themselves, fading effects have to be taken into account.

- Lack of an online centralized management and coordination entity. The fair and efficient use of the available bandwidth of the wireless channel is a hard task in a totally decentralized and self-organizing network. The lack of an entity able to synchronize and manage the transmission events of the different nodes might result in a less efficient usage of the channel and in a large number of packet collisions.

- High mobility, scalability requirements, and the wide variety of environmental conditions. The challenges of a decentralized self-organizing network are particularly stressed by the high speeds that nodes in VANET can experience. Their high mobility presents a challenge to most iterative optimization algorithms aimed at making better use of the channel bandwidth or the use of predefined routes to forward information.

- Security and privacy needs and concerns. There is a challenge in balancing security and privacy needs. On the one hand, the receivers want to make sure that they can trust the source of information. On the other hand, the availability of such trust might contradict the privacy requirements of a sender.

---

[1] The IntelliDrive™ initiative reports that about 25% of all traffic jams are due to crashes.

Figure 1.1 By using vehicle-to-vehicle and vehicle-to-roadside communication, accidents can be avoided (e.g., by not colliding with a traffic jam) and traffic efficiency can be increased (e.g., by taking alternative routes). (Source: Hartenstein and Laberteaux, 2008, reproduced by Permission of © 2008 IEEE.)

- Standardization versus flexibility. Without any doubt, there is a need for standardizing communications to allow VANET to work across the various makes and brands of original equipment manufacturers (OEMs). Yet, it is likely that OEMs will want to create some product differentiation with their VANET assets. These goals are somewhat in tension.

From an application and socio-economic perspective, key challenges are as follows:

- Analyzing and quantifying the benefit of VANET for traffic safety and transport efficiency. So far, relatively little work has been done to assess the impact of VANET as a new source of information on driving behavior. Clearly, the associated challenge in addressing the issue of impact assessment is the modelling of the related human factor aspects.
- Analyzing and quantifying the cost–benefit relationship of VANET. Because of the lack of studies on the benefits of VANET, a cost–benefit analysis can hardly be done.
- Designing deployment strategies for this type of VANET that are not based on a single infrastructure and/or service provider. Owing to the 'network effect', there is the challenge of convincing early adopters to buy VANET equipment when they will rarely find a communication partner.

- Embedding VANET in intelligent transportation systems architectures. VANET will be a part of an intelligent transportation system where other elements are given by traffic-light control or variable message signs. Also public and individual transportation have to be taken into account in a joint fashion. Therefore, truly cooperative systems need to be developed.

As can be seen from the above lists of technical, application, and socio-economic aspects, the field of vehicular application and inter-networking technologies is based on an interdisciplinary effort in the cross section of communication and networking, automotive electronics, road operation and management, and information and service provisioning. VANET can therefore be seen as a vital part of intelligent transportation systems (ITS).

While various projects discussed deployment strategies for VANET, research activities have primarily addressed the technical challenges. Currently, we observe a shift from this classical 'bottom-up' approach to a more 'top-down'-based thinking. Ideally, both approaches will be followed and will finally meet each other: what is identified as a requirement for beneficial deployment can be served by the technological advances. This book is intended to define the current position of the state of the art in VANET research and development.

In Section 1.2 we outline the history of inter-vehicle communications. In Section 1.3 we present an overview of Chapters 2 to 10 and their main contributions.

1.2 Past and Ongoing VANET Activities

The history of the use of radio and infrared communication for vehicle-to-roadside and vehicle-to-vehicle communication is strongly tied to the evolution of intelligent transportation systems. As referenced in Shladover (1989) and in Lasky and Ravani (1993), the basic concepts of roadway automation, i.e., the use of communication and control techniques to make road traffic safe, efficient, and environmentally friendly, were exhibited at the 1939 World Fair. The exhibit, called 'Futurama' by its creator, General Motors, envisioned a peek 20 years into the future, showing both concepts and technology forecasts.[2]

Later, since at least the late 1960s, actual radio-based 'roadway automation' systems were developed and demonstrated. Since this time, one can observe the following facts:

- While safety, efficiency, and environmental friendliness are the key themes, the emphasis consistently changed over time. For example, in the early days route-guidance systems were investigated. Later, for example, tolling systems or research into automatic driving became popular.

- Research and development in various regions, primarily in the USA, in Japan, and in Europe, but also in other parts of the world, were influencing each other.

- The topic has been addressed with a different focus by the automotive and the transportation communities.

---

[2] At the time of writing, video of the 1939 Futurama exhibit can easily be found online.

INTRODUCTION

- A consistent theme over time is 'funding – who will pay?'. Frequently, communication-based tolling and congestion pricing are offered as solutions, although scepticism persists. For example, as cited in Jurgen (1991), Kan Chen and Robert Ervin described the 'chicken and egg standoff' in 1990 as follows: on the one hand, automotive and electronics industries doubt whether the public infrastructure for Intelligent Vehicle-Highway Systems (IVHS) will ever materialize. On the other, highway agencies doubt whether IVHS technologies will deliver practical solutions to real highway problems.

In this section, we present a selection of what we consider important milestone activities with respect to VANET. Owing to the wealth of research and development done in the field of intelligent transportation systems, our overview will be non-exhaustive.

1.2.1 From the beginning to the mid 1990s

In the USA, an Electronic Route-Guidance System (ERGS) was proposed in Rosen et al. (1970):

> 'The system is destination-oriented. The driver enters a code word, representing his intended destination, into the vehicle equipment. Then as the vehicle approaches each instrumented intersection, the destination code is transmitted to the roadside where it is decoded, according to a stored program, and a routing instruction is transmitted back to the vehicle.'

The corresponding communication system was operating at 170 kHz using loop antennas installed at intersections and mounted under the rear of the vehicles. Data transmission rate is reported as 2000 bits per second. According to French (1986, 1987) the ERGS efforts were terminated owing to the expensive roadside infrastructure.

In Japan, the Comprehensive Automobile Traffic Control System (CACS) project was carried out from 1973 to 1979 by the Agency of Industrial Science and Technology of the Ministry of International Trade and Industry (MITI). The objectives of CACS as presented in Kawashima (1990) are still valid after more than 30 years:

- reduction of road traffic congestion
- reduction of exhaust fumes caused by traffic congestion
- prevention of accidents
- enhancement of public and social role of automobiles.

In order to attain this goal, four technical objectives were established:

- to guide drivers along most appropriate routes in order to avoid congestion and air pollution
- to provide useful information in order to assist safe driving

- to give priority of road to public or emergency vehicles

- to provide information promptly to drivers in case of emergency.

The CACS project also carried out a pilot operation with 98 units of roadside equipment and 330 test vehicles (Nakahara and Yumoto 1997). For roadside-to-vehicle communication, loop antennas were used for the roadside units and ferrite core antennas for the vehicle units. The transmission speed is reported as 4.8 kbps.

In Europe, the PROMETHEUS (Programme for European Traffic with Highest Efficiency and Unprecedented Safety) framework initiated in 1986 and launched in 1988 significantly stimulated research and development activities in the area of information technology and mobile communications for motor vehicles and the roads they drive on. PROMETHEUS was supported by 19 European countries and the Commission of the European Communities (Walker 1992; Williams 1988). As for example outlined in Gillan (1989), PROMETHEUS was organized in various sub-programmes including:

- PRO-CAR: driver assistance by electronic systems.

- PRO-NET: vehicle-to-vehicle communications.

- PRO-ROAD: vehicle-to-environment communications.

In PROMETHEUS, vehicle-to-vehicle communication played a prominent role: the report of Dabbous and Huitema (1988) can still be regarded as up to date in many aspects. Dabbous and Huitema analyzed the communication requirements based on typical scenarios, for example for a lane-change maneuver. Assuming a periodic broadcast strategy and collision distance accuracy requirements, they show that a conservative estimate of the transmission rate of the status messages each vehicle sends out periodically might lead to 20 transmissions per second per vehicle. Dabbous and Huitema also indicate that relaxing the requirements on the accuracy of collision distance and introducing prediction methods can significantly reduce the number of transmissions required. Focus was typically on systems operating in the 60 GHz frequency band (see, e.g., Fischer 1991).

Interest in vehicle-to-vehicle communication continued in Japan and the USA. The survey in Kawashima (1990) cites two technical reports by JSK (the Association of Electronic Technology for Automobile Traffic and Driving) published in 1986 and 1988, respectively, on experimental results of vehicle-to-vehicle communication. For the USA, as outlined in Shladover et al. (1991), a main driver appeared to be automatic driving and vehicle platoons. In Sachs and Varaiya (1993), requirements and specifications for vehicle-to-vehicle and roadside-to-vehicle communication for automated vehicles are presented.

An excellent 'snapshot' summarizing project activities of this epoch in the USA, Europe, and Japan as well as looking ahead towards Intelligent Vehicle-Highway Systems is given by Jurgen (1991).

INTRODUCTION

1.2.2 From the mid 1990s to the present

The second half of the 1990s provided remarkable milestones and some major paradigm shifts. Impressive results on cooperative autonomous driving were demonstrated at the San Diego demo of the California Partners for Advanced Transit and Highways (PATH) in 1997, at the Advanced Safety Vehicle (ASV) Phase 2 Demo in 2000 in Tsukuba city, Japan, and within the PROMOTE CHAUFFEUR European project. The focus then shifted from cooperative autonomous driving to cooperative driver assistance systems. In the USA the Intelligent Vehicle Initiative (IVI) in the years 1998 to 2005 (Hartman and Strasser 2005) focused on cooperative active safety. In Europe, the CarTalk and FleetNet projects (Franz et al. 2005) investigated technologies and applications for cooperative driver assistance. In Japan, Phase 3 of the Advanced Safety Vehicle project also acknowledged the role of inter-vehicle communications for cooperative driver assistance.

A game changer: 5.9 GHz DSRC

The concept of VANET has been significantly impacted by the advances in technology and standardization since the mid 1990s. In 1999 a 'game changer' occurred when the US Federal Communication Commission allocated 75 MHz bandwidth of the 5.9 GHz band to Dedicated Short-Range Communication (DSRC). The term 'dedicated short-range communication' was used as a technology-neutral term for short-range wireless communication between vehicles and infrastructure.

A year later, ASTM International established a working group to develop requirements for corresponding DSRC standards. In 2001, the Standards Committee 17.51 of the ASTM selected IEEE 802.11a as the underlying radio technology for DSRC. The corresponding standard was released in 2002 (ASTM E2213-02 2003) and revised in 2003 (ASTM E2213-03 2003). The pressure to make use of the assigned channels and the availability of the IEEE 802.11a technology and standard significantly increased research and development activities. In particular, the mobile networking community's interest in the topic of vehicular networks was revitalized. In 2004, the IEEE started the work on the 802.11p amendment and Wireless Access in Vehicular Environments (WAVE) standards based on the ASTM standard (Jiang and Delgrossi 2008). The Vehicle Safety Communications (VSC) project – backed by the Crash Avoidance Metrics Partnership (CAMP), the US Federal Highway Administration (FHWA), and the US National Highway Traffic Safety Administration (NHTSA) – investigated emerging 5.9 GHz DSRC technology in the years 2002 to 2004 and concluded that the approach based on IEEE 802.11a would be able to support most of the safety applications that VSC had selected. The VSC final report (VSC 2006), however, also explicitly points to challenges of low-latency communication and high availability of the radio channel as well as of general channel capacity-related issues. In 2004, the first ACM International Workshop on Vehicular Ad Hoc Networks took place in Philadelphia, for which the term 'VANET' was coined. The editors of this book served key leadership roles in the early years of this workshop (Ken Laberteaux, General Co-Chair 2004–05; Hannes Hartenstein General Co-Chair 2005, Technical Program Co-Chair 2006).

Current projects and activities

In the following paragraphs we will provide a non-exhaustive overview on ongoing VANET projects and activities in Europe, Japan, and the US.[3] The overview indicates the high level of activity in the field of VANET research and development and demonstrates that the wireless communication technologies covered in this book form the basis of most VANET activities worldwide (see Figure 1.2).

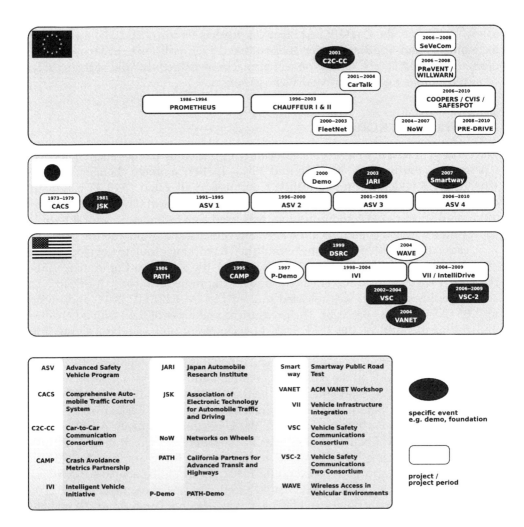

Figure 1.2 A nonexhaustive overview of pioneering VANET activities and milestones as described in this section.

---

[3]We will not provide references to websites featuring those projects and activities since those pages (both content and addresses) change frequently.

Within the Framework Programme 6 of the EU, four integrated projects were started in areas that touch the field of VANET: COOPERS, CVIS, PReVENT, and SAFESPOT. The project Co-operative Systems for Intelligent Road Safety (COOPERS, 2006–10) focuses on innovative telematics applications for cooperative traffic management. From a communication perspective, it therefore primarily addresses vehicle-to-roadside communications and makes use of CALM standards like the CALM infrared communication interface. CALM is the ISO TC 204 (ITS) Working Group 16 (Communication) on 'Continuous Air interface for Long and Medium distance'. CALM aims to support continuous communications for vehicles by making use of various media and communication interfaces. The project Co-operative Vehicle-Infrastructure Systems (CVIS, 2006–10) has a main focus on development and testing of vehicle-to-infrastructure communication and also follows the CALM standards. CVIS makes use of the IEEE 802.11 WAVE-related interface that is denoted as M5 interface (for 'Microwave 5 GHz') within the CALM framework. Project SAFESPOT (2006–10) aims to design cooperative systems for road safety based on vehicle-to-vehicle and vehicle-to-infrastructure communication. The communication technology used in project SAFESPOT is IEEE 802.11a/p. Project PReVENT (2006–08) addressed development of preventive safety applications and technologies. Within the PReVENT Integrated Project, the subproject WILLWARN (Wireless Local Danger Warning) focused on the topic of vehicle-to-vehicle and vehicle-to-infrastructure communication. The WILLWARN system is based on IEEE 802.11a/p and made use of the communication platform developed in the German Network on Wheels (NOW) project. In addition, the project Secure Vehicular Communication (SeVeCom, 2006 to 2008) was dedicated to identifying threats and specifying methods and architecture for securing wireless vehicular communication. Currently, under Framework Programme 7, activities towards field operational tests are funded, in particular the project PRE-DRIVE-C2X that is preparing the building blocks required for successful field operational tests of VANET in Europe.

In Japan, a standard for vehicle-to-infrastructure communication was published in 2001 and denoted as 'Dedicated Short-Range Communication System' (ARIB 2001). The specified system operates in the 5.8 GHz frequency band, is based on time division multiple access (TDMA) and targets a range of about 30 m. The primary use of the system was seen in electronic toll collection but the system was generalized to support various other services (ARIB 2004). In 2008, more than 20 million on-board units for electronic toll collection were deployed in Japan. Based on the success on this 5.8 GHz DSRC system and on infrared-based vehicle-to-infrastructure communication, various ITS projects and activities are currently joining forces to demonstrate and enhance vehicle-to-infrastructure and vehicle-to-vehicle communication under the umbrella of Japan's national ITS Safety 2010 initiative. The Advanced Vehicle Safety initiative, now in its phase 4, is addressing vehicle-to-vehicle communication by a carrier sense multiple access (CSMA) based extension of ARIB STD T-75. The Advanced Cruise-Assist Highway System initiative builds on the 5.8 GHz DSRC system as well as the Driving Safety Support System (DSSS) that is also making use of infrared technology for vehicle-to-roadside communication. The Smartway activity focuses on the ITS services and a common service platform offered on top of the existing networks. In June 2007, it was announced that a 10 MHz channel in the 700 MHz frequency channel will be allocated for safety-related inter-vehicle communication in Japan in 2012.

There are several industry/government projects in the USA ongoing as this book goes to press. Two representative projects are described below. These are chosen for their availability of information and technical scopes that closely track topics in this book. The first, Vehicle Infrastructure Integration (VII), which has been rebranded as IntelliDrive, has recently completed a large proof of concept demonstration. The second, the Vehicle Safety Communication-Applications project, is scheduled to end in late 2009, but several interesting details have been publicly presented, and will be discussed below. Both efforts are substantially funded by the US Department of Transportation and have active participation from several automotive manufacturers and suppliers.

In addition, two other projects deserve a brief mention: an ongoing Integrated Vehicle-Based Safety Systems project explores human-machine interface issues when several safety applications, with potentially overlapping or contradictory advisories, are operated simultaneously (University of Michigan Transportation Research Institute 2007).

A second project, the Cooperative Intersection Collision Avoidance System project, had three components (McHale 2007): a Violation Warning project (demonstrated in Michigan), a Stop Sign Assist project (demonstrated in Minnesota), and a Signalized Left Turn Assist project (demonstrated in California). As equipped vehicles approached a CICAS-V intersection, the vehicles received signal phase and timing information from the intersection light (over DSRC). Each vehicle would then predict the likelihood that it would be in the intersection in violation of a red light, and if appropriate, warn the driver (Maile 2008a). This project was planned to test with naive drivers (Maile 2008a), although the US Department of Transportation has made no such announcement.

1.2.3 Examples of current project results

Vehicle Infrastructure Integration

In 2005, the US Department of Transportation initiated a proof of concept (POC) demonstration. The majority of this testing environment was implemented in the northwest suburbs of Detroit, MI. This system comprised 55 road side equipment (RSE) stations within 45 square miles (see Figure 1.3) and employed 27 vehicles (Kandarpa 2009).

Seven applications were developed and tested:

- In-Vehicle Signage: RSEs trigger displays of advisory messages within the vehicle.

- Probe Data Collection: Vehicles provide historical data on their location/state and share with the RSE, which is then centrally compiled and analyzed.

- Electronic Payments – Tolling.

- Electronic Payments – Parking.

INTRODUCTION

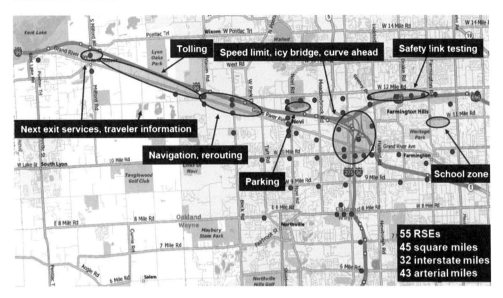

Figure 1.3 VII Proof of Concept Testbed in Michigan. Road Side Equipment locations are shown. Source: Schagrin and Briggs (2008).

- Traveler Information/Off-Board Navigation.

- Heartbeat: RSEs collect periodic (e.g. 100 msec) status messages from vehicles including vehicle speed and location.

- Traffic Signal Indication: Broadcasts traffic light state.

Some key findings (Andrews and Cops 2009; Kandarpa 2009):

- Packet error rates were a strong function of line of sight. When line of sight was maintained, packet error rates remained low even at distances of 100s of meters (e.g. between an RSE with a tall antenna and vehicle). When line of site was lost, error rates increased rapidly.

- IPv6 performed well.

- Low-cost GPS receivers, as used in the POC, did not consistently provide lane-level accuracy.

- End-to-end encryption of packets was successfully demonstrated.

- Heartbeat application worked well for vehicles within approximately 100 m of the RSEs (depending on line of sight).

- Security systems were shown to work, but remained brittle. Also, large-scale tests were not performed.

- Management of network communications resources for multiple simultaneous applications is more complex than expected.
- Installing, configuring and maintaining the RSE was more complex and difficult than expected.

As this phase of the VII program ended, the US Department of Transportation has repositioned and rebranded the VII program (Row et al. 2008). Its new name is *IntelliDrive*. It will emphasize wireless technology in the service of safety applications. Quoting 'The Future of VII' (Row et al. 2008):

> The new VII program will focus research activities in the following areas:
> - technology scanning and research to identify and study a wide range of potential technology solutions
> - research, demonstration, and evaluation of technology-enabled safety applications
> - establishment of test beds to support operational tests and demonstrations for public and private sector use
> - development of architecture and standards to provide an open platform for wireless communications to and from the vehicle
> - study of non-technical issues such as privacy, liability and application of regulation
> - research on ancillary benefits to mobility and the environment.

As an initial step, the Safe Trip-21 project launched in 2008 and has established Connected Traveler test sites in California. Safe Trip-21 'is designed to improve safety and reduce congestion by identifying and harnessing existing technology and adapting it for transportation needs,' and 'will demonstrate that significant advances in solving transportation problems do not have to require large infrastructure investments.' (Safe Trip-21 2008).

No national-scale deployments of VII have been announced in the USA.

Vehicle Safety Communications-Applications

The Vehicle Safety Communications-Applications (VSC-A) is a three year project between five auto makers (Daimler/Mercedes-Benz, Ford, GM, Honda, and Toyota) and the US Department of Transportation. This project is a follow-up to the first Vehicle Safety Communications project (2002–04), and focuses on vehicle communications and relative positioning with the goal of enabling interoperable safety applications (Ahmed-Zaid 2009).

The VSC-A project identified eight crash scenarios as having a 'Top Composite Ranking', based on US Government statistics on crash frequency, cost, and functional years lost. Based on this, VSC-A identified seven safety applications to address these crash scenarios. Those applications are (Maile 2008b):

- Emergency Electronic Brake Lights (EEBL): Warns of sudden braking of vehicles in the forward path.

INTRODUCTION

Table 1.1 Mapping of VSC-A safety applications to crash scenarios of greatest concern (Maile 2008b).

Crash scenarios	V2V safety applications						
	EEBL	FCW	BSW	LCW	DNPW	IMA	CLW
1 lead vehicle stopped		✓					
2 control loss without prior vehicle action							✓
3 vehicle(s) turning at non-signalized junctions						✓	
4 straight crossing paths at non-signalized junctions						✓	
5 lead vehicle decelerating	✓	✓					
6 vehicle(s) not making a maneuver – opposite direction					✓		
7 vehicle(s) changing lanes – same direction			✓	✓			
8 LTAP/OD at non-signalized junctions						✓	

- Forward Collision Warning (FCW): Warns of impending rear-end collision with forward vehicle.

- Blind Spot Warning (BSW)/Lane Change Warning (LCW): Warns during a lane-change attempt if there is another vehicle moving the same direction in (or soon will be in) the blind spot. Secondary advisory whenever there is a vehicle in the blind spot.

- Intersection Movement Assist (IMA): Warns when it is not safe to enter intersection.

- Do Not Pass Warning (DNP): Warns when oncoming vehicle poses collision threat if a lane change is attempted.

- Control Loss Warning (CLW): Self-generated warning when vehicle loses control. Other vehicles will be warned depending on the threat.

The mapping of the eight crash scenarios to the seven VSC-A safety applications is shown in Table 1.1 (Maile 2008b).

In the communications area, the focus is on (Caminiti 2009):

- message composition

- power testing

- message dissemination

- multi-channel operation

- standards coordination and validation.

The message composition subtask has defined a basic safety message which supports all safety applications (Caminiti 2009). This work is strongly influencing the nearly completed negotiations of the SAE J2735 working group. A final standard/recommended practice is expected near the end of 2009.

The power testing work explores whether high power (> 20 dBm) increases safety application performance, and low power (< 20 dBm) addresses congestion control. Comparing packet error rate (PER) around a non-line-of-sight 'closed intersection', the project has shown significantly longer-range low-error communication when using 33 dBm transmission power (over the nominal 20 dBm) (Caminiti 2009). This provides an interesting path for exploration, especially in light of the challenges posed by non-line-of-sight communications reported in the VIIC Report (Kandarpa 2009).

The VSC-A has been actively participating in various DSRC standards, including IEEE 802.11p, IEEE 1609, and SAE J2735 (Caminiti 2009). These standards will be further discussed in Chapter 10. Note that the author of Chapter 10 currently serves as the technical leader for VSC-A in this area.

The message dissemination work will investigate the use of power, rate, and other controls to mitigate network congestion and improve message delivery (Caminiti 2009). The multi-channel operation subtask is investigating channel switching and multi-radio usage (Caminiti 2009). Final results from these subtasks have not been disclosed at the time of this writing. However, results should become available shortly after the completion of the VSC-A project in November 2009.

In addition, VSC-A has a task focused on the security of vehicle-to-vehicle messages. They wish to avoid dedicated security hardware. The project is evaluating elliptic curve digital signature algorithm (ECDSA) (Accredited Standards Committee 2005), timed efficient stream loss-tolerant authentication (TESLA) (Perrig et al. 2002), and several modified versions of the same. Evaluations will be based both on extensive network simulations and on testing of the project's final test-bed of vehicles (Bai 2009). Note that the first-author of Chapter 9 is the technical lead of security work for VSC-A.

1.3 Chapter Outlines

The material contained in this book is organized as follows. Chapters 2–4 identify applications and communication requirements and approaches. Chapter 5 focuses on the mobility models for vehicular traffic, which in turn motivates several aspects and tunings of the protocols described in later chapters. Chapter 6–9 describe the four technical aspects which are most affected by VANET considerations: physical layer, medium access control, middleware, and security. Chapter 10 gives an overview of standards efforts and further describes protocols in the VANET stack.

Cooperative Vehicular Safety Applications (Chapter 2)

This chapter first discusses the enabling technologies for cooperative driving systems. Subsequently, a layered software architecture for cooperative driving systems is described. Following the discussion on cooperative driving architectures, environment mapping, which is the principle architectural component for cooperative safety, is discussed in detail. The stress in this section is on various existing techniques for vehicular path prediction. In the final section of this chapter, several cooperative vehicular safety applications are detailed and illustrated. Each of these applications underlines a particular advantage of vehicle ad-hoc networks (VANET), which is unavailable through other sensing technologies.

Information Dissemination in VANET (Chapter 3)

Vehicles can be seen as probes that locally detect traffic status. Various applications that target transport efficiency could make use of the vast information collected by the vehicles; however, this collection of information needs to be transported over larger distances, for example, city-wide or region-wide. For this purpose, various information dissemination approaches were proposed in the literature and are surveyed in this chapter. The information dissemination 'lifecycle' is structured in four phases: obtaining information, transport of information, summarization of measurements, and aggregation of information. Summarization refers to the process of appropriately combining the measurements of the same observed event of different observers. Aggregation refers to the process of an appropriate reduction of information to deal with the limited capacity of the wireless network.

VANET Convenience and Efficiency Applications (Chapter 4)

VANET convenience and efficiency applications comprise Internet access, service announcements, infotainment, payment services, and most notably collaborative traffic information services. This chapter discusses the suitability of VANET to support this application class. The discussion addresses communication capacity and connectivity limitations as well as the role of competing technologies. In addition, solutions based on centralized client–server systems, on peer-to-peer systems, and on pure vehicle-to-vehicle communications are compared. As the technical basis, data aggregation schemes, as outlined in the previous chapter, are applied to the case of collaborative traffic information systems. Simulation results of these approaches for a city-wide scenario are presented that also indicate the benefit of supporting roadside units.

Vehicular Mobility Modeling for VANET Simulations (Chapter 5)

Since mobility influences the performance of VANET, and the purpose of VANET is to influence mobility, mobility modeling represents a key resource to understand this influencing factor as well as the target domain of VANET research. This chapter surveys various flow and traffic models and presents how these models and corresponding simulators can be used together with communication network simulators. In addition, a design framework for realistic vehicular mobility models is presented that can help researchers and developers to select the right building blocks and the appropriate level of detail depending on their simulation needs. In the chapter, open issues are discussed with respect to the sufficient degree of the level of detail of models as well as on combining mobility and traffic simulators with communication network simulators.

Physical Layer Considerations for Vehicular Communications (Chapter 6)

This chapter begins with an overview of the proposed DSRC standard and the specific parameters of the orthogonal frequency division multiplexing (OFDM) architecture it employs. This is followed by a development of wireless communications channel theory, along with an examination of common metrics used to quantify the performance of wireless channels. The remainder of the chapter describes extensive DSRC channel measurement experiments. The measurement methods are described in detail, followed by a summary and analysis of the results. This shows that the current DSRC standard appears to be sufficient, but not necessarily optimal, for its intended environment.

MAC Layer and Scalability Aspects of Vehicular Communication Networks (Chapter 7)

The vehicles will share a radio channel to exchange safety and control messages without a centralized coordinator for access to the channel. Efficient and effective medium access control (MAC), therefore, represents an essential building block determining the quality of the communication system and its scalability. This chapter first provides an overview of the challenges of medium access control for VANET and surveys the existing fundamental approaches. Then, the carrier sense multiple access based MAC, the IEEE 802.11p draft of standard, is presented. A performance evaluation of IEEE 802.11p for the case of active safety applications is given together with insights into the modeling and simulation methodology. An empirical model for the probability of packet reception is derived from a large number of simulation runs. Finally, various approaches to controlling the load on the radio channel are discussed with a special emphasis on the use of transmit power control.

Efficient Application Level Message Coding and Composition (Chapter 8)

This chapter focuses on message contents for safety applications. The goals are to improve channel utilization by recognizing similarities in transmitted data, and to separate message construction and communication from the application functionality. The chapter begins with an overview of the wireless inter-vehicle communication environment and highlights some desirable features which a system architecture in this environment should possess. Then, the chapter turns to a solution: the message dispatcher (which has subsequently become known as the message handler). This message dispatcher concept is demonstrated by considering some example applications and the resulting message composition. Next, the

INTRODUCTION

concept of efficiency is extended by including linear predictive coding to further reduce bandwidth consumption. The chapter concludes with an examination of the message dispatcher in light of the criteria set out at the beginning of the chapter.

Data Security in Vehicular Communications Networks (Chapter 9)

This chapter begins by presenting the challenges of providing data security in VANET, including attacker and application models. This is followed by an exploration of required supporting infrastructure, including the management and handling of a public key infrastructure (PKI). Protocols for providing secure communication, secure positioning, as well as identification of misbehaving nodes is presented next. This is followed by a detailed exploration of privacy in the VANET context. The chapter concludes with a discussion of implementation aspects, including appropriate key lengths, physical security, organizational aspects, and software updates in the field.

Standards and Regulations (Chapter 10)

This chapter begins with a description of the general protocol stack for DSRC, as well as a description of DSRC regulations in the USA and Europe. The remainder of the chapter describes the most relevant standards activities for VANET, specifically IEEE 802.11p, IEEE 1609, and SAE J2735. The description begins with the lower PHY and MAC/LLC layers, as standardized in IEEE 802.11p. Next, the middle layers, as defined by IEEE 1609 are presented, including multi-channel functioning (IEEE 1609.4) and security services (IEEE 1609.2). While the top-layer applications are described in other chapters, the so-called message sublayer, which provides a message-composition service to other applications, and is defined by SAE J2735, forms the final discussion of this chapter.

References

ANSI X9.62, 2005. *Public Key Cryptography for the Financial Services Industry, The Elliptic Curve Digital Signature Algorithm (ECDSA)*.

Ahmed-Zaid F 2009 Vehicle Safety Communications-Applications (VSC-A) Project Overview. *2009 SAE Government/Industry Meeting*, Washington D.C., February 2009.

Andrews S and Cops M 2009 Final Report: Vehicle Infrastructure Integration Proof of Concept Executive Summary – Vehicle. FHWA-JPO-09-003.

ARIB STD T75, 2001. *Dedicated Short Range Communication System*. Version 1.0 English translation.

ARIB STD T88, 2004. *DSRC Application Sub-Layer*. Version 1.0 English translation.

ASTM E2213-02, 2002. *Standard Specification for Telecommunications and Information Exchange Between Roadside and Vehicle Systems – 5 GHz Band Dedicated Short Range Communications (DSRC) Medium Access Control (MAC) and Physical Layer (PHY) Specifications*. ASTM Committee E17 on Vehicle-Pavement Systems.

ASTM E2213-03, 2003. *Standard Specification for Telecommunications and Information Exchange Between Roadside and Vehicle Systems – 5 GHz Band Dedicated Short Range Communications*

(DSRC) *Medium Access Control (MAC) and Physical Layer (PHY) Specifications.* ASTM Committee E17 on Vehicle-Pavement Systems.

Bai S 2009 Vehicle Safety Communications-Applications (VSC-A) Project: Security for Vehicle Safety Messages. *2009 SAE Government/Industry Meeting*, Washington D.C., February 2009.

Caminiti L 2009 Vehicle Safety Communications-Applications (VSC-A) Project: Communications & Standards Status. *2009 SAE Government/Industry Meeting*, Washington D.C., February 2009.

Dabbous W and Huitema C 1988 PROMETHEUS: Vehicle to Vehicle Communications. Research Report, INRIA-Renault collaboration, August 1988.

Fischer HJ 1991 Digital beacon vehicle communications at 61 GHz for interactive dynamic traffic management. *Eighth International Conference on Automotive Electronics*, pp. 120–124.

Franz W, Hartenstein H and Mauve M 2005 *Inter-Vehicle-Communications Based on Ad Hoc Networking Principles – the FleetNet Project*. Universitätsverlag Karlsruhe.

French R 1986 Historical overview of automobile navigation technology. *36th IEEE Vehicular Technology Conference 1986*, vol. 36, pp. 350–358.

French R 1987 Automobile navigation in the past, present and future. *Proceedings of the International Symposium on Computer-Assisted Cartography*, pp. 542–551.

Gillan W 1989 PROMETHEUS and DRIVE: their implications for traffic managers. *Vehicle Navigation and Information Systems Conference*, Conference Record, pp. 237–243.

Hartenstein H and Laberteaux K 2008 A tutorial survey on vehicular ad hoc networks. *IEEE Communications Magazine* **46**(6), 164–171.

Hartman K and Strasser J 2005 Saving lives through advanced vehicle safety technology: Intelligent Vehicle Initiative Final Report. FHWA-JPO-05-057.

Jiang D and Delgrossi L 2008 IEEE 802.11p: Towards an International Standard for Wireless Access in Vehicular Environments. *IEEE Vehicular Technology Conference Spring 2008*, pp. 2036–2040.

Jurgen R 1991 Smart cars and highways go global. *IEEE Spectrum* **28**(5), 26–36.

Kandarpa R 2009 Final Report: Vehicle Infrastructure Integration (VII) Proof of Concept (POC) Test – Executive Summary.

Kawashima H 1990 Japanese Perspectives of Driver Information Systems. *Transportation* **17**(3), 263–284.

Lasky TA and Ravani B 1993 A review of research related to automated highway systems (AHS). Advanced Highway Maintenance and Construction Technology Research Center, UCD-ARR-93-10-25-01, Dept. of Mechanical and Aeronautical Engineering, University of California, Davis, October, 1993.

Maile M 2008a Cooperative Systems for Intersection Crash Avoidance. *Workshop on VII and CICAS, Transportation Research Board Annual Meeting*, Washington D.C., January 2008.

Maile M 2008b Vehicle Safety Communications-Applications (VSC-A) Project: Crash Scenarios and Safety Applications. *2009 SAE Government/Industry Meeting*, Washington D.C., February 2009.

McHale G 2007 CICAS Program Update. *ITS World Congress 2007*, Beijing, China.

Nakahara T and Yumoto N 1997 ITS development and deployment in Japan. *Proceedings of IEEE Conference on Intelligent Transportation Systems 1997*, pp. 631–636.

Perrig A, Canetti R, Tygar JD and Song D 2002 The TESLA Broadcast Authentication Protocol. *RSA CryptoBytes* **5**, 2–13.

Rosen D, Mammano F and Favout R 1970 An electronic route-guidance system for highway vehicles. *IEEE Transactions on Vehicular Technology* **19**(1), 143–152.

Row S, Schagrin M and Briggs V 2008 The Future of VII. *ITS International*.

Sachs S and Varaiya P 1993 A communication system for the control of automated vehicles. *Path Technical Memorandum*.

Safe Trip-21 2008 Introducing Safe Trip-21: Technology Solutions to Improve Transportation Safety and Reduce Congestion 2008 U.S. Dept. of Transportation, Research and Innovative Technology Administration, newsletter *Horizons*, http://www.rita.dot.gov/publications/horizons/2008_05_06/html/introducing_safe_trip_21.html.

Schagrin M and Briggs V 2008 VII Program Update. *ITS World Congress 2008*, New York, USA.

Shladover S 1989 Research needs in roadway automation. *Vehicle/Highway Automation: Technology and Policy Issues* pp. 89–104.

Shladover S, Desoer C, Hedrick J, Tomizuka M, Walrand J, Zhang WB, McMahon D, Peng H, Sheikholeslam S and McKeown N 1991 Automated vehicle control developments in the PATH program. *IEEE Transactions on Vehicular Technology* **40**(1), 114–130.

University of Michigan Transportation Research Institute 2007 Integrated Vehicle-Based Safety Systems – First Annual Report. Report Number DOT HS 810 842.

VSC 2006 Vehicle Safety Communications Project – Final Report. DOT HS 810 591.

Walker J 1992 Drive, Prometheus & GSM. *Proceedings of the Mobile Radio Technology, Marketing and Management Conference*, London, UK.

Williams M 1988 PROMETHEUS – The European research programme for optimising the road transport system in Europe. *IEE Colloquium on Driver Information*, London, UK, pp. 1/1–1/9.

2

Cooperative Vehicular Safety Applications

Derek Caveney

Toyota Technical Center, Ann Arbor, MI, USA

2.1 Introduction

This chapter provides an overview of cooperative vehicular safety applications and their enabling technologies. We outline appropriate architectures with which to analyze and synthesize cooperative driving applications, the essential components within this architecture necessary for safety applications, and the concepts of operation of some safety applications that can be expected.

2.1.1 Motivation

While the past decade has witnessed a proliferation of active safety systems, this chapter suggests that the next quarter century of active safety systems will be realized by innovative applications of information available through wireless communications. This information may travel between vehicles or between vehicles and fixed infrastructure.

The behavior of current active safety systems is reactive and relies on real-time feedback with small time constants from autonomous sensors. These sensors include 77 GHz and 24 GHz radars, laser radars, and cameras. With the addition of high-accuracy positioning and wireless inter-vehicular communication technologies to a vehicle's sensor suite, cooperative driving applications can be realized to introduce anticipatory, or feedforward, behavior within active safety systems. Anticipatory systems can use longer time constants to discourage drivers from entering risky driving situation, thereby diminishing the need for other active safety systems or passive safety systems. Furthermore, VANET not only promise safety benefits to drivers and those in the surrounding driving environment, but also improved mobility, increased comfort, and reduced environmental impact.

In 2006, 42,642 motor-vehicle-related fatalities were recorded by the National Highway Traffic Safety Administration (NHTSA) Fatality Analysis Reporting System (National Center for Statistics and Analysis 2008). This number was a 2.0% decrease over the number of fatalities (43 510) reported in 2005. However, despite the number of advances in vehicular active and passive safety devices (for example, anti-lock brakes, airbags, crumple zones), the number of fatalities has remained at over 40 000 per year for the past 15 years. One hypothesis for this constant rate is that drivers are willing to undertake more risky driving behaviors with the increased ability of vehicles to protect occupants from severe accidents and injuries. Wireless communications aim to significantly reduce these numbers by providing information to the driver and the vehicle that is unavailable through driver perception and autonomous sensors alone. By leveraging cooperative safety systems, the demand for existing autonomous active safety systems such as pre-crash systems (PCS) as well as the need for passive safety systems such as crumple zones and structure reinforcements would be reduced. Consequently, vehicle weights could be reduced, fuel efficiency could be increased, and emissions could be decreased.

There are four principal benefits of communication-enabled cooperative safety applications over purely autonomous safety systems. Firstly, communication provides an unprecedented field of view and range. Depending on the transmission frequency and power, the radio waves can travel long distances and through obstacles. Line-of-sight is not required. Secondly, the information that can be shared between vehicles, through vehicle-to-vehicle (V2V) communications, or with infrastructure, through infrastructure-to-vehicle (I2V) communications, has greater quantity and higher quality. Vehicle-to-infrastructure (V2I) communications are subsumed by the I2V classification. Information such as the vehicle's predicted route or past trajectory can be best estimated by the vehicle itself, and then shared with others, rather than having each neighboring vehicle estimate this information for every other vehicle. Thirdly, the cost of positioning and communication hardware is significantly less than the equivalent autonomous sensing equipment needed to cover the 360-degree envelope around the vehicle. Finally, communications allow vehicles to coordinate maneuvers for safety goals such as collision avoidance. This coordination can reduce the severity of the maneuvers required by each vehicle to avoid a collision.

2.1.2 Chapter outline

This chapter on cooperative vehicular safety applications first discusses the enabling technologies for cooperative driving systems. The advantages and disadvantages of these technologies are mentioned. Subsequently, a layered software architecture for cooperative driving systems is described and the differences between layers is emphasized. Furthermore, the location of cooperative safety applications within this architecture is highlighted. Following the discussion on cooperative driving architectures, environment mapping, which is the principal architectural component for cooperative safety, is discussed in detail. The stress in this section is on various existing techniques for vehicular path prediction. In the final section of this chapter, several cooperative vehicular safety applications are detailed and illustrated. Each of these applications underlines a particular advantage of vehicle ad-hoc networks (VANET), which is unavailable through other sensing technologies.

2.2 Enabling Technologies

2.2.1 Communication requirements

At the most basic level, the goal of inter-vehicular wireless communications for safety applications is to share current vehicular positions, velocities, and accelerations. Most production vehicles are now outfitted with sensors with which to measure velocities and accelerations. With many vehicles now being sold with positioning systems, this basic information is available. For vehicular safety applications, the task of outfitting vehicles with wireless communications capable of providing high-bandwidth, low-latency messages will belong to car manufacturers, known in the automotive industry as original equipment manufacturers (OEMs). For less time-critical applications involving comfort, mobility, or the environment, it is likely that OEMs will compete with telecommunications companies utilizing WiFi, WiMAX, and cellular wireless technologies. The communication requirements for cooperative safety applications are the most stringent for VANET. Although these requirements are covered in greater detail in subsequent chapters of this book, we mention some to these requirements here.

The first requirement is low-latency dissemination of messages between vehicles and between vehicles and infrastructure. Given the time-critical nature of most safety systems, delayed information is much less useful. The reaction time of the safety system must be of the order of milliseconds. Secondly, all communications must be authentic. Lack of secure communications reduces system trustworthiness, thereby compromising the safety benefit enabled by sharing information. Finally, safety applications require flexible message composition to meet the information requirements of multiple applications running simultaneously.

2.2.2 Vehicular positioning

Accurate autonomous geo-spatial positioning finds itself at the core of most VANET applications, including all safety applications. The lower cost of ground receivers for global navigation satellite systems (GNSS) and the coverage from GNSS satellites make this positioning technology more attractive than other localization techniques using radars, lidars, ultrasonic sensors, or cameras. Furthermore, GNSS satellites provide a common global clock and a common Earth coordinate frame for applications running distributively on multiple vehicles. The Earth coordinate frame is based on the universal transverse mercator (UTM) system grid. UTM is a decimal, rectangular grid to which spherical degree measurements from GNSS signals can be converted using various reference ellipsoids (for example, the world geodetic system (WGS) 84). Additionally, Cartesian UTM positions (x, y, z) are in meters, which allows the positions to be integrated directly into physics-based vehicle models for estimation and control. These benefits make accurate positioning through GNSS a principal enabling technology for VANET safety systems.

The 10–15 m accuracy of a navigation-grade GPS receiver currently in production for the automotive market is only acceptable for route guidance. For such navigation, this rough accuracy is aided by intelligent map-matching algorithms. However, for advanced cooperative driving applications, especially safety systems, this accuracy must be improved and so vehicle positioning is an active field of academic and industrial research (Gao et al. 2006; Petovello 2003).

One approach is differential GPS (DGPS), where a ground-based reference station calculates pseudo-range corrections for each satellite, and wirelessly transmits these corrections to the vehicles' receivers. These corrections account for timing, orbital, and atmospheric errors in the GPS satellite signals, and are applied to the pseudo-ranges observed at the receiver. In the USA, the distribution of corrections is offered by various land-based services, such as nationwide differential GPS (NDGPS), and space-based services, such as the wide area augmentation system (WAAS). The benefit of these corrections is a function of the distance of the receiver from the reference station that is generating its pseudo-range corrections.

As a rule of thumb, the best-case positional accuracy with DGPS remains within approximately 1 meter. For some cooperative safety applications, such as head-on collision warnings, this accuracy is not sufficient. Advanced research into real-time kinematic (RTK) positioning systems for improved relative positioning promises centimeter-level accuracy. RTK positioning uses phase lock loops on the carrier waves of the GPS signals to measure the phase within the cycle. Detecting cycle slips from phase lock loss is critical for RTK positioning. Cycle slips occur when the receiver loses its line of sight to the satellite. Unfortunately, this occurrence is common within the automotive domain when driving under overpasses, through dense foliage, in urban centers, and in deep valleys or canyons. This satellite signal loss is a problem with all GNSS positioning. Its solutions include the fusion of GNSS with inertial measurement units (IMUs) and other vehicle sensors to bridge the gaps during GPS signal outages and also to reduce the IMU position drift errors.

The dynamic models used for vehicular geo-spatial positioning consider the vehicle as a point mass in 3D free space. These equations of motion are frequently used within the mechanization equations employed for GPS/IMU integration (Gao 2006; Petovello 2003). For example,

$$\begin{bmatrix} \dot{\mathbf{r}}^e \\ \dot{\mathbf{v}}^e \\ \dot{\mathbf{R}}^e_b \end{bmatrix} = \begin{bmatrix} \mathbf{v}^e \\ \mathbf{R}^e_b \mathbf{f}^b - 2\Omega^e_{ie}\mathbf{v}^e + \mathbf{g}^e \\ \mathbf{R}^e_b(\Omega^b_{ei} + \Omega^b_{ib}) \end{bmatrix}, \qquad (2.1)$$

where e refers to the Earth-fixed coordinate frame, b refers to the IMU body frame, and i refers to the inertial frame. The state variables are the position vector represented in the Earth frame, $\mathbf{r}^e = [x, y, z]'$, the velocity vector represented in the Earth frame, $\mathbf{v}^e = [\dot{x}, \dot{y}, \dot{z}]' \triangleq [v_x, v_y, v_z]'$, and the rotation matrix between the body frame and the Earth frame, $\mathbf{R}^e_b$. The matrices Ω^a_{bc} are the skew-symmetric forms of the rotation rate vectors $\boldsymbol{\omega}_{bc}^a$, where $\boldsymbol{\omega}_{bc}^a$ represents the rotation rate of the frame 'c' relative to frame 'b', expressed in frame 'a'. The gravity vector in the Earth frame is $\mathbf{g}^e$, while the specific force vector $\mathbf{f}^b$ and the angular rate measurement vector $\boldsymbol{\omega}_{ib}^b$, which are measured by the IMU are the input to the state equations. These equations are perturbed to define the GPS/IMU error equations found within various filtering techniques including extended Kalman filtering, unscented Kalman filtering, and particle filtering. Misalignments between the IMU body frame and the vehicle coordinate frame are often estimated within the GPS/IMU filter.

Several companies[1] now produce commercially available GPS/IMU solutions. At the high-cost end of solutions, the Applanix POS LV 610 PP uses fiber optic gyroscopes in its IMU. It can provide 2 cm horizontal and 5 cm vertical positioning accuracy when a GPS signal is available, while also maintaining 10 cm horizontal and 7 cm vertical positioning accuracy during a 1-minute GPS signal outage. The lower-cost CNS-5000 solution from KVH also uses fiber optic gyroscopes in its IMU, and can provide similar accuracies when a GPS signal is available, but only 8.4 m horizonal accuracy after a 1-minute GPS signal outage. Both these solutions are above the cost suitable for introduction in production automobiles. For production automobiles, MEMS-based IMUs are a viable alternative. Currently, Gladiator Technologies offers the MEMS-based GPS/IMU solution LandMark 20. However, its horizontal accuracy is only 2.5 m when a GPS signal is available. In the LandMark 20, the integration of the MEMS-based IMU aids in GPS reacquisition, which is less than 1 sec, after a GPS signal outage. By compensating for such signal-loss situations, the accuracy of geo-spatial positioning between vehicles increases, and tighter tolerances in cooperative safety applications can be achieved.

2.2.3 Vehicle sensors

As evidenced with the integration of GPS and IMU sensors, future cooperative safety systems will benefit from the fusion of sensors other than just GPS and wireless communications. These sensors include autonomous ranging sensors such as radars, laser radars, and cameras, but also on-board vehicle sensors such as steering wheel angle sensors and wheel speed sensors.

Besides improving vehicle positioning, on-board sensors also aid in vehicle path prediction. As discussed later in this chapter, the combination of vehicle sensor measurements with appropriate vehicle motion models allows accurate path prediction for significant time horizons. These predictions will be the core of many cooperative collision avoidance systems. The challenge facing on-board sensors is the tradeoff between cost and precision. Most production systems use sensors (for example, accelerometers) that meet the minimum specification requirements for their primary application. Low-cost production sensors often provide less accurate or noisier measurements to the feedback system. In vehicular path prediction, these noises can have a more pronounced effect on false alarms and missed detections of a VANET-based safety system, because they are propagated for multiple seconds

---

[1] See Applanix (http://applanix.com), Oxford Technical Solutions (http://www.oxts.com), KVH Industries (http://www.kvh.com), and Gladiator Technologies (http://www.gladiatortechnologies.com).

through a vehicle motion model. Thus, a V2V application designer must consider the noises introduced by the available vehicle sensors. If a certain sensor's noise is too detrimental to the path prediction, the designer may investigate a way to remove the sensor from the prediction routine or insist that this sensor have its precision improved while keeping production costs feasible.

Ranging sensors such as radars, laser radars, and ultrasonic sensors will play a crucial role in detecting and tracking non-communicating vehicles or objects. Low-cost vehicle positioning and wireless communications allow many features of the driving environments to go undetected. As a result, V2V safety applications will find their first implementation in controlled environments such as freeways. When V2V safety applications move into urban environments, the fusion of communication data with autonomous ranging data will be necessary to develop systems with acceptable performance.

Finally, computer vision sensing could supplement vehicle positioning in VANET safety systems. Similar to the goals of fusions between GPS and IMU sensors, computer vision can aid during times of satellite signal loss. For example, when traveling through tunnels, computer vision can be used to lane-match within the tunnel. Furthermore, even during times of sufficient satellite coverage, vision can be fused with navigational maps for map-matching localization and also fused with autonomous ranging sensors for neighborhood mapping.

2.2.4 On-board computation platforms

The confluence of wireless communications, global positioning, and vehicle sensors relies on computational platforms capable of processing large amounts of sensor data, high-bandwidth communications, highly integrated sensor fusion filters, and complicated path prediction and application logic. Similar to on-board sensors, computation platforms in the automotive domain pose a tradeoff between cost and performance. Although one benefit of pursuing cooperative safety system technology is the relatively low cost of integrating global positioning with wireless radios, if the computational platforms necessary to execute the software are too expensive for the automotive manufacturer, the technology might never make it to the market.

2.3 Cooperative System Architecture

The appropriate software architecture for cooperative driving must accommodate applications that can vary dramatically between safety and mobility objectives. The architecture must be modular and flexible enough to allow various degrees of application complexity, communication frequency, sensor capabilities, and control authority. For example, safety applications often require high-frequency or event-driven communications with other vehicles, while mobility applications might only require low-frequency communications with infrastructure. Although vehicular positioning and wireless communications are the fundamental technologies that enable cooperative behavior, the architecture must also accommodate autonomous sensors such as radar, lidar, and computer vision.

A cooperative driving software architecture exhibiting a layered and feedback structure is shown in Figure 2.1. Cooperation and path planning blocks fall under guidance. Localization and mapping blocks undertake navigation duties, while the safety and regulation blocks are issues dealt with under control. For safety applications, the navigation blocks determine the quality of the applications by building the knowledge of the vehicle's own state, the surrounding environment, and the vehicle's position within this environment.

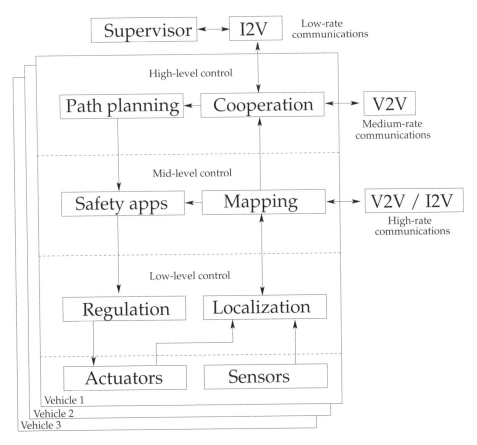

Figure 2.1 A software architecture for cooperative vehicular systems. Cooperation and path planning blocks are aided by low-rate and mid-rate communications. Safety and mapping blocks require high-rate communications and high-rate integration with localization and regulation blocks, which interface with vehicle sensors and actuators.

The importance of the layered architecture lies within the differing time constants of the layers and the respective rates of communication necessary to support each layer. This comparison is given in Table 2.1. Operating with the largest time constant is the cooperation layer. The low-rate I2V communication might occur in

Table 2.1 Differences among the layers of the architecture illustrated in Figure 2.1.

Layer(s)	Time scale	Information span and spatial impact	Application category
Supervisor	Once every 5 minutes	System-wide behavior	Cooperative mobility
Cooperation and path plan	Once every 10 seconds	Between groups	Cooperative comfort
Safety and mapping	Once every second	Local neighborhood of vehicles (group)	Cooperative safety
Regulation and localization	Fraction of a second	Individual vehicle	Autonomous safety and comfort

the order of minutes, while the medium-rate V2V information between cooperating vehicles might occur in the order of seconds. At the cooperation layer, vehicles are sharing information and decision processes to arrive at desired behaviors (for example, speeds, positions, and orientations) within the neighborhood of vehicles. For example, this cooperation might involve formulation, merging, or splitting of groups of vehicles. The nature of cooperative driving applications possible at each time constant is reflected in Table 2.1. Cooperative mobility and comfort applications operate in the higher layers of Figure 2.1.

In general, cooperative safety applications operate within the safety and mapping blocks of the lower layer and require high-rate V2V communications. The constituents of the mapping block, which are path history, path predictions, and target classification, are discussed later in this chapter. While cooperation might occur in the order of seconds, safety and mapping require V2V information from other vehicles at 1–50 Hz in order to update relative maps. Operating beneath the safety and mapping blocks are the regulation and localization blocks. The localization block contains the vehicular positioning components described earlier. It interfaces with the vehicle's sensors, such as the GPS receiver and wheel-speed sensors. The regulation block, which interfaces with the vehicle's actuators at 10–100 Hz, realizes autonomous vehicle behavior, such as vehicle stability control (VSC) and PCS.

2.4 Mapping for Safety Applications

In this section, we investigate the mapping block within the architecture shown in Figure 2.1. For this discussion, each communicating vehicle will be referred to as a 'host' vehicle, and all its neighboring vehicles (both communicating and non-communicating) will be referred to as 'remote' vehicles. The achievable accuracy of the mapping of the surrounding environment, both vehicular and non-vehicular, is directly reflected in the performance of the cooperative safety applications. Vehicular mapping can be broken into three constituents, namely, vehicular path history, vehicular path prediction, and target classification.

Path histories and predictions are most accurately and efficiently constructed by the host vehicle and subsequently communicated to neighboring vehicles. Further, self-predictions eliminate the need of each remote vehicle redundantly estimating paths for all surrounding vehicles. Of course, path histories for remote vehicles can be constructed based upon recording information (for example, positions) received from every other remote vehicle. However, again, the host vehicle is best suited for compressing its past trajectory into a set of points or a path to efficiently transmit to neighboring remote vehicles. Sharing path histories is particularly important during first contact between two vehicles, when neighboring vehicles can immediately gain knowledge of each other's path history.

Path history refers to the recent path that a vehicle has traversed. The path history transmitted between vehicles might be limited in size by time or distance. In particular, a path history might contain position (for example, latitude and longitude) samples over the past 10 seconds or over the previous 200 meters. Sharing a path history primarily benefits those remote vehicles following the host vehicle. These remote vehicles can use the history to classify the lane in which the host vehicle is traveling and also to estimate the relative lateral offset between themselves and the host vehicle. Such classifications allow vehicles to identify only those imminent collisions that are within the same lane of travel. These classifications are only strengthened by fusion with existing road map data and map matching.

An advantage of path histories over path predictions is that they are known, within the uncertainties in past position estimates. As such, path histories can be reduced to representations that contain the trajectory within a desired tolerance. For example, straight-line path histories can be represented by just two time-ordered positions. In Toyota's 'Thin Bird Breadcrumb' proposal (Robinson 2005), a fixed set of past position measurements (or estimates) are condensed into a representative sample for transmission. Various other approaches have been proposed for generating similar representative samples.

The estimation of the future path is a function of vehicle model fidelity, current vehicle state estimates, and expected driver inputs over the prediction horizon. Integration with road map data, navigation systems, and map matching can provide an estimate for the expected future trajectory, but this estimate does not account for future driver decisions such as lane changes, braking, and accelerations. As a result, techniques for path prediction range widely in complexity, accuracy, and applicability. Path prediction for highly dynamic driving situations that invoke vehicle stability systems could be different from path prediction for traffic flow management.

Cooperative safety applications look for the mapping block to output relative information of vehicular and non-vehicular objects from which they can distributively make decisions concerning cooperation with other vehicles or autonomous control actions. Thus, the mapping block updates a map of relative vehicular objects given communicated past- and future-predicted trajectories, while also populating the map with other non-communicating objects detected from autonomous sensors.

In addition, auxiliary information on road geometry, speed limits, and traffic signal phases and timing, might be included in the representation of the map. The map of physical objects is relative to the current localized position and orientation of the host vehicle produced from the localization block of Figure 2.1 and described earlier in this chapter. When vehicles desire to share relative object maps, they can do so in the vehicle-relative coordinate frame (from which a receiving vehicle must transform the map using the current absolute position and orientation of the sender) or in the absolute Earth coordinate frame directly.

The building and updating of such maps relies heavily on the common global clock provided by GNSS. The timestamping of information shared between vehicles through communication is vital to accurate maps. When time discrepancies exist between vehicles, the maps built distributively by different vehicles might produce inconsistent and erroneous behaviors from the cooperative driving applications. Depending upon the nature of the cooperative driving applications, the mapping block might further classify both vehicular and non-vehicular objects into categories reflecting each object's relative position, speed, type, and confidence of classification. In summary, the mapping block and its classifications aim to transform the raw information, which was shared among communicating vehicles or obtained from autonomous sensors, into pertinent summaries upon which the applications can easily operate. The following subsections discuss, in more detail, vehicular path prediction that provides cooperative systems with their anticipatory behavior. These subsections begin with deterministic approaches, and follow with stochastic techniques.

2.4.1 Non-parametric path prediction

Non-parametric prediction refers to those routines that omit the use of vehicle-specific parameters. These routines can be run on any vehicle without any modification or tuning for vehicle model. Consequently, non-parametric routines are often preferred for path prediction. One simple non-parametric prediction approach is to use a route plan as determined by a navigation system. However, using a route plan for prediction only gives road-level accuracy, and this is inappropriate for most cooperative safety applications. A common, shorter-time-horizon prediction approach for vehicular applications is to use the unicycle model. This approach can be implemented easily on current production vehicles without additional sensor cost, because current generation traction control and vehicle stability systems provide the required measurements with sufficient accuracy. At the current time t, the radius R of the road curvature is estimated from the vehicle longitudinal speed v_x, which is measured by wheel speed sensors used for traction control, and the yaw rate ω, which is measured by a gyroscopic device used for vehicle stability control,

$$R = v_x/\omega. \qquad (2.2)$$

This estimate of road curvature is often low-pass filtered because of noisy yaw-rate sensors. The unicycle approach assumes that the vehicle will travel on a constant radius with constant longitudinal acceleration a_x for prediction horizon, T.

The distance d traveled along the arc is

$$d(T) = v_x T + 0.5 a_x T^2, \qquad (2.3)$$

and the angle, $\Delta\psi$, traversed over the prediction horizon is

$$\Delta\psi(T) = \frac{d(T)}{R}. \qquad (2.4)$$

The states x, y, and ψ are with respect to the earth-fixed coordinate frame, and v_x and a_x are with respect to the vehicle-fixed coordinate frame. The state x is the longitude with positive East, y is latitude with positive North, and ψ is the heading of the vehicle, measured positive counter-clockwise from the x-axis.

The relative change in position is

$$\Delta x(T) = R\sin(\Delta\psi(T))$$
$$= R\sin\left(\frac{d(T)}{R}\right), \qquad (2.5)$$

$$\Delta y(T) = R(1 - \cos(\Delta\psi(T)))$$
$$= R\left(1 - \cos\left(\frac{d(T)}{R}\right)\right). \qquad (2.6)$$

So, the predicted position at time $t + T$ is

$$x(t+T) = x(t) + \Delta x(T)\cos(\psi(t)) - \Delta y(T)\sin(\psi(t)), \qquad (2.7)$$
$$y(t+T) = y(t) + \Delta x(T)\sin(\psi(t)) + \Delta y(T)\cos(\psi(t)), \qquad (2.8)$$
$$\psi(t+T) = \psi(t) + \Delta\psi(T). \qquad (2.9)$$

2.4.2 Parametric path prediction

A parametric prediction routine relies on an assumed motion model for the vehicle dynamics. The accuracy of the resultant path predictions will be a function of the accuracy of the assumed motion model. This model will depend on vehicle geometry or other physical models, such as engine or tire dynamics. Depending on the application, parametric prediction models can range from simple two-state systems to complicated 40-state systems. For cooperative safety applications, the dynamics of a moving automobile are sufficiently modeled by a few (3–6) continuous-time, time-invariant, nonlinear equations. Rather than considering engine speeds, manifold air pressures, or suspension displacements, cooperative safety applications consider wheel speeds, wheel angles, and vehicle positions, as state variables. Similar to non-parametric approaches, the measurements necessary for a parametric path prediction are available from sensors already available on many production vehicles for vehicle stability and traction control purposes. A representative vehicle model is the classic bicycle (CB) model (Kiencke and Nielson 2000), which includes as vehicle parameters the mass M, yaw inertia J_z, and the perpendicular distances from the front and rear axles to the vehicle's center of gravity, l_f and l_r, respectively.

The equations of the CB model are

$$\dot{\mathbf{x}}^{CB} = \begin{bmatrix} \dot{x} \\ \dot{y} \\ \dot{\psi} \\ \dot{v}_x \\ \dot{v}_y \\ \dot{\omega} \end{bmatrix} = \begin{bmatrix} v_x \cos(\psi) - v_y \sin(\psi) \\ v_x \sin(\psi) + v_y \cos(\psi) \\ \omega \\ a_x[0] \\ f_1^{CB}(v_x, v_y, \omega, \delta[0]) \\ f_2^{CB}(v_x, v_y, \omega, \delta[0]) \end{bmatrix}, \quad (2.10)$$

where

$$f_1^{CB}(v_x, v_y, \omega, \delta) = \frac{-2(C_{\alpha f} + C_{\alpha r})v_y}{M v_x}$$
$$+ \left(\frac{-2(C_{\alpha f} l_f - C_{\alpha r} l_r)}{M v_x} - v_x \right) \omega$$
$$+ \left(\frac{2 C_{\alpha f}}{M} \right) \delta, \quad (2.11)$$

$$f_2^{CB}(v_x, v_y, \omega, \delta) = \frac{-2(C_{\alpha f} l_f - C_{\alpha r} l_r) v_y}{J_z v_x}$$
$$+ \left(\frac{-2(C_{\alpha f} l_f^2 + C_{\alpha r} l_r^2)}{J_z v_x} \right) \omega$$
$$+ \left(\frac{2 C_{\alpha f} l_f}{J_z} \right) \delta, \quad (2.12)$$

where $C_{\alpha f}$ and $C_{\alpha r}$ are the linear front and rear tire cornering stiffnesses, respectively. The front tire force F_{y_f} is linearly related to the front wheel slip angle α through the gain $C_{\alpha f}$. The lateral velocity state v_y is with respect to the vehicle-fixed coordinate frame. The inputs to the model are the vehicle longitudinal acceleration a_x and the front-wheel angle δ, and are assumed constant over the prediction horizon.

Similar to most parametric vehicle models considered for cooperative safety applications, the CB model is nonlinear, non-stiff, and smooth. As a result, vehicular path predictions employing such nonlinear models can be efficiently computed using numerical integration (NI) rather than linearized, discretized, and propagated using matrix algebra (Caveney 2007a). Table 2.2 shows numerical integration to be more computationally efficient and accurate than a linearization approach over various time horizon predictions. The requirements for the non-parametric unicycle approach is also included for comparison in Table 2.2. The non-parametric approach provides path prediction accuracy similar to the numerically integrated CB model for significantly less computational effort. Obviously, there is a tradeoff between the computational effort and model fidelity. However, one benefit of the model-based approach that is not illustrated by Table 2.2 is that expected driver inputs, such as longitudinal accelerations a_x and steering wheel angles, can be easily incorporated into the predictions. If coarse estimates of the expected steering wheel angles were available and extracted from a map of the upcoming road curvature, then the path prediction accuracy for a parametric approach would significantly increase. This is especially true for long prediction horizons. Unfortunately, map information was not available for the data used in Table 2.2. Furthermore, discrete driver maneuvers such as lane changes could have their expected steering wheel angle profile incorporated into the model-based prediction. Similar driver maneuvers and their associated inputs can be incorporated into non-parametric approaches, but understanding the relationships between steering wheel angles and subsequent yaw rates that must be included in the prediction will nonetheless require some knowledge of a dynamic vehicle model.

Table 2.2 Root mean square (RMS) predicted position error (in meters) and computation time (in milliseconds) comparison for the three deterministic prediction approaches. The vehicle was traveling around a 3 km test course at an average speed of 55 km/h. The non-parametric unicycle model requires the fewest computations. The numerically integrated classical bicycle (CB) is more accurate and requires fewer computations than a linearized CB model.

Metric	Approach	Prediction horizon				
		1 s	2 s	3 s	4 s	5 s
RMS error (m)	Non-para. unicycle	0.31	1.17	3.28	7.05	12.6
	Linearized CB	0.60	2.15	5.31	10.4	17.2
	Integrated CB	0.31	1.12	3.09	6.62	11.8
Comp. time (ms)	Non-para. unicycle	0.003	0.004	0.004	0.004	0.004
	Linearized CB	9.12	17.6	27.8	34.7	42.8
	Integrated CB	7.11	11.1	14.6	18.9	22.2

Successful numerical integration includes choosing a method that is appropriate for the stiffness of the system's dynamics while providing the desired accuracy with the fewest number of function, $\dot{x}(t) = f(x(t), u(t))$, evaluations. Variable step-size methods are attractive because they take large integration timesteps whenever possible. In contrast, a linearization approach might choose a sufficiently small sample time T_s in an attempt to capture any regions with large derivative behavior, but will be slow with unnecessary computations when large steps are possible. In Table 2.2, where T_s was 0.2 seconds, path predictions were less accurate than the NI approach and required more computations than the NI approach. Using a larger value of T_s in the linearization approach could result in fewer computations than the NI approach, but lesser accurate path predictions should be expected (see Caveney 2007a). Furthermore, the average computation time of the linearization approach appears linear with respect to the prediction horizon T. In Table 2.2 the approximate slope is 9 ms per second of prediction horizon. Conversely, the average computation time of the NI approach appears affine with an approximate slope of 4 ms per second of prediction horizon, and a translational offset of 3 ms overhead for any prediction.

In the automotive domain, the variable step-size NI approach can quickly compute a prediction with an arbitrary length prediction horizon when the vehicle is traveling in a straight line, while requiring more computational effort during curving vehicular behavior. For specific situations, such as short prediction horizons or low vehicle speeds, a linearization approach may be appropriate. In the case of short prediction horizons, the number of timesteps (T/T_s) will be small, perhaps resulting in fewer computations than would be necessary with the NI approach. In the case of low vehicle speeds, numerical integrators can take longer to propagate the solution because the vehicle speed often appears in the denominator of vehicle models, which results in more iterations per integration in order to satisfy the specified tolerance. However, cooperative safety applications often require 3- to 5-second path predictions from vehicles traveling at speeds reasonably far from zero and roughly on straight trajectories. Therefore, in general, the NI approach is more efficient than the fixed-step-size approach taken by linearization. Variable-step-size linearization approaches, where T_s is adapted to the nature of recent vehicle behavior, could make linearization approaches more computationally feasible.

Regardless of the approach chosen to propagate a path prediction, the model choice will be dictated by the requirements of the applications that use this prediction. As Table 2.2 demonstrates, the unicycle model produces accurate 1- to 5-second predictions with few computations when the vehicle is traveling under dry road conditions. However, the unicycle approach will be inaccurate if the vehicle is spinning out of control on icy road conditions. This limitation is because the yaw rate measurement is no longer inversely proportional to road curvature, as given in Equation (2.2). A model that incorporates a relationship between the vehicle's tires and the road surface will be necessary if an application is designed to transmit vehicle path predictions in extreme handling scenarios such as this. The wheel slip angle, wheel longitudinal slip, and vehicle slip angle should then be considered. Furthermore, during these extreme scenarios, higher-fidelity models and their state estimates might already be incorporated into autonomous control systems, such as VSC, and therefore available for accurate short-time-horizon predictions. Nevertheless, for cooperative systems in general, the unicycle or similar model will provide sufficiently accurate predictions for most cooperative collision avoidance applications. This is because extreme operating conditions are infrequent and not within the target crash scenarios of cooperative systems. Cooperative safety applications target longer ranges, and subsequently longer time horizons, between vehicles than do autonomous sensors and their respective applications such as PCS.

In summary, the design of a path prediction routine involves various tradeoffs in computational complexity, prediction accuracy, incorporation of a-priori knowledge of driver behavior, sensor quality, and model fidelity. The choices made by the designer must reflect the requirements of the cooperative safety applications under consideration. Only applications that require precise predictions should warrant necessary high-fidelity models and additional computational resources. Furthermore, any limitations imposed by the available sensor quality must be reflected in the model choice and expected prediction accuracy. An accurate model with poor sensors could be less useful and more computationally expensive than a simple model with accurate sensors. Finally, depending on computational complexities and resources, it is possible that different path prediction methods be used to produce multiple predictions for different time horizons, which are all included within a single communication. Thus, vehicles that receive these predictions can use the one that is most appropriate for their application's time horizon.

2.4.3 Stochastic path prediction

The noisy nature of sensor measurements used for path prediction suggests that uncertainty should be incorporated into the prediction. For example, a statistical analysis could be applied to the unicycle model to understand how uncertainties propagate from yaw rate and vehicle speed measurements to the path predictions. A common approach for stochastic prediction involves propagation of the fixed-timestep, discrete, Kalman filter equations without measurement correction. This approach is called Kalman prediction, and produces both a stochastic path prediction and also associated covariance matrices. In terms of vehicular path prediction, the application of Kalman prediction to four different models is shown in Huang and Tan (2006). The Kalman prediction approach can equally be applied to discretized linear systems or linearized nonlinear systems that are then discretized. In the latter case, the Kalman prediction involves the extended Kalman filter (EKF) equations which invoke the Jacobians (first-order EKF) and possibly the Hessians (second-order EKF) of the nonlinear system. The linearization and discretization might take place multiple times over the prediction horizon, depending on the nonlinearities present within the system model and the horizon duration. These techniques are computationally heavy because of the large number of matrix computations required not only in the prediction, but also in the real-time discretization of the continuous-time dynamics. Again, the computational requirements will be a function of the chosen sampling time T_s. An analysis of the computations required to process a Kalman-like parametric path prediction is given in Table 2.3.

To produce both the stochastic path prediction and its associated covariance matrices, numerical integration can be combined with the unscented transform (UT). This combination is termed the UT-NI approach (Caveney 2007b). The unscented transform is a method for calculating the statistics of a random variable which undergoes a nonlinear transformation (Julier and Uhlmann 1997). The intuition behind the UT is that it is easier to approximate a Gaussian distribution than it is to approximate an arbitrary nonlinear function or transformation. In contrast, the EKF approximates a nonlinear function using linearization, and this approximation can be inaccurate when the models have large nonlinearities over short time periods.

Consider propagating an n-dimensional random variable $\mathbf{x}$ with mean $\bar{\mathbf{x}}$ and covariance $\mathbf{P_x}$ through a nonlinear function $\mathbf{y} = f(\mathbf{x})$. To calculate the statistics of $\mathbf{y}$, the UT generates $2n+1$ deterministic points (known as sigma points) $\mathcal{X}_i$ with corresponding weights W_i. The sigma points are defined as,

$$\mathcal{X}_0 = \bar{\mathbf{x}} \tag{2.13}$$

$$\mathcal{X}_i = \bar{\mathbf{x}} + (\sqrt{(n+\lambda)\mathbf{P_x}})_i \quad i = 1, \ldots, n \tag{2.14}$$

$$\mathcal{X}_i = \bar{\mathbf{x}} - (\sqrt{(n+\lambda)\mathbf{P_x}})_{i-n} \quad i = n+1, \ldots, 2n \tag{2.15}$$

Table 2.3 Root mean square (RMS) predicted position error (in meters), 98th percentile predicted position inclusion, and computational time (in milliseconds) comparison for two stochastic prediction approaches. The vehicle was traveling around a 3 km test course at an average speed of 55 km/h. The Kalman prediction approach requires fewer computations, but is less accurate than the UT-NI approach in both mean and covariance.

Metric	Approach	Prediction horizon				
		1 s	2 s	3 s	4 s	5 s
RMS error (m)	Kalman	0.56	2.07	5.26	10.4	17.4
	UT-NI	0.32	1.13	3.07	6.08	8.46
Percentage in 98% ellipse	Kalman	98.6	87.6	76.1	58.7	46.0
	UT-NI	100	98.4	88.9	77.2	73.6
Comp. time (ms)	Kalman	16.5	33.1	47.1	62.1	77.3
	UT-NI	84.6	133	176	217	264

where $(\sqrt{(n+\lambda)\mathbf{P_x}})_i$ is the ith row (or column) of the matrix square root. These sigma points are propagated through the nonlinear function,

$$\mathcal{Y}_i = f(\mathcal{X}_i), \quad i = 0, \ldots, 2n. \tag{2.16}$$

In the case of combining numerical integration and the unscented transform for path prediction, it is this propagation through the nonlinear function that will be performed by numerical integration methods (Caveney 2007b). For the six-state CB model, this will require $2(6) + 1 = 13$ integrations. The mean, $\bar{\mathbf{y}}$, and covariance, $\mathbf{P_y}$, of $\mathbf{y}$ are approximated by a weighted sample mean and covariance of the propagated sigma points, $\mathcal{Y}_i$,

$$\bar{\mathbf{y}} \approx \sum_{i=0}^{2n} W_i^{(m)} \mathcal{Y}_i \tag{2.17}$$

$$\mathbf{P_y} \approx \sum_{i=0}^{2n} W_i^{(c)} (\mathcal{Y}_i - \bar{\mathbf{y}})(\mathcal{Y}_i - \bar{\mathbf{y}})^T \tag{2.18}$$

where the weights are defined by

$$W_0^{(m)} = \frac{\lambda}{n+\lambda} \tag{2.19}$$

$$W_0^{(c)} = \frac{\lambda}{n+\lambda} + (1 - \alpha^2 + \beta) \tag{2.20}$$

$$W_i^{(m)} = W_i^{(c)} = \frac{1}{2(n+\lambda)}. \tag{2.21}$$

Here $\lambda = \alpha^2(n+\kappa) - n$ is a scaling parameter. α determines the spread of the sigma points around $\bar{x}$ and is typically set to $1e-3$. κ is a secondary scaling parameter which is usually set to 0, and β is used to incorporate prior knowledge of the distribution of $\bar{x}$ (for Gaussian distributions, a value of 2 for β is optimal) (Wan and Van der Merwe 2000).

Table 2.3 shows a comparison between the UT-NI approach and the Kalman prediction approach. The superiority of the UT-NI approach is particularly apparent during long prediction horizons. A more accurate prediction is a result of maintaining the nonlinear dynamical equations and only applying linearizing approximations on the state derivatives within the integration technique. It is particularly interesting to note that the means of the (stochastic) predicted position using the UT-NI approach can be more accurate in terms of RMS errors than the results shown in Table 2.2, where the nonlinear CB model is simply integrated ahead using the (deterministic) initial condition. The superior prediction capabilities of the UT-NI approach over the Kalman prediction approach are further illustrated by the second metric included in Table 2.3. This metric shows the percentage of actual positions that fall within the 98-percentile ellipse, which is a function of the prediction covariance, of the predicted position.

In terms of computation requirements, Table 2.3 shows a significantly longer execution time using the UT-NI approach over the Kalman prediction approach. However, a particular property of the UT-NI approach, which was not exploited in this comparison, is that the individual sigma points could be propagated in parallel. In Table 2.3, it can be seen that, for a given prediction horizon T, the UT-NI approach takes approximately 13 times as long as the NI approach in Table 2.2. Again, this is expected because the six-state CB model will require $2(6) + 1 = 13$ integrations. Conversely, the Kalman prediction must be executed sequentially because each linearization is a function of the previous state estimate, which is updated every T_s seconds during the prediction horizon T. A parallel-processing architecture for propagating the sigma points could result in computation times for the UT-NI approach that are significantly shorter than those of the Kalman prediction approach. These shorter times should be in the vicinity of the computation times shown in Table 2.2, with some additional time required for the sigma point generation and combination steps.

2.5 VANET-enabled Active Safety Applications

The combination of global navigation satellite systems (GNSS) and wireless communication technologies has stimulated research by automotive manufacturers and academia into cooperative driving applications. These applications include safety applications, which aim to mitigate vehicular collisions, such as rear-ending; mobility applications, which look at increasing traffic flow through information sharing, such as road conditions; and comfort applications, which aim at reducing the driver's workload, such as cooperative adaptive cruise control.

Several industry/government consortiums strive to identify which vehicular safety applications (and related technologies) will provide the greatest safety benefits. These organizations include the Crash Avoidance Metrics Partnership (CAMP) in the USA, the Car2Car Communication Consortium in Europe, the Advanced

Safety Vehicle (ASV) Project in Japan. For example, the deliberations between the NHTSA, the US Department of Transportation (USDOT), and the Vehicle Safety Communications Consortium (VSCC) of CAMP have identified eight such applications (CAMP Vehicle Safety Communications Consortium 2006). The near-term applications are traffic signal violation warning, curve speed warning, and emergency electronic brake lights. The mid-term applications are pre-crash warning, cooperative forward collision warning, left turn assistant, lane change warning, and stop sign movement assistance. Improved positioning accuracy is the key differentiator between the mid-term applications and those that will be introduced first to market.

The communication requirements of these eight safety applications are shown in Table 2.4. Note that communication frequency is in the range of 1–50 Hz, and the maximum communication range is 50–300 meters. Further, high-level data element requirements are specified. Several of these data elements, such as position and heading, are needed by multiple applications. These messages can be efficiently composed using a message dispatcher (Robinson et al. 2007), which results in messages whose payload size ranges from 25 bytes to several hundred bytes. However, the small 25-byte messages, which are often called heartbeat messages, are the most common and comprise a significant fraction of all messages sent.

Cooperative safety applications consist of three phases when interfacing with the vehicle and its driver. These phases are informational, warning, and automated control. The informational and warning phases include the driver in the control feedback loop. The informational phase must not distract the driver while providing information about the current driving situation, whereas the warning phase must alert the driver of an impending critical situation. The automated control phase removes the driver from the loop and directly controls the vehicle's actuators to avoid the critical situation. Automated control may be appropriate in cases when the driver's response time would inhibit mitigation of the critical situation. Safety applications that incorporate autonomous control over the vehicle should substantially reduce vehicular collisions and fatalities, but face greater scrutiny from driver acceptance and litigation standpoints.

When a designer must implement a set of cooperative safety applications, path prediction routines must be chosen that satisfy the minimum requirements of all the applications. It is desirable that the software architecture detailed above contains only one path predictor within its mapping block, from which each application gets its vehicle's path prediction, if necessary. However, as mentioned previously, it is possible that different predictions be used for different time horizons. Thus, if one application requires a road-level path prediction and another application requires a lane-level path prediction, the system can choose to perform only the lane-level path prediction and use it for both applications or do two separate predictions. This choice depends on the available computational resources. In terms of path history, its deterministic nature should allow for a single representation to meet the requirements of all applications sufficiently. Finally, multiple applications could require target classifications of remote vehicles. Again, this target classification should only be performed once, and its results made available to all applications.

Table 2.4 Eight high-priority cooperative vehicular safety applications as chosen by the National Highway Traffic Safety Administration and the Crash Avoidance Metrics Partnership (CAMP Vehicle Safety Communications Consortium 2006).

Application	Comm. type	Freq.	Max. latency	Data transmitted	Range
Traffic signal violation	I2V	10 Hz	100 ms	Signal phase, timing, position, direction, road geometry	250 m
Curve speed warning	I2V	1 Hz	1000 ms	Curve location, curvature, slope, speed limit, surface	200 m
Emergency brake lights	V2V	10 Hz	100 msec	Position, heading, velocity, acceleration	200 m
Pre-crash sensing	V2V	50 Hz	20 msec	Vehicle type, position, heading, velocity, acceleration, yaw rate	50 m
Forward collision	V2V	10 Hz	100 msec	Vehicle type, position, heading, velocity, acceleration, yaw rate	150 m
Left turn assist	I2V or V2V	10 Hz	100 msec	Signal phase, timing, position, direction, road geometry	300 m
Lane change warning	V2V	10 Hz	100 msec	Position, heading, velocity, acceleration, turn signal status,	150 m
Stop sign assist	I2V or V2V	10 Hz	100 msec	Position, velocity, heading, warning	300 m

The remainder of this chapter presents the concepts of operation for various applications enabled through VANET systems. These applications fall within three categories; vehicle-to-vehicle (V2V), infrastructure-to-vehicle (I2V), and pedestrian-to-vehicle (P2V). Although most prototype implementations of V2V applications only warn the driver of imminent collisions, the use of autonomous vehicle actions such as braking and steering are increasingly likely in future production systems.

In the applications below, the I2V application is expanded upon in more detail than others because it relies heavily on the path prediction of the host vehicle. Also, it is expected to be one of the first cooperative safety applications to be introduced to production vehicles because it requires low penetration rate to be effective in reducing vehicle-related fatalities. Finally, it is one application where automated vehicle control is likely to be accepted by drivers.

2.5.1 Infrastructure-to-vehicle applications

Intersection violation warning

The intersection violation warning (IVW) application warns the driver when violating a red light seems imminent, as shown in Figure 2.2. The application uses I2V communications and path prediction of the host vehicle. Acquiring dynamic information such as the light phase and light timing is the principal advantage of the communication link.

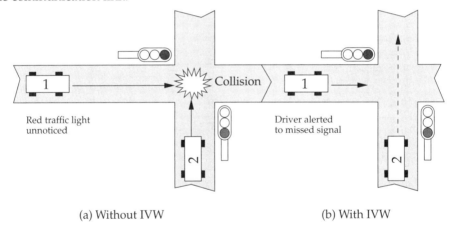

(a) Without IVW (b) With IVW

Figure 2.2 (a) Severe collisions result from red lights being violated by inattentive drivers. The intersection violation warning (IVW) application aims to reduce such violations. Without IVW, the inattentive driver of vehicle 1 runs a red light and collides with vehicle 2. (b) With IVW, the driver of vehicle 1 is alerted of the impending violation and able to bring their vehicle to a stop before the intersection.

In the USA, a roadside unit co-located with a traffic light controller will broadcast traffic light information including its location, light phase, light timing, and intersection geometry. Vehicles approaching the intersection compare this information with their projected trajectories and determine whether a signal violation is imminent. If so, the driver is alerted. A message is also sent from the vehicle to the traffic light and surrounding vehicles indicating that a warning has been issued.

Positioning requirements for the IVW application require lane-level accuracy at a minimum because traffic signal information varies over the various lanes arriving at an intersection. Each approach can have different phases and timing. Thus, to avoid false alarms and missed detections, accurate absolute positioning is necessary to map match the vehicle's position with an approach. If available, this map matching could be enhanced by differential GPS information included in messages sent from the infrastructure.

In terms of path prediction, the IVW application also requires lane-level minimum accuracy. The application will use 3- to 5-second path predictions to determine whether the vehicle is predicted to enter the intersection after the light for that particular approach has turned red. Fortunately, most intersections have straight approaches, making unicycle or bicycle models appropriate. In previous work, a 4-second path prediction was performed with the CB model (Caveney 2007a). This allowed naive drivers approximately 3.5 secs to react to violation warnings and bring their vehicles to a stop before the intersection. Recently, automated braking by prototype vehicles has been implemented when a violation was detected and the driver did not respond to audible and visual warnings (15th World Congress on Intelligent Transport Systems 2008).

Difficulties in path prediction arise when additional turn lanes appear as the vehicle nears the intersection. Transitional maneuvers to enter these lanes should be incorporated into the path prediction when a change of approach is detected. In Japan, the communication of upcoming traffic signal information will be made by approach-specific infrared beacons. This makes the map matching and path prediction significantly easier, and so a simple linear kinematic path prediction can be used.

The quality of vehicle sensors necessary will be a tradeoff between cost and false alarms or missed detections. The size of the intersection geometries allows for relatively large path prediction errors concerning whether the vehicle will be in the intersection after the signal for that approach has turned red. Therefore, accelerometer and wheel speed measurements available on current production vehicles will likely be sufficient. Conversely, efforts on improving the GPS absolute positioning will be necessary for map matching. Furthermore, accurate modeling of driver braking behavior is necessary to determine when drivers intend to brake in the future versus when they misinterpret or misjudge the upcoming signal status and should be warned.

2.5.2 Vehicle-to-vehicle applications

Electronic brake warning

The electronic brake warning (EBW) application alerts the driver when a preceding vehicle performs a severe braking maneuver. The application is particularly useful for the host vehicle when the view of the braking vehicle is blocked by other vehicles. The application operates by augmenting the rear brake lights with a system that leverages the non-line-of-sight advantage of VANET systems to help prevent rear-ending accidents.

In Figure 2.3, when vehicle 1 brakes sharply, it produces an event-based message that is broadcast indicating it is undergoing severe braking. Surrounding vehicles that receive the message must then discern whether the event is relevant. They can ignore warnings from vehicles traveling behind, far ahead, or in the opposite direction.

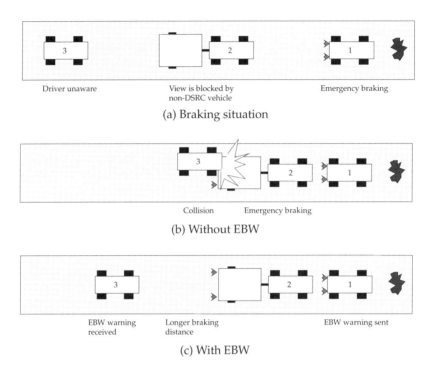

Figure 2.3 (a) Brake lights are often difficult to see if there is a blocking vehicle. The electronic brake warning (EBW) application informs following vehicles of severe braking maneuvers of preceding vehicles within their lane. (b) Without EBW, the driver of vehicle 3 has no time to react to the delayed braking of vehicle 2. (c) With EBW, vehicle 3 is informed of the braking of vehicle 1 before the driver of vehicle 2 has begun braking.

The EBW application operating on a host vehicle needs path histories with lane-level minimum accuracy from remote vehicles. These path histories can be sufficiently computed from DGPS receiver measurements. The host vehicle must then compare its current and future path with the path history provided by the braking vehicle. Depending on the length of the path history, the host vehicle may use various combinations of its current position or its predicted position to determine whether the position of the braking vehicle poses a potential threat.

Although the past position of the preceding braking vehicle may be co-located with the current position of the host vehicle, it does not necessarily mean that both vehicles will be co-located in the future. Its own path prediction also requires lane-level minimum accuracy, and as such, the quality of vehicle sensors necessary, besides the DGPS receiver, may be low and a vehicle model may or may not be necessary. If a linearization approach is chosen for path prediction, a sample time (T_s) of 1 second should be sufficient for moderate (2- to 4-second) prediction horizons.

The event-based nature of the EBW application requires reliable communications between vehicles during all times. If the emergency-braking message is not received from the remote vehicle by the host vehicle, the application is ineffective and the traffic situation reverts to the default behavior of the drivers reacting only to visible braking vehicles. Furthermore, a long (greater than 200 m) reliable communication range is beneficial, but not necessary for the EBW application to be effective.

On-coming traffic warning

The on-coming traffic warning (OTW) application alerts the driver of on-coming traffic during overtaking maneuvers that require the driver to cross into the opposite lane of travel. The application operates through accurate relative positioning between vehicles. In particular, the lateral (across) distance between two vehicles that are approaching each other must be predicted. This OTW application leverages the long range advantage of VANET.

In Figure 2.4(a), vehicle 1 misjudges the distance to vehicle 2 while overtaking vehicle 3. Without OTW, Figure 2.4(b) shows that vehicle 1 may have a near collision with vehicle 2, due to this misjudged distance. However, with OTW, vehicle 1 is informed of a potential collision with vehicle 2 and chooses to delay its overtaking maneuver until after vehicle 2 has passed.

The OTW application requires accurate path predictions from one or both the host and remote vehicles. For example, the host vehicle can use the path prediction of the remote vehicle to determine the relative lateral distance between its current position and the predicted path of the oncoming remote vehicle. Similarly, the host vehicle can use its own path prediction and the current position of the oncoming remote vehicle to determine this relative lateral distance. Therefore, precise positioning and path predictions are required for such an application. The potential positioning errors from a DGPS receiver might lead to a false head-on collision indication. GPS/INS integrated systems or RTK positioning systems should be incorporated. The path prediction horizons have to be long, preferably up to 10 seconds. If a model-based prediction approach is used, it should require little computational effort while still providing sufficient accuracy over long duration. Furthermore, the communication range for the OTW application should be as long as possible to incorporate the lengths of the two approaching path predictions. Finally, the OTW application could benefit from accurate map matching of one or both of the host and remote vehicles, to eliminate false alarms and missed detections due to movements made by the host vehicle.

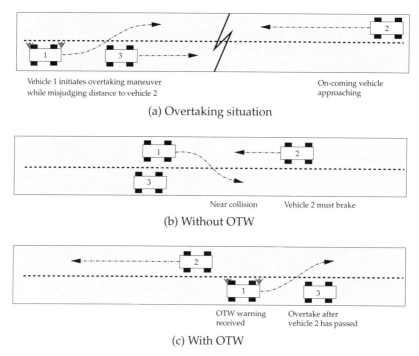

Figure 2.4 (a) Drivers often misjudge the distance to, and speed of, on-coming traffic when initiating an overtaking maneuver. The on-coming traffic warning (OTW) aims to mitigate head-on collisions and near collisions. (b) Without OTW, the driver of vehicle 1 causes a near collision with both vehicles 2 and 3, with vehicle 2 having to brake sharply. (c) With OTW, vehicle 1 can assess the time to collision (TTC) with vehicle 2 when the overtaking maneuver is initiated and inform the driver of vehicle 1 if the maneuver is ill-advised.

Vehicle stability warning

The vehicle stability warning (VSW) application alerts the driver of preceding vehicles that have recently required intervention by the vehicle stability control (VSC) system. The application leverages the range of the communication link to allow following vehicles to prepare for upcoming potentially hazardous driving conditions, such as ice or oil, through adaptation of their own VSC system parameters. Alternatively, the application can alert drivers to prepare for the upcoming road conditions by suggesting eyes forward or both hands on the steering wheel. To illustrate the VSW application, Figure 2.5(a) shows vehicle 1 encountering an icy patch of road that requires activation of its stability control system to maintain its curved path.

Figure 2.5(b) shows that without knowledge of the icy patch, vehicle 2 is unable to negotiate the curved path because it is traveling at a higher speed and subsequently collides with vehicle 3. However, with the VSW application, vehicle 2 is informed of

COOPERATIVE VEHICULAR SAFETY APPLICATIONS

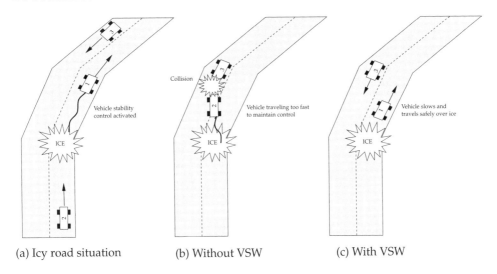

(a) Icy road situation (b) Without VSW (c) With VSW

Figure 2.5 (a) Predicting upcoming road surface conditions is difficult, especially in rapidly changing weather conditions. The vehicle stability warning (VSW) application shares information from preceding remote vehicles to following host vehicles to reduce the number of instances of vehicle control loss. Here, vehicle 1 encounters an icy patch of road that requires activation of its stability control system to maintain its curved path. (b) Without VSW, the driver of vehicle 2 is unable to negotiate the same curved path because it is traveling at a higher speed and subsequently collides with vehicle 3. (c) With VSW, the driver of vehicle 2 is informed of the activation of the stability control system of vehicle 1 and slows to maintain control of their vehicle through the curve.

activation of the stability control of vehicle 1 and slows to maintain control of the vehicle, as shown in Figure 2.5(c).

The positioning, path history, and path prediction requirements of the VSW application are similar to those of the EBW application. It is also an event-based application. However, the VSW application does not require path histories or path predictions to be implemented. The geo-spacial information alone indicating where a VSC system of a remote vehicle was activated can benefit a surrounding host vehicle. Furthermore, the geo-spacial information does not require DGPS-level accuracy. Navigation grade positioning systems are sufficient.

Lane change warning

The lane change warning (LCW) application warns the driver who is intending to perform a lane change when it is unsafe to do so. The application uses V2V communications and path prediction of remote vehicles. The application operates by leveraging the range of the communication link to predict immediate future conflicts should the driver complete the lane change, as shown in Figure 2.6. Path predictions are used to determine whether the area in the lane to the side to which the driver

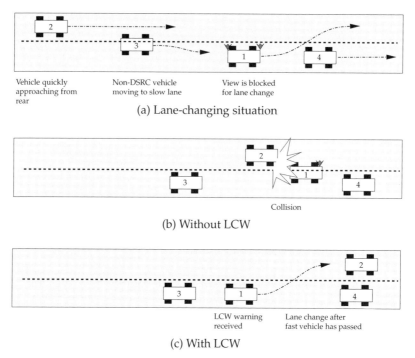

Figure 2.6 (a) The lane change warning (LCW) application warns drivers intending to lane change if their blind spot zone will be occupied by any following vehicles in the near future, thereby creating a possible rear-ending or side-swiping collision. (b) Without LCW, vehicle 1 proceeds with a lane-change without noticing vehicle 2 quickly approaching. (c) With LCW, when vehicle 1 initiates its lane change, vehicle 1 is warned of the existence of vehicle 2, and waits for vehicle 2 to pass before completing the lane change.

desires to move will be occupied during a 3- to 5-second time horizon. If so, the driver is alerted to the danger.

For the LCW application, shorter (1- to 3-second) prediction horizons are required. Relevant vehicles are traveling in the same direction and are in close proximity to each other. Moderate 1-meter accuracy of positioning methods and path predictions are necessary to eliminate both false alarms and missed detections. DGPS-level positioning should be sufficient, but precise vehicle accelerometers are necessary for accurate path predictions. In particular, because the relevant vehicles are in close proximity and traveling in the same direction, drivers will be sensitive to false alarms from poor path prediction because an attentive driver will be able to see the scenario for multiple seconds before warnings are falsely given.

2.5.3 Pedestrian-to-vehicle applications

Pedestrian in roadway warning

The pedestrian in roadway warning (PRW) application allows vehicles to perceive pedestrians that are potentially out of sight with respect to the vehicle. The concept relies on either personal transmission devices that are perhaps incorporated into individual cell phones or infrastructure-based pedestrian detection devices that communicate pedestrian positions or paths to the vehicle. With a communication link between pedestrians and vehicles, vehicles can obtain enhanced knowledge of the upcoming road, such as when to turn corners in busy urban centers. If the upcoming road will be occupied with pedestrians, the vehicle can warn the driver to pay particular attention. Similarly, pedestrians may receive broadcasts from vehicles that can be leveraged to avoid misjudged entries into the roadway.

Figure 2.7 illustrates a scenario for the PRW application. When the driver of vehicle 1 decides to execute its left turn between vehicles 2 and 3, they don't see the pedestrian crossing the road. Without the PRW application, a collision with the pedestrian is possible or, if the driver stops during the left turn, a collision with vehicle 3 is possible. With PRW, the driver of vehicle 1 is alerted of the pedestrian crossing the roadway and delays its left turn until after vehicle 3 and the pedestrian have both cleared the intersection.

Stringent requirements exist for pedestrian positioning and pedestrian path prediction within the PRW application. Pedestrians are highly erratic in behavior and could lead to many false warnings within the vehicle if pedestrian positions and paths are inaccurately computed and transmitted to the vehicle. On the vehicle side, positioning and path predictions with a lane-level minimum accuracy are adequate,

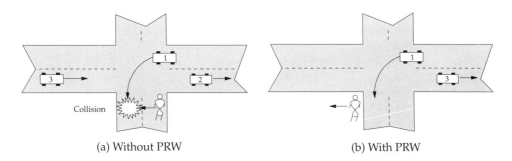

(a) Without PRW (b) With PRW

Figure 2.7 (a) Collisions with pedestrians often result when drivers rush their left and right turns. The pedestrian in roadway warning (PRW) application allows vehicles to predict collisions with pedestrians outside the field of view of the driver. Vehicle 1 decides to execute its left turn between vehicles 2 and 3 but, in its haste, doesn't see the pedestrian crossing the road. (b) With PRW, the driver of vehicle 1 is alerted of the pedestrian crossing the roadway. Vehicle 1 then delays its left turn until after vehicle 3 and the pedestrian have cleared the intersection.

but subdecimeter accuracy is preferred. Again, having subdecimeter accuracy might be unnecessary when the pedestrian behavior itself contains large uncertainties.

References

15th World Congress on Intelligent Transport Systems 2008 *11th Ave. Theater Toyota demo*. Technical report. Toyota Motor Corporation.

CAMP Vehicle Safety Communications Consortium 2006 *Vehicle safety communications project – final report*. Technical Report DOT HS 810 591, National Highway Traffic Safety Administration, Washington, DC.

Caveney D 2007a Numerical integration for future vehicle path prediction. *Proceedings of the American Control Conference*, pp. 3906–3912, New York, NY.

Caveney D 2007b Stochastic path prediction using the unscented transform with numerical integration. *Proceedings of IEEE Intelligent Transportation Systems Conference*, pp. 848–853, Seattle, WA.

Gao J 2006 GPS/INS/g sensors/yaw rate sensor/wheel speed sensors integrated vehicular positioning system. *Proceedings of ION GNSS*, pp. 1427–1439, Fort Worth, TX.

Gao J, Petovello MG and Cannon M 2006 Development of precise GPS/INS/wheel speed sensor/yaw rate sensor integrated vehicular positioning system. *Proceedings of ION National Technical Meeting*, pp. 780–792, Monterey, CA.

Huang J and Tan HS 2006 Vehicle future trajectory prediction with a DGPS/INS-based positioning system. *Proceedings of the American Control Conference*, pp. 5831–5836, Minneapolis, MN.

Julier SJ and Uhlmann JK 1997 A new extension of the Kalman filter to nonlinear systems. *Proceedings of Signal Processing, Sensor Fusion, and Target Recognition VI*, vol. 3068, pp. 182–193, Orlando, FL.

Kiencke U and Nielson L 2000 *Automotive Control Systems: For Engine, Driveline, and Vehicle*. Springer, New York, NY.

National Center for Statistics and Analysis 2008 *Traffic safety facts – 2006 data*. Technical Report DOT HS 810 809, National Highway Traffic Safety Administration, Washington, DC.

Petovello MG 2003 *Real-Time Integration of a Tactical Grade IMU and GPS for High-Accuracy Positioning and Navigation*. PhD thesis, University of Calgary.

Robinson C 2005 Toyota 'thin bird breadcrumb' proposal. *Toyota Internal Report*.

Robinson C, Caveney D, Caminiti L, Baliga G, Laberteaux K and Kumar P 2007 Efficient message composition and coding for cooperative vehicular safety applications. *IEEE Transactions on Vehicular Technology* **56**(6), 3244–3255.

Wan E and Van der Merwe R 2000 The unscented Kalman filter for nonlinear estimation. *Proceedings of IEEE Adaptive Systems for Signal Processing, Communications, and Control Symposium*, pp. 153–158, Lake Louise, Canada.

3

Information Dissemination in VANETs

Christian Lochert, Björn Scheuermann and
Martin Mauve

Heinrich Heine University, Düsseldorf, Germany

3.1 Introduction

While VANET safety applications typically require information only from a relatively limited geographical area – the vicinity of the car – this is not the case for many convenience and driving comfort applications. For example, navigation systems could make use of information on the current traffic situation in a larger surrounding. Another example is VANET-based parking guidance systems: these distribute information about the parking situation in an entire city. In this type of application, information is cooperatively collected and shared within a large area, such as a highway network or the road network of a city. The information might be on traffic intensity and travel times, free parking places, or any other parameter that can be observed by individual participants, and collected and distributed by a cooperative mechanism in the vehicular ad-hoc network.

Distributing information over long ranges in a VANET is a very challenging task. It can be fulfilled only when the inherent properties and limitations of VANETs are considered in all components of the system. Particularly relevant are (1) connectivity and (2) capacity constraints. Limited initial market penetration results in a very sparse network. Even once a higher penetration ratio is achieved, network connectivity will often be constrained, e.g., during low-traffic hours. Among other effects, limited connectivity directly affects the possible speed at which information can be transported over a VANET, and hence directly limits the up-to-dateness of the shared information that can be achieved.

Capacity constraints in VANETs are mainly due to the limitations of the shared wireless channel. It is intuitively evident that the capacity of the network will not suffice to provide detailed and continuously updated information on every small-scale geographical entity to all participants of the network. This is also underlined by known results on the capacity limits of wireless multihop networks, first and foremost the seminal results by Gupta and Kumar (2000). The limited capacity necessitates mechanisms to reduce the amount of information that needs to be transported through the network, by summarizing and aggregating individual data items.

Four central challenges have been identified in the literature on long-range data dissemination in VANETs:

- how to *obtain* the information, i.e., how to make local observations that can form a data basis for the application

- how to *transport* the information, that is, given that information is available at some point in the VANET, how to deliver it to interested parties in other parts of the network

- how to *summarize* measurements of the same or very similar parameters made by distinct VANET participants, to reduce redundancy when it is not possible to distribute all individual observations

- how to *aggregate* information on larger geographical entities, considering the fact that network capacity constraints do not allow for an arbitrarily detailed picture of distant regions.

A basic sketch of a dissemination-based system is shown in Figure 3.1. Information is collected from internal and external sources (i.e., from sensors in the car and potentially also from roadside infrastructure) and is stored in a local knowledge base. The contents of this knowledge base are made available to the application, and are furthermore shared and exchanged with other cars by means of wireless communication. The amount of data exchanged between cars is reduced by applying summarization and aggregation techniques to the data in the knowledge base.

Our discussion will roughly follow the structure given by the four questions posed above. The next section will look at issues of, and proposals for, local data generation, before we then consider protocol issues for data transport in Section 3.3. Sections 3.4 and 3.5 subsequently turn towards data representation and transformation aspects, for summarizing and aggregating the collected information, respectively. We conclude this chapter with a summary in Section 3.6.

3.2 Obtaining Local Measurements

Before a dissemination-based VANET application can start to propagate and process data, local observations (i.e., measurements of one or more local parameter(s)) need

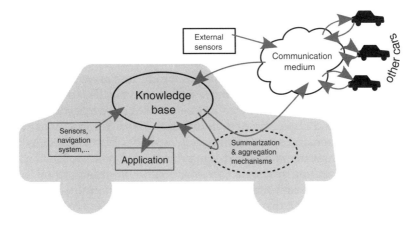

Figure 3.1 A dissemination-based VANET application.

to be made. In the discussion of data dissemination protocols and dissemination-based applications, the aspect of how these observations are made is often neglected – their existence is simply assumed, and the parameters to be disseminated are chosen in a more or less straightforward way. However, a closer look reveals that generating local observations that are suitable for specific applications is often not that straightforward, and the choice of parameters to be transmitted in the network deserves careful consideration. It heavily depends on the specific purpose, i.e., on the aim of the application, and it is intimately interrelated to other design parameters concerning the protocol, aggregation mechanisms, etc.

Sensors in vehicles The easiest way to obtain information is to use data provided by integrated sensors in the vehicles. Many such sensors are already available and are used for other electronic subsystems in the car. They provide information on environment parameters such as temperature, light intensity, position (GPS), speed, or road surface condition (ice, water, etc.). More and more often, sophisticated devices and processing logic are also able to provide information such as the distance to preceding and successive vehicles (e.g., from radar distance measurements), visibility distance, or traffic signs and traffic lights (through camera pictures and image recognition techniques).

Reading the speedometer or rev counter alone can already provide important information and may allow conclusions to be drawn about the current traffic situation. In order to improve the accuracy and granularity of the information, it is possible to combine information from multiple sources, a process often called *sensor fusion*. Nowadays, many vehicle sensors are already connected to the Controller-Area Network (CAN), specified by the International Organization for Standardization (2003, 2004), and sensor readings can be obtained in a standardized way via this channel. A system that integrates and combines data from different sources in order to gain insight into the current road situation is discussed by Ozbay et al. (2007). In the context of accurate localization, sensor fusion is discussed by Boukerche et al. (2008).

The navigation system as an information source The majority of the VANET convenience applications discussed so far are more or less directly related to a navigation system. Prime examples are again a distributed traffic information system for finding routes with short travel times based on the current traffic situation and a system for finding free parking places. From the perspective of information generation in VANETs, the fact that more and more vehicles are equipped with a navigation system means that more and more vehicles have a particularly powerful and sophisticated kind of 'sensor' at their disposal: a navigation system not only has quite accurate position and speed information available, but also detailed map data and information about the intended driving direction. Unlike, for instance, a 'bare' GPS receiver, a navigation system is therefore able to provide position information on a higher abstraction level: instead of geographical coordinates, a navigation system knows the road ID, driving direction, or even the lane (e.g., regular versus turning lane).

Distributed traffic information systems like SOTIS by Wischhof et al. (2003b), TrafficView by Nadeem et al. (2004), or CASCADE by Ibrahim and Weigle (2008) often use vehicle position and driving speed as the underlying primitive (which is then exchanged in more or less aggregated form). If the aim of the application is to optimize the route choice based on the current traffic situation in the road network, an alternative approach is to instead use travel times along road segments as the local observations, as suggested by Goel et al. (2004), Xu and Barth (2006a), and Lochert et al. (2008) – with the aid of a navigation system this parameter is straightforward to obtain locally. While individual (and quickly outdated) vehicle position information needs further, non-straightforward processing before it can serve as a basis for a road network routing algorithm, travel times along road segments can directly be used as an edge weight in the road network representation of a VANET-supported navigation system.

Both approaches may exhibit their individual benefits and drawbacks. However, aggregation mechanisms, dissemination protocols, the interpretation of the values by the application, etc., differ largely depending on whether highly dynamic, short-lived individual vehicle positions and speeds or more time-stable, cooperatively estimated average travel times are shared in the network – the impact of the choice on other system components is evident in this example.

External sensors It is conceivable to complement the built-in sensors in the vehicles with additional, external information sources. These external sensors might, for example, reside in the road pavement and count the number of passing-by vehicles and thereby generate raw data for an analysis of the current traffic situation. Parking spaces can be equipped with devices to sense whether they are currently occupied or not, an approach discussed by Piorkowski et al. (2006) and Markoff (2008). Often, it will not even be necessary to install new sensing equipment, because the data is already available. For example, parking meters may already know which parking places are occupied; induction loops are often present to sense passing-by cars; the traffic light controllers are aware of the current traffic light status.

Regardless of whether new equipment is installed in order to supplement information collected by the cars themselves or whether readily available, external information sources are used, a way to communicate this data to the VANET is necessary. Typically, this will happen by integrating external devices into the wireless network, such that passing-by cars can communicate with them and thereby obtain the respective information. Concrete scenarios of such applications are, for example, discussed by Caliskan et al. (2007; 2006).

The VANET device as a 'sensor' An interesting idea to complement the so far mentioned car-internal and -external information sources is to draw conclusions about the traffic situation from observations concerning the VANET itself. Many proposed VANET safety applications build upon periodic beacons. Such information messages, single-hop broadcast to all cars in the surrounding, can provide a fine-grained picture of the local environment, and therefore serve as a basis for an estimation of the local traffic density. Mechanisms in this direction are suggested, for instance, by Jerbi et al. (2007). Using VANET-based local information exchange to derive information about the local environment will, of course, be easier to implement once a substantial fraction of cars are equipped with VANET devices; until then, information acquired through periodic local presence announcements of VANET nodes will necessarily be incomplete.

Prediction of parameters All above-mentioned sensed parameters take only the current situation into account. However, Caliskan et al. (2007) point out that this will often lead to suboptimal results. They motivate this using a distributed application guiding cars to parking sites with free capacity. Due to a limited information propagation speed in the VANET, there will often be significant delay between the measurement and the arrival of the respective information at an interested car. Furthermore, a significant time span will typically elapse until the interested car is able to arrive at the parking place. A driver will, however, not be interested in the number of free parking places at some time in the past, but instead intends to maximize their individual probability that at least one parking place is free upon arrival at the parking site.

In their approach, Caliskan et al. assume that each parking lot consists of a number of parking places and has a central instance that is able to keep track of arrivals and departures (e.g., a fee payment terminal). They propose to not only monitor the current occupancy of each parking site, but also the current dynamics, expressed through an arrival rate and the average parking duration. With these additional parameters available, cars can make predictions about the probability of finding a free parking place at the respective parking lot at the estimated time of arrival.

The central insight is of broader interest than just for a parking guidance application; similar observations of course also hold for other applications. The aspect discussed by Caliskan et al. therefore generally underlines that a holistic design of a VANET dissemination application makes sense: the nature of the information, the means by which it is obtained, the way it is used by the application, and the way it is communicated from its point of origin to interested parties are all interrelated.

3.3 Information Transport

After obtaining information from different sensors, a local view of a car's surrounding can be created. However, not only the car that obtained this information will need this data, but also other vehicles should be able to adapt their behavior based on this information. It is thus important to distribute the information to the cars interested in it. In the following paragraphs we deal with the different methods for the *transport* of information. In order to avoid misconceptions we define the terms *broadcasting*, *flooding*, and *beaconing* as follows. *Broadcasting* is a *single-hop* transmission of a packet to all nodes within radio range of the sending node. It is often used as a primitive to implement *flooding*, which means distributing a packet over a range spanning multiple wireless hops. The receiving nodes rebroadcast the packet and thereby deliver it to all nodes in the network or to a subset thereof, for instance the nodes within a limited geographical region. *Beaconing* is the periodic transmission of information to all neighbors within radio range. In a sense beaconing is periodic broadcasting. One single packet transmitted by beaconing is called a *beacon*.

3.3.1 Protocols for information transport

Flooding and geocasting Well known from its frequent use in MANETs, one way to propagate information very fast is to use flooding. In a naive implementation every node that receives this information will simply rebroadcast it. To avoid infinite packet duplication, each node will broadcast a given packet at most once. In addition a time to live (TTL) counter may be used to limit the area where the packet is distributed. This naive approach will transmit a large amount of redundant packets, potentially leading to severe congestion. This is known as the 'broadcast storm problem' originally identified in a study by Ni et al. (1999). Many approaches have been proposed to deal with this problem.

In vehicular networks flooding is often used to disseminate traffic information messages to other vehicles. The general procedure is that a vehicle – the source – detects an incident or situation which should be communicated to other vehicles. The source thus starts flooding this information. In the following paragraphs we deal with different approaches that aim to limit the number of concurrent packets within the network. The common idea of all these methods is to influence the forwarding behavior of the vehicles, either by adapting the time when to forward the packet or by introducing rules on whether a given vehicle should forward the packet at all.

In the approach of Tonguz et al. (2007) it is proposed to adapt the broadcasts used to realize flooding depending on the density of the traffic. When the traffic is dense not every car needs to retransmit the message. Alternatively, the broadcasting of information can also be adapted to the covered distance. In the closer vicinity of the source the message is repeated by many nodes. By overhearing, ideally all or almost all nodes in close proximity to these broadcasters will be informed. With increasing distance from the source the frequency of broadcasts is decreased. Similar to this approach Brønsted and Kristensen (2006) propose to use flooding *only* in the surrounding of the message's source. This allows vehicles close by to be informed very fast without having to flood a packet into a large area. Dornbush and Joshi (2007) in their study rely on techniques called 'rumor spreading', 'gossip', or 'epidemic' dissemination. Here, a node which receives a message stores it locally and will forward it later to other nodes. Nevertheless, these techniques still use a flooding-like approach. In order to limit the number of transmitted messages, nodes will decide to disseminate only the latest information. Also, approaches combining these two ideas are possible. Wegener et al. (2007) propose to adapt flooding not only to the density of nodes in the vicinity but also to the novelty of the flooded information. Thus, a new event will trigger the creation of a new message. To meet the density rate adaption criterion only a few vehicles will further broadcast this newly generated information. While being forwarded the message gets older which results in a reduced broadcasting frequency. By limiting the number of concurrent packets this scheme aims to avoid the broadcast storm problem.

A slightly different scheme to limit the flooding of information is to let the receiver decide whether to rebroadcast the packet at all. Chisalita and Shahmehri (2004), for instance, consider such an approach, albeit that they primarily look at safety applications. They assume that when a dangerous situation is detected a warning message will be broadcast. In contrast to simply rebroadcasting this message, a receiver of this data analyzes whether it is interested in the warning. This could be the case if the receiver is driving on the same road and in the same direction as the source vehicle of the message. The assumption is that if it itself is interested in the data then other cars in its vicinity might also profit from the message. The information is thus broadcast again. This approach may be leveraged by including additional information in the message to ease the decision on which nodes are interested in the data. Ducourthial et al. (2007) present an approach to add, e.g., geocast addresses to the message in order to facilitate the analysis performed by a receiving node. By evaluating this condition it may then decide whether to forward the message. Similar to these schemes Briesemeister et al. (2000) propose taking into account the distance to the previous forwarding node. The further away a node is, the more likely it will be to rebroadcast the received message. Nodes that overhear this transmission in between these two nodes do thus not need to broadcast. So as to determine the furthest node explicitly, Korkmaz et al. (2004) apply an RTS/CTS scheme for broadcasts. After the transmission of a 'request to broadcast' all nodes in transmission range send a jamming sequence called 'black-burst' with a duration dependent on the distance between the sender and the node. The furthest node thus sends the longest burst and indicates by sending a 'clear to broadcast' that it wants to receive this broadcast packet. In order to reduce the overhead of such a scheme, Zhao et al. (2007) use a position-based geocast routing approach. Here, the receiver decides – based on a local neighbor table – to which node the message will be forwarded. Due to the usage of 'directed' broadcasts, nodes within the radio range can overhear the message.

In an early study, Kosch et al. (2002) propose to extend the reactive routing protocol AODV of Perkins and Royer (1999) by some geocasting functionality. Messages containing information on, for instance, current road conditions are disseminated within a given region. The same authors also propose an approach where information is disseminated according to a local interest rate. If a node is interested in this data it will also forward the message. Although this approach is appealing because existing protocols can be used, it might have scaling limitations in large networks where a lot of nodes are the source of information. In order to deal with the distinct mobility patterns of vehicular networks Sormani et al. (2006) add a store-and-forward mechanism to the receiving nodes. Thus, the nodes contribute to transporting the information by wireless transmissions as well as by their own locomotion.

It should be noted that message spreading schemes where nodes decide whether to forward information based on their own interest miss the opportunity to use vehicles that are themselves not interested in the information for improved dissemination. This may turn out to be a severe performance issue. For instance, it has often been observed that using the oncoming traffic for information transport results in significantly better performance – the oncoming traffic will, however typically not be interested in data about the region it is coming from.

Williams and Camp (2002) compare the different flooding approaches according to their performance and overhead. Although this study is done in the context of generic MANETs the results can be helpful for VANETs as well. Their results and conclusions underline the expectation that flooding approaches show very poor performance in terms of congestion. The study considered only one flooding 'stream' starting at one single source. In a vehicular information dissemination scenario, information flooding might be started by multiple nodes concurrently which would most likely deteriorate the presented results further.

Request/reply The transport of information in the context of vehicular applications is not limited to solely proactive flooding approaches. If the information is not needed by many users it is worthwhile thinking about reactive or on-demand algorithms. In this case a vehicle will ask explicitly for specific information by transmitting a request message. The final destination of this packet may be known, for instance if the user requests the current gas prices from the next gas station or asks a parking meter for a free parking spot. The forwarding of this message may be implemented in different ways.

Zhao and Cao (2008) propose forwarding request messages in a unicast fashion by constructing an explicit route, while Basu and Little (2002, 2004) use position-based routing approaches. The reply message will then be returned similarly to the request. Instead of asking for the information directly at the source, vehicles may also ask other vehicles passing by whether they know about it. According to Sago et al. (2007) the waiting time at a red traffic light may last long enough to fill up the local knowledge base by querying other cars. In most studies requesting is performed in an indirect fashion by announcing the entries of the local knowledge base. A node that receives this message can analyze if it is aware of information that is new or unknown to the requester. Different techniques are proposed to implement these request mechanisms. In Figure 3.2 an autonomous update process is depicted. In addition to this procedure it is also possible to announce the local knowledge while other vehicles request explicitly for a specific entry. In contrast to a unicast request-reply mechanism, Wegener et al. (2007) propose a periodic single-hop broadcast requesting technique based on a data structure similar to the one of

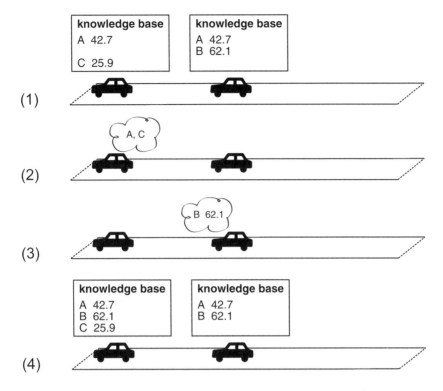

Figure 3.2 Implicit information exchange. In (1) two cars with different knowledge bases meet. In (2) the left car propagates its own knowledge. By comparing this announcement the right car can differentiate which entries it owns and are thus needed by the left car. In (3) it transmits the missing entry. Finally, in (4) the left car has updated its knowledge base.

the study mentioned above. Furthermore, by using techniques to prevent overload of the network they suggest adapting the frequency of these requests depending on the traffic density.

In order to overcome the partitioning in VANETs Fujiki et al. (2007) propose making use of delay-tolerant routing approaches. In this context request packets do not need to follow a prebuilt unicast route but are geocast to a certain region and can be transmitted implicitly by the locomotion of vehicles. By doing this, different partitions can be spanned. The transport of the reply message follows the same rules as the request packets did.

It is worth stating that request/reply schemes face serious scalability problems. Due to the well known capacity constraints in wireless multihop networks, as stated by Gupta and Kumar (2000), the capacity usable by any source converges to zero as the size of the network increases to infinity. Thus these approaches must take special care to limit the request/reply exchange to vehicles that are close to each other or to employ some other way to limit the load on the network.

Clustering In some situations it becomes obvious that flooding a message into a whole network is not appropriate and will cause a high level of congestion. At the same time request/reply schemes might not be appropriate, e.g. since the information is needed by many nodes at the same time. It has therefore been proposed to replace the unstructured flooding of packets by some sort of hierarchical distribution using clustering.

Agarwal et al. (2007, 2008), and Little and Agarwal (2005) present the directional propagation protocol (DPP). In this approach, the authors consider information dissemination along a highway. When the traffic is dense, nodes within some region form a cluster. Within this cluster messages are shared with all nodes of this cluster. In order to propagate a message in one direction, overlapping clusters are used to transmit the message. Clusters in the opposing direction are considered as well to propagate the message. DPP adopts procedures of a delay tolerant network (DTN). If there is no overlapping cluster the propagation of messages is done simply by the locomotion of vehicles.

A dissemination approach that also makes use of clustering techniques is proposed by Chang et al. (2008). If a car needs some information, it forms a cluster. This initiates the clustering process. In order to transmit data to other clusters relay vehicles connect two clusters. In a first phase, information within one cluster is gathered in a time division multiple access (TDMA) fashion. After the first – collection – phase, all cluster heads own the information of all vehicles within their cluster. In the second – retrieval – phase this information is then sent back to the initiator of the first cluster, i.e., the requesting node.

Another cluster-based approach is presented by Chennikara-Varghese et al. (2007). Here cars within a small region autonomously form a cluster. Owing to the close vicinity, direct communication between those cars is possible. However, communication to other vehicles or clusters has to be performed by relaying vehicles that are part of several clusters. In the context of parking meter status dissemination Basu and Little (2002, 2004) propose using clustering approaches only within the group of the static parking meters.

Sharing Leontiadis and Mascolo (2007) present an approach which is based on the publish/subscribe paradigm. Messages are disseminated and kept in a predefined area. If a node is interested in any type of message it can subscribe to this event locally. If a messages is received by that node it can easily check if this matches its subscription or not. Information is duplicated and these replicas are transmitted to some nodes within this area. If a node leaves this area the information is then transferred to a node which moves into the area or remains within this area. The authors show in their study that the number of duplicates can be dynamically adjusted and is in general lower than the number of subscribers. The main difference of this approach compared to the approaches above is that information should remain in some area.

The approach proposed by Shinkawa et al. (2006) follows the same general idea. Vehicles possess information about their current area. If they leave this area the information is discarded. But in order not to lose this information globally, especially if the density of equipped nodes is low, the authors propose to use buses as 'message ferries' which will not drop any received information. The motivation behind using these vehicles is that they follow a regular and recurrent route covering multiple regions and are thus able to supply other vehicles with information.

Beaconing In many VANET applications, time-varying data is announced periodically by each node in form of a beacon. In this way the network load is limited to a fixed level determined by the density of the nodes and the beaconing frequency. When receiving such a one-hop broadcast, nodes do not react to it with a new transmission directly. They instead integrate the content of the message into their local knowledge base. With the next beacon that they will send this information, or a part thereof may then be further distributed.

In order to maximize the information dissemination with the restriction of only few vehicles equipped and without causing congestion in the network Xu and Barth (2006a) propose adapting the rate of the beacons. With increasing distance and depending on the average velocity of vehicles along a road the frequency of beacons to be transmitted is adapted. Similarly Fujiki et al. (2007) also present a mechanism that adapts the rate of the periodic beacons whereas Wischhof et al. (2003a) recommend taking into account whether a different view of the traffic situation has arisen from the received message. Additionally, Wu et al. (2004) analyze the environment of a vehicle. If new neighboring vehicles appear, a beacon is sent to update their local knowledge base.

Saito et al. (2004) use a TDMA approach where one second is divided into ten slots. The used slot depends on the current speed of a node. The idea is that if there is a traffic jam the current speed will be low and thus only few beacons need to be sent.

Torrent-Moreno (2007), and Torrent-Moreno et al. (2005, 2006) present an approach to controlling the channel load used by safety beacons and to provide a more reliable information transmission to receivers close to the sender. Instead of varying the frequency of beacons as in the above-mentioned studies they propose adjusting the transmission power of beacons. The goal is to diminish the risk of packet collisions through congestion and thereby increasing the likelihood that beacons are successfully received.

3.3.2 Improving network connectivity

Store and forward Since VANETs are not only highly dynamic, but also highly partitioned networks, continuous connectivity may not be assumed in a protocol design. To allow for long-range information dissemination beyond the extension of a single network partition, concepts from delay-tolerant networking can be applied. In particular, store-and-forward approaches for information transport are a common construction principle of VANET dissemination protocols. In a store-and-forward approach nodes do not immediately forward messages, but carry the information along with their movement. When opportunities arise, e.g., by meeting other vehicles, the information is transmitted to forward it further. In the simplest case, it is spread epidemically by beaconing, but more elaborate schemes have also been discussed. Approaches that make use of these ideas have, for example, been proposed by Caliskan et al. (2006); Leontiadis and Mascolo (2007); Lochert et al. (2007a, 2008); Sormani et al. (2006); Wischhof et al. (2005); Wu et al. (2004); Xu et al. (2004); Zhao and Cao (2008), to name just a few.

When it comes to bridging network gaps between groups of cars on the same road, it has often been observed that making use of oncoming traffic for information transport brings large performance benefits. Especially when information about traffic or road conditions is exchanged, it is vital to communicate the data to vehicles driving far behind the region where the observations are made. Transportation by the locomotion of the creator of the information and by other vehicles driving in the same direction is thus not helpful. However, using oncoming traffic to forward the information can greatly improve the performance of the system. For instance, Yang and Recker (2005) as well as Agarwal et al. (2007, 2008) demonstrate this effect in their studies. Nadeem et al. (2006) even conclude that it is efficient to use *only* cars going in the opposite direction to rapidly disseminate information.

Roadside units Yang and Recker (2005) aim to answer two questions on information propagation: what is the minimum necessary penetration ratio, and which other requirements need to be met in order to disseminate (warning) messages faster than by the pure locomotion of vehicles? The main focus here is whether it is possible to transmit data quickly to distant regions with only few vehicles equipped. They conclude that at least 20% should use VANET technology. However, the results depend significantly on the communication range. Generally, as we discuss in Lochert et al. (2007b), specifying a definitive limit on the necessary penetration ratio seems difficult or impossible: the traffic density varies heavily depending on the specific road, the time of day, etc. – thus, a penetration ratio that is sufficient during rush hour in a city center may be entirely insufficient at night time or in a more rural region. We therefore argue that we need to specify the *equipment density* instead: the density of VANET-equipped vehicles on the road.

Independent of any specific penetration ratios or equipment densities, however, Yang and Recker make the – later often confirmed – observation that especially during rollout, when very few vehicles are equipped, extra measures for ensuring connectivity will to some extent be necessary for many applications. They suggest analyzing the impact of additional infrastructure – roadside units. Such infrastructure devices have appeared in many different flavors in the literature. Goodman et al. (1997), for instance, suggest that external information such as gas prices may be provided via 'information kiosks', or simply by a WiFi access point. Other terms that have been used in the literature include '(stationary) supporting units' and 'roadside units (RSUs)'. Here, we will use the latter term.

RSUs may generally be either stand-alone devices that communicate only with vehicles via wireless communication, or they may be interconnected via a backbone network, which in turn could be realized either by wired or infrastructure wireless (GSM, UMTS, ...), or via a mesh network of RSUs. Essentially, their purpose is always to connect the VANET to external information sources, or to increase connectivity and/or capacity of the network. Zhao et al. (2007) for instance propose the use of supporting units at road junctions. When installed at traffic lights, these devices are able to receive information from passing cars and will inform newly approaching vehicles with this data.

The idea of using this kind of infrastructure device is not new. Back in 1964, Bowes (1964), and Covault and Bowes (1964) presented results from a testbed using roadside radio communications. They analyzed whether it was feasible to install RSUs in order to provide the drivers with additional information and to exert traffic control. Besides the technical challenges of such a system, user acceptance and cost were questioned. Although the technical prerequisites were quite different, the goals of the roadside units were comparable to the ones discussed today. Different incidents were tested that had some influence on safety or traffic flow. Radio devices installed in the cars received the messages from the roadside units when passing by. In contrast to digital data messages used today, they used spoken messages output via a loudspeaker.

Banerjee et al. (2008) theoretically analyze the performance of different kinds of supporting units. In particular, they consider backbone-interconnected RSUs or base stations, a mesh network of RSUs, and stand-alone RSUs. They compare the number of stations necessary to keep the packet delay low. One conclusion is that a lot (5–7 times) more stand-alone RSUs are needed in order to achieve the same performance as with interconnected base stations. Banerjee et al. use an abstract, random node movement model. For stand-alone RSUs, node movement is important for delivering information to more distant regions. In real VANETs, nodes often move preferentially in one direction (e.g., towards the city center during morning rush hour). As mentioned above, typical long-range dissemination applications require that information travels *against* the main movement direction. Interconnected RSUs can meet these requirements and quickly deliver information to distant regions – much more effectively than stand-alone RSUs can. Therefore, in real VANETs, the performance benefit of networked RSUs can be expected to be even larger.

This matches the results from our study in Lochert et al. (2007b), where we assess the impact of RSUs on a beaconing-based information dissemination approach. The RSUs use the same application as ordinary vehicles. We show that, in order to have a substantial effect, the supporting units should be interconnected via a backbone network, such that they can use a common knowledge base: as soon as one unit receives some information, all other supporting units can use it and distribute it further. In Lochert et al. (2008) we examine the impact of the number of required RSUs and their placement within a city on the travel time savings gained with a VANET-based traffic information system, and we propose a methodology for finding good RSU positions using genetic algorithms.

The concept of RSUs is of course not restricted to use with beaconing-based dissemination schemes. Chennikara-Varghese et al. (2007), for example, analyze how stand-alone or interconnected RSUs provide easier communication between clusters of vehicles. Here, RSUs are basically access points placed at the road side, which participate in the clustering protocol.

3.3.3 What to transport

While discussing how to transport information within the network, many studies also consider which parts of information or which messages should be transmitted.

Relevance functions – a concept introduced by Kosch et al. (2006) – are used to compare the available information with respect to its novelty or importance. Caliskan et al. (2006) use the age and the distance to a resource for their relevance function. Dornbush and Joshi (2007) classify the received messages according to their 'significance'. When transmitting local knowledge to other vehicles, a message is filled until the maximum packet size is reached. This keeps the message size at a fixed level and avoids network congestion. Other metrics to decide which part of the local information should be transmitted are proposed as well. Saito et al. (2007) suggest different approaches. In a probabilistic selection method no ordering of the information is performed but the data to be sent is determined in a purely random fashion. They also propose ordering the information based on its age. Where other authors only make use of the newest and thus most current information, this study also considers propagating older information in order not to lose it.

Fujiki et al. (2007) assume that information about the vicinity of a vehicle is required more often than information about a distant region. It is thus useful to transmit these data more frequently. Adler et al. (2006a,b) suggest computing a global message benefit of a message to be sent. This benefit is used to parameterize the medium access procedure. Since a highly relevant message should be sent first, the contention phase is modified in order to prioritize the message. This is performed by adapting the size of the contention window. However, in an environment with many transmissions and considering a more realistic (probabilistic) radio propagation model, Torrent-Moreno et al. (2004) showed that such a prioritization method would most likely not achieve the desired effects.

An implicit or soft-state approach to sharing information only in a bounded area is presented by Xu et al. (2004). The authors assume that if two vehicles meet they will exchange their local knowledge. In order to limit the size of the information, only the newest entries are transmitted. On the other hand, by adding new information to the local knowledge base, existing data will age and die out after some time.

Apart from being able to map position information to logical entities in the road network, the (static) parameters of the road network known by a navigation system (such as, for example, typical speeds on individual roads) may also be of great use when generating data that is helpful in VANET applications. It is a design option to inform other vehicles only in case of significant deviations from 'typical' values. For example, a traffic information application might report travel times only if they are significantly different from what would usually be expected on the respective road. This has, for example, been proposed by Goel et al. (2004), as well as Dornbush and Joshi (2007). Such an approach can potentially save significant bandwidth. In order to judge locally whether such a deviation currently exists, the respective travel times from the navigation system's static map data may serve as a reference.

However, there is a general problem with such 'minimalist' approaches that transmit information only under specific circumstances: in a case where no information is received, it is not clear whether this is due to the fact that the situation is 'normal', or whether there are simply no current measurements, so that nothing is known. Again, depending on the way the information is used and interpreted by the application, this may be acceptable or not.

3.4 Summarizing Measurements

So far, we have looked at how local measurement data can be obtained, and we have seen quite a number of approaches as to how this information can be transported to interested receivers. However, distributing all individual measurements made by all cars to all recipients who are interested in the respective road or area may be theoretically conceivable – but it is evident that this cannot scale. Therefore, mechanisms are required that reduce the amount of data dynamically and within the network – while the environmental parameters monitored by the system vary, while the positions of the cars and network topology continuously change, and while data is being exchanged wherever opportunities arise. In the remaining part of this chapter, we thus now focus on the problems of *summarizing* and *aggregating* measurement data.

In VANET dissemination applications, it is often the case that multiple measurements on the same or on closely related observed entities are available in the network. For instance, multiple vehicles may have traversed the same road segment, each of them measuring its own individual travel time. Or multiple vehicles may have obtained information on the current occupancy level of a parking site, but they may have done so at slightly different points in time. When such data – referring to the same entity, but to different observations – 'meet' in the network, capacity limitations suggest that they are 'merged' in some way. If they are combined into one single value, it is not necessary to store two individual values locally. Much more importantly, however, it is then also not necessary to transmit two values when information is exchanged between network participants. The aim of the combination operation is to generate one single data item which reflects the known information as well as possible. We call this process of combining multiple observations *summarization*.

A close relative of summarization is *aggregation*, and the distinction between those two is not always perfectly clear in the literature. In our terminology, aggregation means that data concerning *different* entities are combined into one value. The intention of aggregation is to produce coarser data representations of areas in larger distances, which can save even more bandwidth than combining multiple measurements of the same parameter through summarization. We will discuss aggregation in the next section.

Blind averaging In the simplest form of the data summarization problem, a device participating in the VANET is confronted with two distinct values for one and the same parameter (e.g., for the number of free parking places on a specific parking site, or for the time needed to travel along some specific road segment): one of the values may be stored in its local knowledge base, another value is obtained from some information source. For example, the car may have previously made an observation of the parameter itself and now receives a respective value from another car.

Unless additional information such as, for instance, a timestamp is stored and communicated along with the data, it is not straightforward to decide if the current value in the knowledge base should be updated or not. Xu and Barth (2006b) discuss the problem of data summarization in the context of measuring travel times along road segments. They establish fixed ten-minute time slots, and summarize the travel time measurements on the same road segment falling into the same time slot. In the first approach they suggest, when the travel time along a road segment is received from another car, the received and local values are averaged, and the average value is stored in the knowledge base. However, it turns out that this approach does not perform well. The problem is that measurements made early during a time slot will dominate the dynamically updated average. When additional measurements are made later on, the early values are already known to many cars, so that they are often received. Therefore, the new measurements are quickly canceled out through repetitive averaging with the old values.

Timestamp-based comparison This problem points to the idea of complementing the measurement values by timestamps. When a car makes an observation, it stores not only the observed value in its knowledge base, but also the current time. When the data is transmitted to other network participants, the timestamps are sent along with the values. A straightforward 'summarization' mechanism – actually one that is often used in the literature – is to replace older measurements by newer ones, based on comparing the timestamps. Such an approach assumes that contributing entities have sufficiently accurately synchronized clocks available. This is not unreasonable since VANET-enabled cars will typically use some geopositioning system such as GPS anyway. These systems can serve as a highly accurate time source, and therefore the requirement of synchronized clocks is not too severe a constraint.

A timestamp-based comparison mechanism certainly avoids the problem that old measurements persist in the network and dominate newer, more up-to-date data. However, it also means that the value in the knowledge base of each car will always stem from one single observation. Whether this is a problem or not depends on the application; in particular it depends on the specific data source: in a system like the one proposed by Caliskan et al. (2006), where fixed infrastructure (here, at parking sites) generates virtually exact measurements, newer values will indeed always describe the current situation better than older ones. Thus, in this case, replacing old measurements based on newer ones is certainly appropriate, and is an easy way to achieve good data quality.

In other situations, however – especially when the measurements are noisy – keeping only the most up-to-date value is not a good idea. When measuring travel times through direct observations by the cars, influencing factors such as individual driving style, traffic light phases, etc., may cause large variations. One individual measurement may therefore not reflect the actual situation very well.

Timestamp-based averaging Xu and Barth (2006b) discuss two improved alternatives to the blind travel time averaging scheme discussed above, in which they avoid the averages being dominated by old, but widespread, measurements on the one hand, but also avoid discarding old data completely upon availability of new observations. Both algorithms assign timestamps to travel time values, essentially in the way discussed above. The first approach is based on assigning the older value a lower weight upon averaging: when a new observation is made or when a value is received from another car (regardless of whether its timestamp is newer or older), a weighted average is calculated, where the older of the two values is assigned a lower weight than the newer one. Xu and Barth found that good results are achieved when the two weights are chosen as 0.8 and 0.2.

As their third approach they propose updating the travel times in the local knowledge base only when the car itself makes a newer observation; other cars of course still assist in distributing the averaged values. In their scheme, vehicles store the number of samples in the current time slot so far, denoted by n, the timestamp T of the latest observation that contributed to the value, and the average travel time value v itself. A newly observed value o of the parameter can then – in the observing car itself – be integrated by an appropriately weighted average, generating the new value v', in the following way:

$$v' = \frac{1}{n+1}(nv + o).$$

When such an update is performed, T is updated to the current time, and n is increased by one. The parameters n and T are communicated to other cars along with the current average value. When data is received from other vehicles, based on the timestamp T either the local value or the received value is kept; the older of the two is discarded.

This scheme ensures that each sample is included only once in the summarized values, and that all the measurements contributing to the average value have the same weight. It may still happen, though, that observations made by a car and locally included in the value are lost in later steps. This occurs in the following situation: assume that a car A makes an observation and includes it in its local knowledge base. In the course of this operation it also updates the associated timestamp. A short time later, car B has not yet received the data updated by A, but itself makes an observation of the parameter, and integrates it into its respective local value. The data in B's local knowledge base does not contain the observation made by A, but it bears a newer timestamp. When these two data items are subsequently exchanged in the network and compared for up-to-dateness, B's data will 'win' and replace A's data, thereby overriding the update operation made earlier by A. Consequently, A's contribution will be lost. Despite this effect, Xu and Barth find that out of the three schemes they propose, this approach yields the most accurate representations of the exact travel time values.

Road-segment averaging A different variant of an averaging scheme is used in SOTIS by Wischhof et al. (2005). The general mechanism applied there for summarizing information from multiple sources is called Segment-Oriented Data Abstraction and Dissemination (SODAD). SODAD is based on a subdivision of longer roads into segments. The road segments are hard-coded, so that all vehicles agree on the same set of segments, and each segment can be addressed with a unique ID. Locally, individual measurements are exchanged between the vehicles, so that a

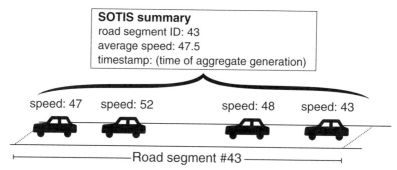

Figure 3.3 Per-segment vehicle velocity summarization in SOTIS.

typical vehicle will be able to collect multiple samples from its immediate vicinity. For disseminating information to further away network areas, vehicles combine the samples that they have for the road segment they are currently driving on. Upon generation, the per-segment values are assigned a timestamp – the time when the summarization has been performed. When they are disseminated in the network, the timestamp determines whether the value currently in the knowledge base or a newly received value are kept.

Wischhof et al. (2005) state that the function determining how exactly the individual samples are combined into a per-segment summary is application-dependent. In the concrete case of SOTIS, the application is based on exchanging information on the velocity of vehicles. The measured parameter is therefore the current driving speed. For summarizing the velocity information of multiple vehicles into a per-segment value, the average value is calculated. This is schematically shown in Figure 3.3.

A concept exhibiting some parallels with SODAD is outlined by Brønsted and Kristensen (2006). The Zone Diffusion Protocol described there also divides the roads into so-called 'cells', and uses an application-specific data combination policy to merge observations made on the same road cell. As an example application, the paper considers road condition dissemination, and suggests the use of a conservative estimation policy: in case the information on the road condition for a cell diverges, the worst case is assumed, i.e., if one measurement classifies the road as icy while some other car considers it dry, it is conservatively assumed to be icy.

3.5 Geographical Data Aggregation

The summarization techniques discussed in the previous section enables comparison and merging of data from multiple observations concerning the same parameter, like the current driving velocity on the same road segment in SOTIS or the number of free parking places on the same parking place in the parking guidance system by Caliskan et al. (2006). Thus, summarization can keep the amount of data per measured parameter (parking lot, ...) constant. In larger road networks – for instance in a whole city or a network of highways – there will, however, still be a huge amount of data left: the number of road segments in a city, for example, is high, as is the number of parking places. Generally, if observed parameters are homogeneously distributed over the area of the network, then their number will increase quadratically with the covered radius. The limited bandwidth available for dissemination applications in VANETs will therefore not allow the continuous spread of such detailed information over long distances, always maintaining a fine-grained picture in the whole network.

To overcome this limitation, it has been proposed to use in-network data aggregation techniques, which aim to combine the current values of multiple parameters into one single aggregate value. Such an aggregated value could be the average speed on a longer part of a highway, or the total number of free parking places in a larger part of a city. In this section, we will discuss data aggregation schemes for VANETs.

Clustering groups of similar vehicles The TrafficView system by Nadeem et al. (2004) aims to facilitate distributed cooperative traffic monitoring of the highway ahead of the vehicle. While this goal is similar to that of SOTIS, TrafficView takes a different approach with regard to the type of information made available via the network. TrafficView is based on data on the current position and speed of individual vehicles rather than on speed averages for road segments as in SOTIS. Because the amount of data soon exceeds reasonable limits, an adaptive aggregation mechanism is used. Locally each car stores a list of vehicle IDs, positions, and speeds. These are transmitted to neighboring vehicles via beacons. The size of these transmissions is limited, and therefore mechanisms are introduced to reduce the amount of data while at the same time conveying approximate information about as many vehicles as possible. For this purpose, TrafficView nodes use so-called aggregate records in their transmissions. Aggregate records describe not single vehicles, but groups of vehicles with similar properties.

An aggregate record in TrafficView consists of one single speed, position, and timestamp value, along with a list of vehicle IDs. This saves bandwidth for transmitting separate values for each individual vehicle. The speed and position values in the aggregate record are weighted averages over the respective parameters of the vehicles in the record. An aggregate is schematically shown in Figure 3.4. TrafficView proposes aggregation algorithms that are parameterized for regions of the road ahead. The most important input parameters are an aggregation ratio, indicating how many vehicles should be combined into one aggregate, and the amount of data that may be transmitted for a given region. Based on these parameters TrafficView then aggregates appropriate values, aiming to keep the introduced errors small.

The CASCADE scheme by Ibrahim and Weigle (2008) shows some similarities to TrafficView's aggregation approach. It also distributes information on individual vehicle positions and velocities with the aim of supporting dynamic route planning by networked navigation systems. CASCADE uses syntactic compression on clusters of vehicles that are similar with respect to position and speed. Groups of

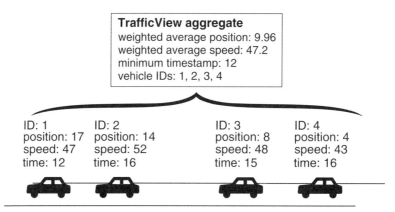

Figure 3.4 An aggregate over four vehicles in TrafficView; the calculation of the weighted averages here assumes that the aggregating vehicle is at position 0.

vehicles are summarized in one record, where the amount of per-vehicle data within the record is minimized.

CASCADE vehicles have detailed information about individual vehicles in their neighborhood, so called 'primary records'. The entirety of the detailed knowledge of a vehicle is termed 'local view'. The vehicles on which information is available are grouped into clusters depending on the distance from the local vehicle, their heading, and their altitude. The per-vehicle data in the cluster records is then compressed by expressing vehicle coordinates and speed relative to the cluster center and the cluster's median speed. Consequently, an aggregated cluster record contains the (absolute) cluster position and speed, and one 'compact record' (with coordinates relative to the cluster) per vehicle in the cluster, as shown in Figure 3.5. Aggregated information is distributed over longer distances. When aggregate records are received, they are incorporated into the 'extended view' of the receiving vehicle.

SOTIS, TrafficView, and CASCADE show a number of similarities. All three systems collect information on position and speed of vehicles, and all of them combine information on vehicles with similar parameter values. None of these systems supports hierarchical aggregation of information on a whole road network, i.e., information on distinct roads is never combined. Thus, they are borderline cases between summarization and aggregation. Here, we decided to classify SOTIS' mechanisms for data reduction as summarization, and TrafficView's and CASCADE's as aggregation, because SOTIS considers vehicle data as samples of the same parameter – the velocity on a road segment – and then summarizes these samples. In TrafficView and CASCADE, in contrast, distinct vehicles are consequently treated as distinct observed entities.

Distributed data clustering In their StreetSmart system, Dornbush and Joshi (2007) tackle the problem of a distributed traffic monitoring system from a different

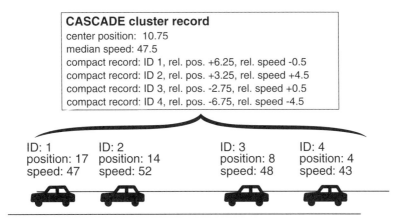

Figure 3.5 An aggregated cluster record in CASCADE.

angle. In contrast to the approaches discussed so far, StreetSmart does not exchange any data about individual vehicles or road segments at all. In StreetSmart, each vehicle records samples – essentially position and speed – along its own movement path. Individual position data in StreetSmart is represented by a road ID and an offset (i.e., position) along the road. These samples are *not* transmitted to other vehicles. They are locally subjected to data mining techniques, so as to generate a more abstract view. This aggregated view is transmitted by the network.

The speed/position samples are locally aggregated using a data clustering mechanism with an application-tailored similarity measure. This measure depends on (1) the road that the measurement was made on, (2) the position along the road, and (3) the recorded speed. It is used to identify groups of samples which exhibit unexpectedly low speeds on the one hand, and are otherwise similar in terms of the above-mentioned criteria. If clusters of samples exhibit a speed which is not typical for the respective road, traffic congestion is assumed to exist. Clusters for which this applies are then communicated to other network participants, abstractly described through their cluster centroids.

So, while StreetSmart also uses clustering of similar data sets, these data sets are position/speed samples collected over time from the *same* vehicle, while the previously discussed systems TrafficView and CASCADE aggregate position/speed records from different vehicles.

Aggregation over hierarchical areas The approaches discussed so far aggregate data from multiple vehicles, or multiple measurements consecutively made by the same vehicle. They are therefore able to significantly reduce the amount of data to be exchanged. However, they alone are not sufficient to fully overcome the scalability problem for two reasons. First, they can sensibly only combine data from the same road and the same driving direction. So, the total amount of data in the network still quickly increases with the size of the covered road network.

The second and probably even more important reason is that the amount of data for a complete picture still increases with the number of vehicles: in TrafficView and CASCADE, the aggregates contain lists of the IDs of included vehicles. So, there is a reduction of the number of bytes used per vehicle, but the size of the aggregates still grows linearly with the number of included data items. In StreetSmart, local measurements are clustered, but each car distributes its centroids in the complete network, without further aggregation; therefore, there, too, the amount of data increases with a larger number of vehicles (or, equivalently, a larger covered area). We will thus now look at proposals where the size and count of aggregated data representations is independent from the number of measuring vehicles, and where the amount of data does not increase linearly with the covered area (or with the number of roads in the covered area).

The first approaches which achieve such a behavior have been proposed by our group in Caliskan et al. (2006) and Lochert et al. (2007a). In these works, area-based aggregation models are used: with increasing distance, measurements from larger and larger areas are combined into aggregated values. The decentralized parking guidance scheme described in Caliskan et al. (2006) uses such an aggregation scheme. Local infrastructure at each parking site is used to locally summarize information on the overall occupancy of the parking site. This is founded on the assumption that larger parking sites will typically have infrastructure for payment and/or admission installed anyway. Such infrastructure already has an up-to-date overview of the occupancy level of the site, so potentially noisy measurements conducted by the vehicles themselves are unnecessary.

The 'atomic' information on the current occupancy status of a single parking site is handed over to passing cars, and is subsequently passed on between vehicles. Within the network, aggregates are formed. They serve as a compact representation of the parking situation in larger parts of a city, by describing the total number of free parking places on all parking sites within some larger area. A fixed subdivision of the city is used, defined by a quadtree over the two-dimensional plane. A quadtree is a hierarchical subdivision of the plane into smaller and smaller squares, as indicated in Figure 3.6. Within a certain radius, information on all individual parking sites is

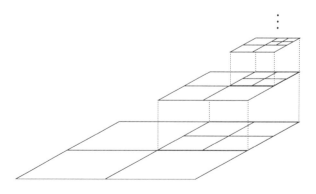

Figure 3.6 A quadtree.

exchanged in the network; at larger distances, only aggregate values are used. These aggregate values become coarser and coarser the more distant the described area is. The motivation for hierarchical aggregates is clear from the application focus: at a larger distance, it is sufficient to know about the approximate situation and the coarse direction where parking places can be found. It is not necessary to know the detailed occupancy level of all parking sites. The closer a car gets to its destination area, the more fine-grained the information gets, so that a decision for one specific parking site can finally be made.

Vehicles form aggregates of the lowest level (i.e., the cells in the smallest quadtree grid) by adding up the number of free parking places known in the area. That is, if a vehicle has received atomic information from a number of parking places in the same quadtree cell, it may generate a lowest-level aggregate. Aggregates for use at again larger distances use the higher levels of the quadtree hierarchy. They are generated from aggregates of the next-lower level, by adding the respective numbers of free parking places from the four component subaggregates. So, larger aggregates are hierarchically created from smaller aggregates.

An aggregate essentially comprises the ID and hierarchy level of the respective quadtree cell, a timestamp, and the total number of free parking places. Thus, each aggregate has the same size, regardless of the size of the area it covers and of the number of parking sites within that area. Consequently, for an increasing aggregation level – and thus with increasing geographical distance from the described area – the amount of data per covered area decreases.

Comparing and merging hierarchical area aggregates In the previous section we saw that by using timestamps one can easily keep only the most up-to-date value observed for some parameter, discarding older measurements. This is possible because the up-to-dateness can be compared based on the timestamps. Unfortunately, this cannot be transferred in a straightforward way to aggregates describing more than one parameter. The reason is that an aggregate generator – a car in the network – will typically not have 'perfect' (i.e., complete and fully up-to-date) information available for the generation of the aggregates. As a concrete example, consider an application for disseminating information about the parking place situation. Assume that three parking sites – A, B, and C – lie within an aggregated area. Car x knows that at A there were 10 free parking places at time t_1, and that at time t_2 there were 15 free parking places at B. x has no information about C. The aggregate generated by car x will state that within the area a total of 25 free parking places is known. Car y has more up-to-date information on B than x, and is aware of 19 free parking places there at some point in time later than t_2. y also knows about 13 free parking places at C at time t_3, but does not know anything about A. y's aggregate will thus state 32 free parking places in the area. Assume now that a third car z (directly or indirectly) receives the respective aggregates from both x and y. Which one should it keep? Which one is 'better'? Clearly, the answer to this question is not straightforward, if there is a definite answer at all. The information in the aggregates is partially overlapping, and none of them clearly dominates the other.

The central insight from this example is that aggregates will not be 'perfect' representations of the current situation in the area they cover, but only the best possible approximation based on the current knowledge of the generating node. If different nodes generate aggregates for the same area, these aggregates will typically be based on different, but partially overlapping information. The fundamental issue that therefore arises is that the completeness and up-to-dateness of aggregates can not be expressed through a single timestamp. A data structure describing the set of contained information and the respective timestamps is obviously also not an option, because it would compromise the aim of small aggregates, the size of which should ideally not increase with the amount of data contained.

Summarization and aggregation mechanisms deal differently with this issue, but typically they resort to using some kind of 'best guess' heuristic timestamp. For example, SOTIS uses the time of aggregate generation as a timestamp for the aggregate; obviously, however, it is well possible that a node with old and incomplete information generates an aggregate, which is then considered more up-to-date than a slightly earlier generated aggregate with a more solid and (at least for the most part) more up-to-date basis. The aggregate timestamp generation schemes for TrafficView (where the timestamp of the oldest observation is used) and Caliskan et al.'s parking guidance system (which calculates the average timestamp of the components used for aggregation) exhibit similar artifacts; there, too, situations are easily constructed where aggregates with quite comprehensive and up-to-date coverage are assigned an older timestamp than aggregates with a much weaker information basis.

In our work, Lochert et al. (2007a), we tackle this problem of non-comparability of aggregates. We propose the use of a special form of data representation to overcome this problem: by storing the observations in a duplicate insensitive probabilistic data structure called 'soft-state sketches'. Duplicate insensitivity means that information which is already present in both source aggregates will not have a higher impact on the result. Returning to our example above, the soft-state sketch scheme does not *need* to answer the question of which of the two aggregates is the better one, because it can merge the aggregates of x and y in a duplicate insensitive way. From the two source aggregates, z can generate an aggregate which contains information on A, B, and C. This merged aggregate will then incorporate all the information and will represent the total number of $10 + 19 + 13 = 42$ free parking places.

Using such a mechanism, it becomes entirely unnecessary to compare aggregates regarding their up-to-dateness: there is no need to discard one of the aggregates, because the information contained in both of them can be kept, by merging wherever and whenever they 'meet' in the network. There is no need for one single car to first collect the underlying information as completely as possible, and then to calculate an aggregate, which can subsequently only be updated by a complete recalculation. Instead, aggregates can incorporate new information from the underlying parameters whenever it becomes available, while they are being passed around in the network.

Soft-state sketches are based on a data structure proposed by Flajolet and Martin (1985); their FM-sketches were originally designed to estimate the number of distinct elements in a multiset in linear time. We adapted them for VANET applications, so that they can be used to collect and distribute measured parameters. We also devised an extension which allows for the automatic removal ('soft-state') of old observations, if they are no longer backed up by recent measurements. The drawback of a sketch-based data representation is that it is probabilistic – i.e., it does not store the exact values of the parameters, but probabilistic approximations. The accuracy of soft state sketches can be adjusted, trading off accuracy against aggregate size. Generally, an approximate representation of measurements seems reasonable in typical VANET convenience applications. The soft-state extension for the removal of old data can also result in a certain time lag between a changing value and the change being reflected in the aggregates. This time lag is, however, of the same order of magnitude as the delay caused by the delay-tolerant data transport over the VANET – a delay which can generally not be avoided. An extensive example of how this idea can be used to distribute the number of free parking places will be given in the chapter on VANET convenience applications.

We also generalize the concept of hierarchical aggregation areas in Lochert et al. (2007a). The quadtree-based approach used in Caliskan et al. (2006) allows for an easy way of addressing areas and their hierarchy. But it has the drawback of not considering the topology of the underlying city: the same quadtree field may include areas which are barely connected in the road network, and the properties of which may largely differ – for instance, areas on two sides of a river or a railway. We argue that aggregation areas should follow the topology of the underlying road network, and give a generic formulation of an aggregation hierarchy, along with some hints on how it should be designed to allow for effective sketch-based aggregation.

Hierarchical aggregation of travel times in road networks The aggregation algorithms described in Caliskan et al. (2006) and Lochert et al. (2007a) serve well in the case where the application requires *area-based* summaries, i.e., when the intention is to obtain an overview of the (total, average, maximum, . . .) value of some parameter within larger and larger areas. A parking guidance system – disseminating the total number of free parking places within an area – is indeed a prime example of such an application. Let us now, however, consider the case of travel times in a road network. While it *is* helpful to know the total number of free parking places on all roads within some part of the city, it is *not* particularly useful to know, for example, the total sum of the travel times along all these roads; this value does not convey useful information for route planning through that particular area. Therefore, for a scalable decentralized road navigation system, it is vital to devise a way in which travel times through a road network can be represented in an aggregated, hierarchically coarser and coarser way.

In Lochert et al. (2008) we consider such a system which disseminates travel times along road segments as a basis for navigation systems. There we propose the use of hierarchical approximations of the road network with increasing distance. Travel times along all links between all road junctions are locally observed, and are exchanged within the closer vicinity. The more important junctions are then used as 'aggregation landmarks', and an approximation of the road network is constructed, which consists of these landmarks and a set of virtual links between them. That is, the approximate road network in some sense simplifies the true topology, by

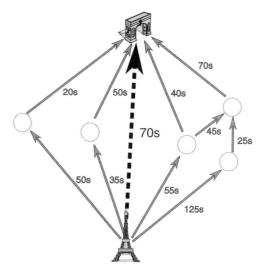

Figure 3.7 The shortest possible travel time from Tour Eiffel to Arc de Triomphe is 70 s.

leaving out smaller streets and junctions and summarizing travel time information over all possible routes between pairs of landmarks. Only the more central points and abstract connections between them are considered. In Figure 3.7, for instance, cars that are close enough would be provided with detailed information on the travel times between all shown lower-level landmarks (represented by circles), while cars beyond a certain distance only know that the shortest possible travel time between the high-level landmarks 'Tour Eiffel' and 'Arc de Triomphe' is 70 s. Similar abstractions of the road network are also used in some modern algorithms to quickly determine shortest paths through road networks.

In the proposed scheme, aggregates are travel times along these virtual links between landmarks. In the true road topology, there may be many possible routes between two landmarks connected by a virtual link. The aggregated value essentially only states that 'it is possible to drive from landmark A to landmark B within t seconds'. In order to calculate the aggregate value, a car uses the locally available small-scale information on road segment travel times to calculate the travel time along the currently shortest route from A to B. The key point is that it is not necessary to disseminate this shortest route itself nor do the travel times along the individual road segments along the shortest route need to be known by a car at a larger distance: the abstract landmarks and the travel times between them are sufficient for route planning. Because more detailed information becomes available as the car comes closer to the respective region, the precise path which actually enables travel from A to B within the given time can then be determined.

The very same concept is applied on higher hierarchy levels that represent the road network at even larger distances. Just as a subset of the junctions are denoted as first-level landmarks, a subset of these first-level landmarks are at the same time

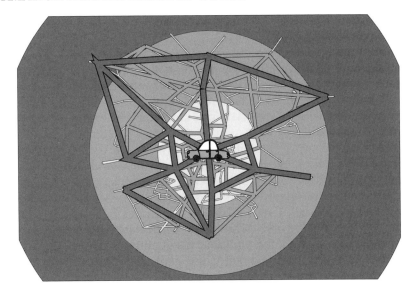

Figure 3.8 A hierarchical road network abstraction.

a second-level landmark, and so on. Higher-level landmarks are connected through higher-level virtual links, forming coarser and coarser approximations of the road network with increasing hierarchy level. Figure 3.8 shows this trait: the car in the center of the figure has detailed information about the individual road segments in the inner circle available. For regions further away on the map, coarser information in the form of virtual links between higher- and higher-level landmarks are used. The travel times along the higher-level virtual links constitute the respective higher-level aggregates; they are calculated from the travel times along the virtual links of the next-lower hierarchy levels, and again represent the travel times along the respective shortest paths. That is, where travel times along road segments are used to calculate travel times between landmarks on the first hierarchy level, these first-level virtual links serve as a basis for calculating travel times between landmarks of the second hierarchy level, etc. We will take up this idea again in more detail in the chapter on VANET-based convenience applications, where we will also investigate the impact that such a system might have on travel time savings.

3.6 Conclusion

In this chapter, we have considered the communication characteristics of VANET convenience applications. These applications often require data to be distributed in a larger surrounding. We have started by surveying the different sources of information, and looked at how the data can be obtained in the first place. We then looked at dissemination protocols and algorithms, which provide the means to deliver the data from their points of origin to the regions of the network where they are demanded and useful.

We finally turned towards mechanisms for summarizing and aggregating data. We saw how multiple observations can be combined into more compact representations, and how aggregated data can be maintained and updated cooperatively in the network. Because of the inherent limitations of wireless multihop networks, especially in terms of capacity, such mechanisms are a vital cornerstone of dissemination-based VANET applications.

In the future, we expect that it will turn out to be more and more vital to consider the interplay of all the aspects covered here and their interrelation to the specific applications to be supported. Instead of targeting isolated aspects, our focus should move towards a more integrated view of data acquisition, data transport, and in-network data processing. Questions such as the prediction of future developments based on past observations, dealing with noisy measurements, and automatically assessing and verifying the quality of the available data are still far from being solved. Likewise, real-world experiences are still very scarce, and many assumptions and conjectures still demand to be backed up by experimental verification in real environments. Addressing these issues in the context of emerging – and undoubtedly useful – applications will certainly raise a plethora of interesting new aspects and challenging research questions.

References

Adler C, Eichler S, Kosch T, Schroth C and Strassberger M 2006a Self-organized and context-adaptive information diffusion in vehicular ad hoc networks. *ISWCS '06: Proceedings of the 3rd International Symposium on Wireless Communication Systems*, pp. 307–311.

Adler C, Eigner R, Schroth C and Strassberger M 2006b Context-adaptive information diffusion in VANETs: Maximizing the global benefit. *CSN '06: Proceedings of the 5th IASTED International Conference on Communication Systems and Networks*.

Agarwal A, Starobinski D and Little TDC 2007 Exploiting downstream mobility to achieve fast upstream message propagation in vehicular ad hoc networks. *MOVE '07: Proceedings of the 2007 Workshop on Mobile Networking for Vehicular Environments*, pp. 13–18.

Agarwal A, Starobinski D and Little TDC 2008 Analytical model for message propagation in delay tolerant vehicular ad hoc networks. *VTC '08-Spring: Proceedings of the 67th IEEE Vehicular Technology Conference*, pp. 3067–3071.

Banerjee N, Corner MD, Towsley D and Levine BN 2008 Relays, base stations, and meshes: Enhancing mobile networks with infrastructure. *MobiCom '08: Proceedings of the 14th Annual ACM International Conference on Mobile Computing and Networking*.

Basu P and Little TD 2002 Networked parking spaces: architecture and applications. *VTC '02-Fall: Proceedings of the 56th IEEE Vehicular Technology Conference*, vol. 2, pp. 1153–1157.

Basu P and Little TDC 2004 Wireless ad hoc discovery of parking meters. *WAMES '04: Proceedings of the MobiSys 2004 Workshop on Applications of Mobile Embedded Systems*, pp. 8–11.

Boukerche A, Oliveira HABF, Nakamura EF and Loureiro AAF 2008 Vehicular ad hoc networks: A new challenge for localization-based systems. *Elsevier Computer Communications* **31**, 2838–2849.

Bowes RW 1964 *A study of the feasibility of using roadside radio communications for traffic control and driver information*. PhD thesis, Georgia Institute of Technology Atlanta, GA, USA.

Briesemeister L, Schäfers L and Hommel G 2000 Disseminating messages among highly mobile hosts based on inter-vehicle communication. *IV '00: Proceedings of the IEEE Intelligent Vehicles Symposium*, pp. 522–527.

Brønsted J and Kristensen LM 2006 Specification and performance evaluation of two zone dissemination protocols for vehicular ad-hoc networks. *ANSS '06: Proceedings of the 39th Annual Simulation Symposium*, p. 12.

Caliskan M, Barthels A, Scheuermann B and Mauve M 2007 Predicting parking lot occupancy in vehicular ad-hoc networks. *VTC '07-Spring: Proceedings of the 65th IEEE Vehicular Technology Conference*, pp. 277–281.

Caliskan M, Graupner D and Mauve M 2006 Decentralized discovery of free parking places. *VANET '06: Proceedings of the 3rd ACM International Workshop on Vehicular Ad Hoc Networks*, pp. 30–39.

Chang WR, Lin HT and Chen BX 2008 Trafficgather: An efficient and scalable data collection protocol for vehicular ad hoc networks. *CCNC '08: Proceedings of the 5th IEEE Consumer Communications and Networking Conference*, pp. 365–369.

Chennikara-Varghese J, Chen W, Hikita T and Onishi R 2007 Local peer groups and vehicle-to-infrastructure communications. *GLOBECOM '07: Proceedings of the IEEE Global Telecommunications Conference – Workshops*, pp. 1–6.

Chisalita I and Shahmehri N 2004 A context-based vehicular communication protocol. *PIMRC '04: Proceedings of the 15th IEEE International Symposium on Personal, Indoor, and Mobile Radio Communications*, vol. 4, pp. 2820–2824.

Covault DO and Bowes RW 1964 A study of the feasibility of using roadside radio communications for traffic control and driver information. *Highway Research Record* **49**, 89–106.

Dornbush S and Joshi A 2007 StreetSmart Traffic: Discovering and disseminating automobile congestion using VANET's. *VTC '07-Spring: Proceedings of the 65th IEEE Vehicular Technology Conference*, pp. 11–15.

Ducourthial B, Khaled Y and Shawky M 2007 Conditional transmissions: Performance study of a new communication strategy in VANET. *IEEE Transactions on Vehicular Technology* **56**(6), 3348–3357.

Flajolet P and Martin GN 1985 Probabilistic counting algorithms for data base applications. *Journal of Computer and System Sciences* **31**(2), 182–209.

Fujiki T, Kirimura M, Umedu T and Higashino T 2007 Efficient acquisition of local traffic information using inter-vehicle communication with queries. *ITSC '07: Proceedings of the 10th International IEEE Conference on Intelligent Transportation Systems*, pp. 241–246.

Goel S, Imielinski T and Ozbay K 2004 Ascertaining viability of WiFi based vehicle-to-vehicle network for traffic information dissemination. *ITSC '04: Proceedings of the 7th International IEEE Conference on Intelligent Transportation Systems*, pp. 1086–1091.

Goodman DJ, Borràs J, Mandayam NB and Yates RD 1997 INFOSTATIONS: a new system model for data and messaging services. *VTC '97: Proceedings of the 47th IEEE Vehicular Technology Conference*, vol. 2, pp. 969–973.

Gupta P and Kumar PR 2000 The capacity of wireless networks. *IEEE Transactions on Information Theory* **46**(2), 388–404.

Ibrahim K and Weigle MC 2008 Optimizing CASCADE data aggregation for VANETs. *MoVeNet '08: Proceedings of the 2nd International Workshop on Mobile Vehicular Networks*.

International Organization for Standardization 2003 Road vehicles – controller area network (CAN). *ISO 11898*.

International Organization for Standardization 2004 Road vehicles – diagnostics on controller area networks (CAN). *ISO 11765*.

Jerbi M, Senouci SM, Rasheed T and Ghamri-Doudane Y 2007 An infrastructure-free traffic information system for vehicular networks. *VTC '07-Fall: Proceedings of the 66th IEEE Vehicular Technology Conference*, pp. 2086–2090.

Korkmaz G, Ekici E, Özgüner F and Özgüner Ü 2004 Urban multi-hop broadcast protocol for inter-vehicle communication systems. *VANET '04: Proceedings of the 1st ACM International Workshop on Vehicular Ad Hoc Networks*, pp. 76–85.

Kosch T, Adler C, Eichler S, Schroth C and Strassberger M 2006 The scalability problem of vehicular ad hoc networks and how to solve it. *IEEE Wireless Communications* **13**, 22–28.

Kosch T, Schwingenschlögl C and Ai L 2002 Information dissemination in multihop inter-vehicle networks. *ITSC '02: Proceedings of the 5th International IEEE Conference on Intelligent Transportation Systems*, pp. 685–690.

Leontiadis I and Mascolo C 2007 Opportunistic spatio-temporal dissemination system for vehicular networks. *MobiOpp '07: Proceedings of the 1st international MobiSys workshop on Mobile opportunistic networking*, pp. 39–46.

Little TDC and Agarwal A 2005 An information propagation scheme for VANETs. *ITSC '05: Proceedings of the 8th International IEEE Conference on Intelligent Transportation Systems*, pp. 155–160.

Lochert C, Scheuermann B and Mauve M 2007a Probabilistic aggregation for data dissemination in VANETs. *VANET '07: Proceedings of the 4th ACM International Workshop on Vehicular Ad Hoc Networks*, pp. 1–8.

Lochert C, Scheuermann B, Caliskan M and Mauve M 2007b The feasibility of information dissemination in vehicular ad-hoc networks. *WONS '07: Proceedings of the 4th Annual Conference on Wireless On-demand Network Systems and Services*, pp. 92–99.

Lochert C, Scheuermann B, Wewetzer C, Luebke A and Mauve M 2008 Data aggregation and roadside unit placement for a VANET traffic information system. *VANET '08: Proceedings of the 5th ACM International Workshop on VehiculAr Inter-NETworking*, pp. 58–65.

Markoff J 2008 Can't find a parking spot? Check smartphone. *The New York Times*.

Nadeem T, Dashtinezhad S, Liao C and Iftode L 2004 TrafficView: traffic data dissemination using car-to-car communication. *ACM SIGMOBILE Mobile Computing and Communications Review* **8**(3), 6–19.

Nadeem T, Shankar P and Iftode L 2006 A comparative study of data dissemination models for VANETs. *MobiQuitous '06: Proceedings of the Third Annual International Conference on Mobile and Ubiquitous Systems: Networking & Services*, pp. 1–10.

Ni SY, Tseng YC, Chen YS and Sheu JP 1999 The broadcast storm problem in a mobile ad hoc network. *MobiCom '99: Proceedings of the 5th Annual ACM/IEEE International Conference on Mobile Computing and Networking*, pp. 151–162.

Ozbay K, Nassif H and Goel S 2007 Propagation characteristics of dynamic information collected by in-vehicle sensors in a vehicular network. *IV '07: Proceedings of the IEEE Intelligent Vehicles Symposium*, pp. 1089–1094.

Perkins CE and Royer EM 1999 Ad-hoc on-demand distance vector routing. *WMCSA '99: Proceedings of the 2nd IEEE Workshop on Mobile Computing Systems and Applications*, pp. 90–100.

Piorkowski M, Grossglauser M and Papaioannou A 2006 Mobile user navigation supported by WSAN: Full-fledge demo of the SmartPark system. *MobiHoc '06: Proceedings of the 7th ACM International Symposium on Mobile Ad Hoc Networking and Computing*. Technical Demonstration.

Sago H, Shinohara M, Hara T and Nishio S 2007 A data dissemination method for information sharing based on inter-vehicle communication. *AINAW '07: Proceedings of the 21st International Conference on Advanced Information Networking and Applications Workshops*, vol. 2, pp. 743–748.

Saito M, Funai M, Umedu T and Higashino T 2004 Inter-vehicle ad-hoc communication protocol for acquiring local traffic information. *Proceedings of the 11th World Congress on Intelligent Transport Systems*.

Saito M, Tsukamoto J, Umedu T and Higashino T 2007 Design and evaluation of intervehicle dissemination protocol for propagation of preceding traffic information. *IEEE Transactions on Intelligent Transportation Systems* 8(3), 379–390.

Shinkawa T, Terauchi T, Kitani T, Shibata N, Yasumoto K, Ito M and Higashino T 2006 A technique for information sharing using inter-vehicle communication with message ferrying. *MDM '06: Proceedings of the 7th IEEE International Conference on Mobile Data Management*, pp. 130–134.

Sormani D, Turconi G, Costa P, Frey D, Migliavacca M and Mottola L 2006 Towards lightweight information dissemination in inter-vehicular networks. *VANET '06: Proceedings of the 3rd ACM International Workshop on Vehicular Ad Hoc Networks*, pp. 20–29.

Tonguz O, Wisitpongphan N, Bai F, Mudalige P and Sadekar V 2007 Broadcasting in VANET. *MOVE '07: Proceedings of the 2007 Workshop on Mobile Networking for Vehicular Environments*, pp. 7–12.

Torrent-Moreno M 2007 Inter-vehicle communications: assessing information dissemination under safety constraints. *WONS '07: Proceedings of the 4th Annual Conference on Wireless On-demand Network Systems and Services*, pp. 59–64.

Torrent-Moreno M, Jiang D and Hartenstein H 2004 Broadcast reception rates and effects of priority access in 802.11-based vehicular ad-hoc networks. *VANET '04: Proceedings of the 1st ACM International Workshop on Vehicular Ad Hoc Networks*, pp. 10–18.

Torrent-Moreno M, Santi P and Hartenstein H 2005 Fair sharing of bandwidth in VANETs. *VANET '05: Proceedings of the 2nd ACM International Workshop on Vehicular Ad Hoc Networks*, pp. 49–58.

Torrent-Moreno M, Santi P and Hartenstein H 2006 Distributed fair transmit power adjustment for vehicular ad hoc networks. *SECON '06: Proceedings of the 3rd Annual IEEE Communications Society Conference on Sensor and Ad Hoc Communications and Networks*, vol. 2, pp. 479–488.

Wegener A, Hellbrück H, Fischer S, Schmidt C and Fekete S 2007 AutoCast: An adaptive data dissemination protocol for traffic information systems. *VTC '07-Fall: Proceedings of the 66th IEEE Vehicular Technology Conference*, pp. 1947–1951.

Williams B and Camp T 2002 Comparison of broadcasting techniques for mobile ad hoc networks. *MobiHoc '02: Proceedings of the 3rd ACM International Symposium on Mobile Ad Hoc Networking and Computing*, pp. 194–205.

Wischhof L, Ebner A and Rohling H 2005 Information dissemination in self-organizing intervehicle networks. *IEEE Transactions on Intelligent Transportation Systems* 6(1), 90–101.

Wischhof L, Ebner A, Rohling H, Lott M and Halfmann R 2003a Adaptive broadcast for travel and traffic information distribution based on inter-vehicle communication. *IV '03: Proceedings of the IEEE Intelligent Vehicles Symposium*, pp. 6–11.

Wischhof L, Ebner A, Rohling H, Lott M and Halfmann R 2003b SOTIS – a self-organizing traffic information system. *VTC '03-Spring: Proceedings of the 57th IEEE Vehicular Technology Conference*, pp. 2442–2446.

Wu H, Fujimoto R, Guensler R and Hunter M 2004 MDDV: A mobility-centric data dissemination algorithm for vehicular networks. *VANET '04: Proceedings of the 1st ACM International Workshop on Vehicular Ad Hoc Networks*, pp. 47–56.

Xu B, Ouksel A and Wolfson O 2004 Opportunistic resource exchange in inter-vehicle ad-hoc networks. *MDM '04: Proceedings of the 5th IEEE International Conference on Mobile Data Management*, pp. 4–12.

Xu H and Barth M 2006a An adaptive dissemination mechanism for intervehicle communication-based decentralized traffic information systems. *ITSC '06: Proceedings of the 9th International IEEE Conference on Intelligent Transportation Systems*, pp. 1207–1213.

Xu H and Barth M 2006b Travel time estimation techniques for traffic information systems based on inter-vehicle communications. *TRB '06: 82nd Annual Transportation Research Board Meeting*.

Yang X and Recker W 2005 Simulation studies of information propagation in a self-organizing distributed traffic information system. *Elsevier Transportation Research Part C* **13**(5-6), 370–390.

Zhao J and Cao G 2008 VADD: Vehicle-assisted data delivery in vehicular ad hoc networks. *IEEE Transactions on Vehicular Technology* **57**(3), 1910–1922.

Zhao J, Zhang Y and Cao G 2007 Data pouring and buffering on the road: A new data dissemination paradigm for vehicular ad hoc networks. *IEEE Transactions on Vehicular Technology* **56**(6), 3266–3277.

4

VANET Convenience and Efficiency Applications

Martin Mauve and Björn Scheuermann

Heinrich Heine University, Düsseldorf, Germany

4.1 Introduction

Convenience applications may prove to be a key enabler of VANET technology since they allow for competition between vehicle manufacturers. Safety-related applications such as intersection warning require that the functionality is standardized in order to be of use. An application that guides a driver to an appropriate parking place or that avoids a traffic jam may, in contrast, very well be manufacturer-specific and provide a distinct advantage over other manufacturers. This is a very powerful motivation to invest into non-safety-related VANET applications.

At the same time convenience applications require very careful consideration when deciding whether it makes sense to built them on top of direct communication between cars. The three key reasons why direct communication may not always be an optimal choice for a given application are: the limited capacity of the network, the lack of connectivity, and, finally, the presence of competing technology such as cellular networks.

The contrast between the fundamental limitations of convenience applications on the one hand and the potentially high reward for successfully providing those applications on the other hand will be the guiding theme of this chapter. We will start our discussion with a presentation of the fundamental limitations that VANET application developers face. This is followed by an overview of non-safety applications that have been proposed for VANETs. Then we will turn our attention to collaborative traffic information systems (TIS) as one particularly popular class of applications in VANETs. These systems collect observations of individual vehicles on the current traffic situation and then distribute them to other participants where they will typically be used for route planning. The main challenge in this class of applications is distributing a large amount of dynamic information to many distinct recipients. We discuss three different ways in which this challenge can be met: a solution using a centralized server connected via a cellular network, employing peer-to-peer technology over cellular networks, and a VANET-based approach. After the advantages and drawbacks of each approach have been highlighted we will dwell on the VANET-based solution, and show how data aggregation – as discussed in the chapter on data dissemination – can be used to overcome the core problem of the limited network capacity.

4.2 Limitations

When developing an application for use in VANETs it is vital to be aware of the limitations that are present in this environment. The main challenges are capacity restrictions, limited connectivity, and competing alternative technologies. In this section we describe these challenges in detail and provide hints on how to overcome them. The focus of this discussion will be on the limitations of pure vehicle-to-vehicle communication.

4.2.1 Capacity

In their landmark paper Gupta and Kumar (2000) showed that the transmission capacity of an individual node in a wireless multihop network converges to zero as the number of nodes in the network increases. Since VANETs *are* wireless multihop networks, the most fundamental limitation that developers for non-safety-related applications face is the limited capacity of the network.

Unfortunately, the specific topology of VANETs may amplify the capacity limitations even further. The key problem is that, in VANETs, nodes are not evenly distributed over a given two-dimensional area, they are restricted to move along roads. Consider the extreme case of a single highway where a message is transmitted from a sender to a receiver. A highway, generally, is of such limited width that any wireless transmission covers its whole breadth. Therefore, transmitting messages from the sender to the receiver de facto amounts to flooding the highway between those two vehicles no matter how smart or efficient the employed routing algorithm is. The medium will be occupied at least once around each single vehicle between sender and receiver.

Assume now that all vehicles should, on average, receive the same amount of bandwidth for transmitting their own data to a receiver of their choice. Any such transmission will burden all vehicles between the sender and the receiver. Their number grows as the average density of cars multiplied by the average distance between sender and receiver. Consequently, following our considerations it is straightforward to see that the capacity available to each sender will decrease antiproportionally both with the average density of vehicles and with the distance between the sending and the receiving vehicle.

There are multiple ways to deal with the capacity limitation of VANETs. The simplest approach is to make sure that the constants are chosen such that the capacity of the network is not exceeded for a given application. One could, for example, design applications that require only a small data rate over relatively short distances. Or one could aim to improve the single-hop data rate to match the communication requirements of the application.

A more sophisticated way to deal with the capacity constraints in VANETs is to design the applications so that they become inherently scalable in this environment. One frequently used approach is inspired by Grossglauser and Tse (2002) where the mobility of the participating nodes is used to increase the capacity of the network. For many VANET-related applications information, such as a road congestion notification, needs to be transmitted to cars at some distance behind the sender. The sender can simply transmit the message to a car driving in the opposite direction. This car will carry the message until it can be delivered directly to the intended receiver. In this fashion the discussed capacity limitations can be alleviated.

Another approach is to eliminate the density of senders from the capacity limitation. For many VANET-related applications the transmitted data is not specific to one sender–receiver pair. Data such as the current congestion situation or the amount of free parking space are of general interest to many vehicles. Thus, there is no need for all cars to report on a given piece of information. It is sufficient that one representative transmits a message towards those areas where the information may be useful. As outlined in the chapter on data dissemination, this can be combined with an aggregation scheme. Such a scheme may also eliminate the distance aspect in addition to the density aspect of the capacity restriction, yielding fully scalable VANET applications.

4.2.2 Connectivity

The second key challenge that non-safety VANET applications face is the limited connectivity of the network. The network formed by direct communication between cars consists of many clusters that are not interconnected at any given point in time. The reason for this is twofold. First, the initial deployment density of VANET technology will be low for a long time after the first vehicles have been equipped. A study by Matheus et al. (2005) projects that even if all middle class cars and above were equipped with VANET technology it would take three years to reach 10% market penetration and about 16 years until 45% is reached.

The second reason why the network between vehicles is highly partitioned is that the probability for the formation of an uninterrupted chain of cars in radio range decreases exponentially. Lochert et al. (2007b) explains this in detail and derives an approximation for the likeliness of chain formation. Figure 4.1 shows the resulting probability of uninterrupted connectivity depending on the number of equipped vehicles per radio range (ρ) and the covered distance. It is important to realize that this is an optimistic estimate, because the analytical model assumes that cars are

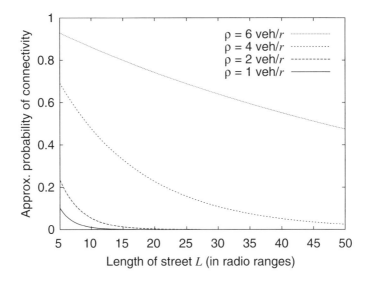

Figure 4.1 Approximated probability of radio connectivity depending on the equipment density and the covered distance.

homogeneously distributed over the street. In reality, they tend to form clusters, which will reduce the probability of chain formation even further.

Owing to the limited connectivity that can be expected from VANETs, most proposed applications either rely on local communication only (such as almost all safety-related applications) or use the mobility of the vehicles to bridge connectivity gaps. Handing data to oncoming traffic or simply carrying the data for a while not only reduces the required capacity of the application but also enables the dissemination of data beyond a single partition of the VANET.

In order to get a better understanding of the performance that can be achieved by combining direct forwarding along a chain and physically carrying data to bridge partitions, Lochert et al. (2007b) simulate the dissemination of data in the medium-sized German city of Brunswick. The simulated scenario consists of more than 500 km of roads and up to 10 000 vehicles. The network simulator ns-2 is combined with the microscopic traffic simulator VISSIM, as introduced in Lochert et al. (2005). ns-2 uses the IEEE 802.11 MAC, two-ray ground propagation, a communication range of 250 meters, and an obstacle model that does not allow radio signals to propagate through the walls of buildings.

An information source was placed in the center of the city. Each car that passes by the information source would learn about its current 'value' (in this case, simply a timestamp). Cars also send periodic beacons, so that information can also be handed over via wireless transmissions. In this simulation, a penetration ratio of 100% with a radio communication range of 250 meters corresponds to an average equipment density ρ between 2.25 and 5 vehicles per radio range, depending on the simulated time of day.

VANET CONVENIENCE AND EFFICIENCY APPLICATIONS

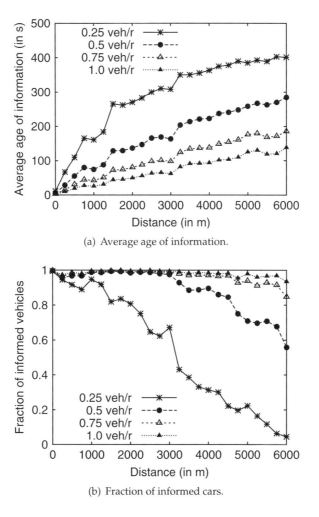

Figure 4.2 Dissemination of data using chains and the mobility of vehicles.

Figure 4.2(a) shows the average age of information available in a vehicle as a function of the distance from the information source. It can be seen that the age grows approximately linearly with the distance. The propagation speed rises significantly with increasing equipment density. This is because the probability of the formation of long chains increases with growing equipment density. This in turn means that data transport via wireless communication becomes more and more predominant, while the importance of transport via locomotion decreases.

At a low equipment density, the average duration for the dissemination of information to the outer areas can reach a value as high as 400 s. But a low equipment density has an even more serious influence on the probability of a vehicle knowing anything about the data source *at all*. This statistic, after 500 s of simulation time, is depicted in Figure 4.2(b). The probability of obtaining information at an increasing

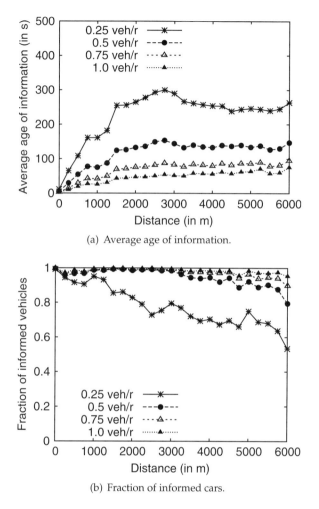

(a) Average age of information.

(b) Fraction of informed cars.

Figure 4.3 The impact of networked roadside units.

distance decreases rapidly. The main reason for this trait is that it takes some time after a vehicle starts its trip from a residential area, a parking lot, etc., until it meets another car from which it can obtain information. From the results it can be seen that a relatively high equipment density is necessary for sufficiently reliable information dissemination. Consequentially, the roll-out phase of a VANET application, where only a low equipment density is available, is highly problematic.

In order to specifically support this roll-out phase, it has been proposed to install a small number of roadside units (RSUs) at key road intersections. These RSUs participate in the network and are all interconnected by a backbone network, they thus share a common knowledge base and are able to bridge large distances. The impact of using seven interconnected RSUs in the scenario described above can be seen in Figure 4.3.

The main lesson learned from these experiments is that even a relatively small amount of additional infrastructure may be able to achieve substantial improvements. Nevertheless, significant delays and a notable fraction of uninformed cars remain, so, depending on the application's characteristics, connectivity remains an important issue also with RSUs.

4.2.3 Competition

Direct communication between cars faces strong competition from current and future cellular communication. For example, 3G systems already offer affordable and widely available mobile Internet access. It is reasonable to assume that low-cost cellular Internet access will soon be common, long before VANET technology becomes a reality. The main advantage of cellular communication is that only one hop is wireless. Once the data has reached the base station of a cellular network, neither the capacity nor the connectivity constraints outlined above apply. This is of particular importance if the communication spans longer geographic distances. Furthermore, cellular networks exist today while VANET technology may not be available for several years to come. Therefore, a large number of applications that requires communication between a car and the Internet or between two specific cars are likely to use cellular networks.

On the other hand, VANET-style communication is very well suited for fast and high volume local data exchange. It is simply faster and more efficient to exchange data between two cars in radio range directly than to use the detour via a cellular network. Also, there are applications, such as a traffic information system, where the data is of interest to many or even all cars in the vicinity of the transmitting car. This, too, is a situation where direct communication between cars may have an edge over cellular networks, because the local broadcast property of wireless transmissions can be exploited. Finally, VANET communication is inherently free of charge, making it very attractive for any applications, if they can deal with the restricted capacity and limited connectivity.

4.3 Applications

The idea of letting cars exchange information with each other has given rise to a vast number of proposals for applications. In this section we will highlight some application classes that are frequently mentioned when discussing VANET convenience and efficiency applications.

One of the earliest applications proposed for inter-vehicle communication was Internet access. In several projects VANETs were considered a mobile extension of the Internet. Data transmission between cars would then be handled in a very similar way to routing packets in the Internet with vehicles being the equivalent to Internet routers. Given the capacity constraints outlined above, the perspective on Internet access has changed quite considerably. Currently VANETs are considered to be able to extend the radio range of wireless access points that might be located, for instance, at gas stations by a small number of wireless hops. Cars that drive by could then be granted Internet connectivity for a longer period of time compared to direct contact with a single base station. Commonly it is expected that this kind of connectivity will be provided free of charge and that the owner of the access point will get some non-monetary benefit in return. He might, for example, be able to advertise the prices of gas or other products. Such a service could be of interest for applications that require only intermittent Internet connectivity. One example could be push-based email access. But, in any case, free, intermittent Internet access will have to compete with the much more reliable service offered through cellular networks. Their service might become very cheap in the near future.

An application class that is in some aspects related to Internet access is the distribution of advertisements by means of inter-vehicle communication. For example, Lee et al. (2007) propose a system where vehicles receive an incentive from forwarding and carrying advertisements in a VANET. Often free Internet access, as described above, is combined with some form of advertisement distribution. Since the distribution of advertisements is mostly single-hop communication it is unlikely to be hindered by capacity or connectivity issues.

The ability to communicate with other passengers in the close vicinity has inspired researchers to look at novel ways of entertainment (at least for passengers in the back seat); see, for instance, Palazzi (2007). Quizzes, paper chase, rally support and mixed reality games are just some examples of entertainment applications that might be supported through inter-vehicle communication. Most of these games explicitly target local communication so that capacity and connectivity of the network do not pose serious constraints. However, all things considered, there has been remarkably little work in this area, so far. At this time, it does not seem as if this application class will be a driving force for VANETs in the immediate future.

A very straightforward application for vehicle-to-roadside systems would be payment services, targeting for example the collection of toll fees. Such an application could replace the multitude of heterogeneous systems in use today. For this low-volume, single-hop application class neither capacity nor connectivity are expected to be a challenge.

The desire to get to a destination as fast as possible while consuming minimal resources has given rise to research on VANET-based traffic information systems (TIS) – where 'traffic information' may generally be understood in a broad sense. The core idea of these applications is to use vehicles as mobile sensors which monitor parameters such as traffic density, road and weather conditions, or parking space occupancy. The information is then distributed to other vehicles where it can be used for tasks such as route optimization or other adaptations of driving behavior. VANET-based traffic information systems have received a lot of attention over recent years; see, for instance, Caliskan et al. (2006); Dornbush and Joshi (2007); Ibrahim and Weigle (2008); Lochert et al. (2007a, 2008); Nadeem et al. (2004); Wischhof et al. (2003); Xu and Barth (2006a,b). Furthermore, they are an excellent example for a challenging application with respect to the limitations outlined above: a large volume of dynamically changing data is transmitted by many cars over long distances. At the same time, cellular networks are a strong competitor to VANETs for distributing traffic-related information. In the remainder of this chapter, we will therefore take a closer look at traffic information systems. Our goal is to use them as an example for the decision and design process of VANET-based convenience applications. We start by investigating potential design alternatives, in particular regarding the use of cellular networks. After this discussion we will focus on the challenges of capacity for distributing information on the current traffic situation and parking space occupancy over a VANET.

4.4 Communication Paradigms

The inherent limitations of VANET communication, the limited initial VANET equipment density, and the competition with cellular communication systems emphasize that VANETs should not be the only, unreflected focal point of our attention. When we think about vehicular convenience applications, it is always worthwhile to discuss the benefits and drawbacks of different communication paradigms in a little more detail. We do so in this section, considering the specific case of a traffic information system (TIS). With a focus on this particular application, we will contrast VANET-based communication with two infrastructure-supported approaches: the straightforward way of building them as a client/server system, and a peer-to-peer approach where the cars form an overlay network via cellular communication.

4.4.1 Centralized client/server systems

The usage of cellular communication networks reduces the problem of implementing a working traffic information system to the question of how to make collectively gathered data available to all interested parties. One way of achieving this is to use a client/server architecture. In this approach one server (or a server farm) on the Internet stores a central knowledge base consisting of all the data that has been collaboratively gathered and contributed by the cars. Cars make observations, e.g., on the current traffic situation, while driving. These observations are sent directly to the server.

A car may then request this information. For the purpose of identifying the best out of many alternative routes towards some destination, information on substantial parts of the road network is required. To fulfill this task two approaches are conceivable: (i) The car downloads (and regularly updates) traffic data for all possible alternative routes. It can then compute the fastest route to the destination based on current traffic information. (ii) As the downloads require a high communication effort, the route computation might alternatively be performed by the server. This, however, requires substantial computational resources in a central location.

When using a client/server architecture combined with an infrastructure-based communication approach the TIS application can avoid the network layer problems that typically arise in VANETs. The greatest advantage of infrastructure-based communication is the fact that the density of the equipped cars needed for a working application is much smaller than in the case of a VANET.

The major technical challenge of a centralized system is to deal with the huge amount of simultaneous updates and queries, as pointed out by Hull et al. (2006); Rybicki et al. (2007) – recall that each car is a source of queries and sends its own measurements regularly. Furthermore, owing to the high degree of the centralization, the server can become a bottleneck or even a single point of failure. However, the main reason not to use a centralized system for managing traffic information could very well be non-technical: it simply does not seem to be desirable to hand the control of this data over to one central authority, potentially limiting the access to data collected conjointly by all traffic participants.

4.4.2 Infrastructure-based peer-to-peer communication

To avoid these drawbacks of a client/server solution Rybicki et al. (2007) have shown how infrastructure-based cellular communication can be leveraged to build a distributed peer-to-peer network for traffic information systems. This approach combines robustness resulting from decentralization and independence from a central authority on the one hand with good connectivity provided by infrastructure-based communication on the other hand.

In such a system, each participating car is a peer. The data access interface of the central server is substituted by a distributed hash table (DHT). Similar to its non-distributed counterpart, the DHT associates keys with values. For traffic information systems, one may use geographical coordinates as keys and the respective observations as values. Each peer is thus responsible for a specific subset of keys (e.g., streets) and associated data (observations). The unambiguous mapping of the keys to peers is done by means of a random and uniform hash function. The main part of the DHT is the implementation of a lookup algorithm in a decentralized fashion. For a given key, the lookup returns the responsible peer. This is done in the following way. Participants form an overlay network, where each peer is connected to a number of neighbors following a carefully designed structure. Either a peer possesses a given key or it knows a peer offering a progress towards the responsible one in the overlay.

In Figure 4.4 the considered scenario is divided into multiple areas with unique IDs used as keys. In this example, the DHT Chord by Stoica et al. (2003) is used as the lookup protocol. In Chord, both peers and keys are hashed to an identifier. These Chord IDs form a ring structure. Each peer has a link to its direct successor (the peer with the next higher Chord ID). The keys are also hashed and stored on their successors. In the example, area 16 is hashed to Chord ID 14. So, the peer with Chord ID 35 (successor of Chord ID 14) is responsible for the data concerning area 16.

To find a peer responsible for a given key one has to send a lookup query along the ring. By passing it on to a neighbor it will eventually reach its destination. In order to perform such queries more efficiently, each peer maintains a finger table with fingers pointing to peers that are at least 2^i IDs away with increasing i (indicated by the dotted lines). With this extension, the complexity of locating the responsible peer becomes logarithmic in the number of network participants.

Peers can join and leave the network freely. Upon a join, the new participant takes the responsibility for some set of keys, reducing the average workload in the network. When leaving, the peer hands over the stored data to some other responsible peer.

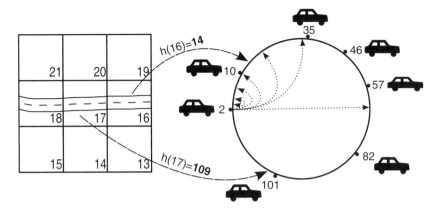

Figure 4.4 Distribution of street segments among peers.

In order to use the peer-to-peer system as a basis for a TIS application, participants send their observations to the responsible peers. Thereby, the information is made accessible to all other drivers. To gain information about current traffic conditions on a planned route, the inquiring application has to perform a number of lookups. For each segment of the route the responsible peer needs to be found and queried for the data. Owing to the random allocation of the hashed keys, a typical usage pattern involves many queries for neighboring segments that form a route.

Rybicki et al. (2007) propose a possible solution that aims to deal with this problem. If the observations of all segments of a route are stored 'close together' in the system, the number of independent lookups can be reduced. In order not to harm the Chord structure, a second finger table that consists of *semantic* fingers could be used. Such fingers point at peers responsible for street segments adjacent to the ones stored at the participant. Therefore the requesting node will only have to find a peer responsible for the first segment of the planned path and then 'follow' the route in the peer-to-peer structure without performing further lookups – in some sense, the road network is reflected in the overlay structure.

A standard challenge in peer-to-peer networks is to ensure fair load balancing. This becomes even more vital when dealing with data exhibiting a high skew, i.e., when some data are more demanded than others, which is also the case for TIS applications. For example, the responsibility for a segment on a highway means much more data to store and more queries to serve than managing a small street in the suburbs. The simplest way to address this problem and to obtain a fair load balancing is to increase the number of peers responsible for each segment beyond one, thereby introducing some redundancy and resulting in a fairer workload distribution.

The system as outlined thus far can offer the same functionality as the client/server architecture, while it does not require dedicated, centralized infrastructure. However, up to now, each car needs to obtain the full information about planned and alternative routes. In order to keep track of changes in the traffic situation on the planned route and to react to possible changes, the application has to perform periodic queries to other peers. This proactive solution will cause a substantial amount of redundant network traffic.

A possible solution would be to use communication paradigms for group communication like, for example, publish/subscribe. In this approach, the application, instead of periodically requesting new information, can register its interest for given data on the responsible peer ('subscribe') once and will be informed about any important changes on the planned route. The peer that is responsible for this route notifies all interested parties of substantial changes in the traffic conditions.

Probably the most compelling feature of the peer-to-peer approach is its self-scalability. As opposed to the client/server architecture, the amount of resources available in the network is proportional to the number of participants. Hence, a peer-to-peer network scales gracefully with an increasing number of users. Since the data are distributed among the participants there is no single point of failure and the financial and organizational costs of dedicated servers can be avoided.

A peer-to-peer overlay is based upon existing infrastructure-based communication networks. So, similar to the client/server case, the approach is not as dependent on a high equipment density as a VANET application. However, a drawback of the peer-to-peer approach is the higher overall bandwidth usage compared to a client/server architecture. This consists of the costs of relaying queries and updates of others and to maintain the overlay.

4.4.3 VANET communication

Considering the characteristics of a TIS application, it turns out that many vehicles are interested in similar information, especially when the vehicles themselves and their driving directions or destinations are similar. Neither client/server nor peer-to-peer networks based on cellular communications can make use of this fact in a straightforward way – VANETs, in contrast, can: because of its broadcast nature, the wireless multihop environment supports communication patterns where the same information is delivered to many or all the nodes within close vicinity very well. Hence, for this type of application, it is not unreasonable to use VANETs as a basis.

A common approach would be to use periodic update beacons in the following way: each car makes observations such as the traffic density, the number of free parking places, etc., related to a position in space (i.e., a road segment or a small area) and a point in time when the observation has been made. All or part of the locally stored information is periodically single-hop broadcasted. Upon reception of such a broadcast, a node incorporates the received data into its local knowledge base. By comparing the timestamps of observations, it can ensure that always the most up-to-date value for each position is stored and redistributed. A deeper discussion of this approach and alternative dissemination protocol designs can be found in the chapter on information dissemination in this book.

Because of the inherent scalability problems discussed above, disseminating all detailed measurement data within a large area is prohibitive. To overcome this problem, the use of hierarchical data aggregation has been proposed: with increasing distance, observations concerning larger and larger areas (or road segment lengths) are combined into one single value. Such an aggregated value could, for example, be the average speed on a longer road segment, or the percentage of free parking places in a part of a city. Coarse aggregates are made available at greater distances, more detailed data is kept only in the closer vicinity.

The quality of the TIS information provided to the driver depends fundamentally on the quality of the aggregation technique. In the subsequent sections, we will therefore look at two specific aggregation mechanisms in more detail. Both target an environment where cars take measurements of time-varying and location-dependent parameters, and this observation data is disseminated in a VANET via periodic beacons. We first describe an approach that employs a probabilistic data representation technique for aggregating information in an area-based fashion: with increasing distance, measurement values concerning larger and larger areas are combined into one single data item. This is very useful for aggregating parameters such as the number of free parking places within the area, the vehicle density, or the road condition.

We will then see that an area-based approach has limitations when it comes to the question of aggregating travel times in a road network for a dynamic road navigation system. To determine the driving route towards a destination that currently has the shortest travel time, summaries over areas alone are not sufficient: the average travel time along all road segments within some part of a city is not particularly useful for the purpose of identifying the optimal route. In Section 4.6 we will therefore continue with an approach that is able to aggregate route information.

4.5 Probabilistic, Area-based Aggregation

A key problem of data aggregation in VANETs is the rating of the quality or up-to-dateness of aggregated information: in contrast to single observations, aggregated values cannot be directly compared based on timestamps. The summaries – for instance, the total number of parking places within some part of a city – are created by cars that will typically not have the most up-to-date version of all underlying data available. Therefore, multiple aggregates for the same area may exist, based on different, but likely overlapping knowledge. To decide which one is based on 'better' underlying data is hard, if not impossible. So, how should a car which is provided with two different 'versions' of the same aggregate decide which one it should keep (and disseminate further)?

In order to answer this question we will expand the idea first outlined in the chapter on information dissemination and discuss a scheme, originally proposed in Lochert et al. (2007a), in more detail. This approach shows how the problem can be overcome if both single observations and aggregates do not carry the observed value directly, but instead contain an approximation of this value in form of a so-called soft-state sketch. While soft-state sketches still do not provide a way to compare the quality of two aggregates directly, they allow for something even better: multiple aggregates for the same area can be merged, yielding a new one that incorporates all the information contained in any one of the aggregates. Hence, there is no need to decide which aggregate contains more up-to-date information since the resulting aggregate comprises all the information from all aggregates that have been merged.

4.5.1 FM sketches

Soft-state sketches are an extended and modified variant of FM sketches. FM sketches were originally introduced in the context of databases, by Flajolet and Martin (1985). An FM sketch represents an approximation of a positive integer by a bit field $S = s_1, \ldots, s_w$ of length $w \geq 1$. The bit field is initialized to zero at all positions. To add an element x to the sketch, it is hashed by a hash function h. This hash function has geometrically distributed positive integer output; the probability that the hash value $h(x)$ of x is equal to i is given by

$$P(h(x) = i) = 2^{-i}. \tag{4.1}$$

The entry $s_{h(x)}$ is then set to one. (With probability 2^{-w} we have $h(x) > w$; in this case, no operation is performed.) A hash function with the necessary properties can easily be derived from a common hash function with uniformly distributed bit string output by using the position of the first 1-bit in the output string as the hash value.

The central result of Flajolet and Martin (1985) is that an approximation $C(S)$ of the number of distinct elements added to the sketch can be obtained from the length of the initial, uninterrupted sequence of ones, given by

$$Z(S) := \min(\{i \in \mathbb{N}_0 \mid i < w \wedge s_{i+1} = 0\} \cup \{w\}), \tag{4.2}$$

by calculating

$$C(S) := \frac{2^{Z(S)}}{\varphi}, \tag{4.3}$$

with $\varphi \approx 0.77351$.

The (quite substantial) variance of $Z(S)$ can be reduced by using multiple sketches in parallel, using a technique called probabilistic counting with stochastic averaging (PCSA). For details, see Flajolet and Martin (1985). When m sketches are used, PCSA yields a standard error of approximately $0.78/\sqrt{m}$.

FM sketches can be merged to obtain the total number of distinct elements added to any of them by a simple bit-wise OR. Important here is that, by their construction, repeatedly combining the same sketches or adding already present elements again does not change the results, no matter how often or in what order these operations occur.

4.5.2 Using sketches for data aggregation in VANETs

For the purpose of simpler discussion, let us consider a specific kind of traffic information system: assume that we are interested in disseminating the number of free parking places. For the moment, we ignore the fact that the measured quantities change over time. As a first step, we use a sketch (or, with PCSA, a set of sketches) for each road segment. We assume that a car is able to observe the current number of free parking places while passing a road segment, e.g., by collecting data from

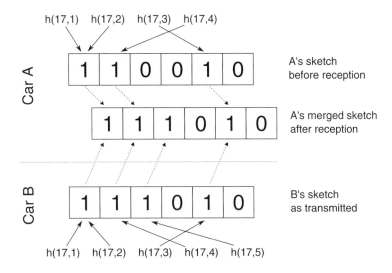

Figure 4.5 Generation and merging of FM sketch aggregates.

sensors on the parking places, as proposed by Piorkowski et al. (2006). After passing a road segment with ID r and observing x free parking places on it, a car may add the tuples $(r, 1), \ldots, (r, x)$ to the sketch for r, by hashing them and setting the respective bits. The same principle could be used if x were, for instance, the number of cars on the road segment as a measure of the traffic density.

The locally stored sketches for the road segments are periodically broadcast to the node's one-hop neighbors. Upon reception, received and local sketches are merged by calculating the bit-wise OR. Figure 4.5 exemplifies this procedure. Two cars, A and B, make independent observations on the same road segment (with ID 17). A observes four free parking places and thus hashes the tuples $(17, 1), \ldots, (17, 4)$ into its sketch for road segment 17. B observes five free parking places, and consequently adds $(17, 1), \ldots, (17, 5)$. If A and B meet later, and A receives a transmission containing B's sketch, A merges it by bit-wise OR and obtains a new sketch, replacing its previous one. Obviously, this can be repeated whenever cars meet, so that information from further cars can be incorporated, exchanged, and passed on, thereby accumulating 'knowledge' in the aggregate.

Note that the hashed tuples (r, i) are identical for different observers, the observed value determines only how many tuples are added. If all observers use the same hash function (something that could easily be standardized), the same number of free parking places on the same road segment will set the same bits, a lower number will set a subset thereof. Of course, in the current basic algorithm, bits that have once been set will never get unset again, and the sketch is therefore not able to follow decreasing values. We will therefore now discuss how to extend the data structure in order to overcome this limitation.

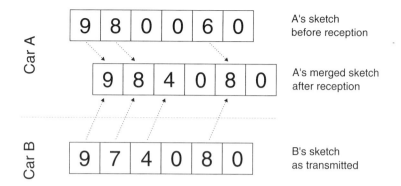

Figure 4.6 Merging of soft-state sketches.

4.5.3 Soft-state sketches

Since bits will never get unset, the sketches will always represent the maximum of all ever observed values for each road segment. To overcome this problem, a method is needed to remove old observations. This is accomplished by modifying the original FM sketches. We use small counters of n bits length instead of single bits at each index position. These counters represent a time to live (TTL) in the range $0, \ldots, 2^n - 1$ for that bit. The operation of setting a bit to one after an observation is replaced by setting the corresponding counter to the maximum TTL, to $T := 2^n - 1$. Beacons containing the sketches are sent at regular intervals. Just before sending such a beacon with information from the local knowledge base, all counters in the locally maintained sketches are decremented by one, if they are not yet zero.

When incorporating a received sketch into the local knowledge base, the bit-wise OR is substituted by a position-wise maximum operation. This yields a soft-state variant of FM sketches, in which previously inserted elements essentially die out after their TTL has expired, unless they are refreshed by a newer observation. The merging is visualized in Figure 4.6. Car A receives an aggregate from car B, and updates its own soft-state sketch accordingly.

For obtaining the current value from a soft-state sketch, the algorithm remains essentially unchanged; still, the smallest index position with value zero is identified and used. Note that this incurs some delay if a bit position is no longer set in newer observations. Coming back to our above example of observing parking places, assume that no further observations are made which set a particular bit position (e.g., because there is no longer a free parking place being hashed to it). If the position had previously been set in an aggregate, then the TTL value will decrease over time until it arrives at zero.

4.5.4 Forming larger area aggregates

Based on soft-state sketches, hierarchical aggregation can be accomplished in the following way. First, we need a hierarchy of areas over which the aggregation is performed. An area of a higher hierarchy level will typically consist of two or more areas of the lower level. (For a more general and more formal definition see Lochert et al. (2007a).) For each area on each hierarchy level, we make arrangements for one soft state sketch representing the (aggregated) value for the area. Any observation made for some location can immediately be incorporated into each aggregate that contains the location. Merging information about one subarea into a larger aggregate can be done in exactly the same way as merging two sketches for the same area: by calculating the position-wise maximum of the soft-state sketches.

From the properties of the sketches, it is easy to see that aggregates formed in this way will include the measurements from all their component areas, that they remain duplicate insensitive, and that therefore new information can be added dynamically at any time. Any received sketch can immediately be incorporated into any larger aggregate that fully contains the area of the received sketch. This finally gives us all the desirable properties mentioned above: we do not need to throw away one of multiple alternative aggregates for the same area – we can combine them and will obtain a new aggregate, which can be expected to show better coverage and to provide more up-to-date information of the area than any individual component from which it was formed.

Coming back to our example application of counting free parking places, these aggregates will contain the total counts in their respective areas. Information on small-scale areas will then typically be kept in the closer vicinity, while cars farther away will preferably maintain and distribute larger-scale aggregates of the region.

4.5.5 Application study

To complement the theoretical discussions of the scheme above, we show one example result from the simulation study conducted in Lochert et al. (2007a), using a simulation environment similar to the one above in Section 4.2.2, with 20% VANET equipment penetration ratio in the same city scenario.

We compare the results achieved with probabilistic aggregation to an artificial optimal reference protocol. In this protocol, too, information is disseminated by periodic beacons. But the optimal protocol does not care about practical bandwidth limitations. In a simulator the packet size and the amount of information actually exchanged between the nodes are independent. The reference protocol exploits this fact and does *not* aggregate information at all. Instead, a received 'aggregate' contains the sending node's most up-to-date measurement values of all the locations. Obviously, implementing this optimal reference protocol is easily possible in a simulator, but not in practice. However, it is well suited as a benchmark: with a practical protocol based on beacons and a knowledge base, the cars can never have better information.

We evaluate both our scheme and the optimal reference protocol with an idealized application. We subdivide the city area into 256 small areas, which we use as single locations. A simple stochastic process produces a time-varying value for each of these areas, which can be 'measured' by the cars the respective area. This could be interpreted as 'counting' the number of free parking places in this area, to stick with the example used above. It may likewise be considered an abstract model

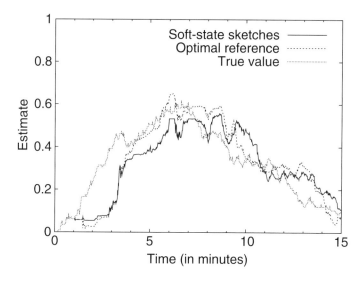

Figure 4.7 Accuracy of distant aggregate.

for any other time-varying, measurable quantity. In relation to the relatively short simulated time of 15 minutes, the changes of the simulated parameter are very rapid. The setting is thus very challenging for a dissemination protocol that needs to keep track of them. In addition to the time-varying parameter itself, one reference value per area is disseminated – e.g., the 'total number of parking places'. This allows us to calculate relative representations (the 'fraction of free parking places') and to distinguish between situations where no information is available on the one hand and where the disseminated value is zero on the other hand.

Figure 4.7 shows an instance of a region which is an aggregate over four of the areas in the simulation. The figure depicts how the aggregate value of this area is seen by cars in another area at a distance of about 3 km beeline. This is compared to the optimal reference protocol and the true simulated value in the observed area. For both the probabilistic aggregation scheme and the optimal reference protocol there is a clearly visible time lag for the information transport through the VANET. For the soft-state sketch aggregation scheme, there is also a certain delay before it follows a decreasing value, corresponding to the time needed for the soft-state decaying of the no longer set bits. It is also not astonishing that even the optimal reference protocol does not always have complete information. If up-to-date data only on parts of the total aggregation area is present and this data is not typical for the whole aggregate, effects like the overestimation around simulation minute six are the unavoidable consequence. Nevertheless, the estimates reflect the true situation in the modeled region well; this is also visible in other instances. For more details, we refer the reader to Lochert et al. (2007a).

In summary, sketch-based probabilistic aggregation can be used to create aggregates that come close to what can theoretically be achieved with the considered kind of system. This is a very encouraging result, which underlines that the proposed algorithm is a suitable way to overcome the often observed general difficulties of distributed, uncoordinated data aggregation in dissemination schemes.

4.6 Travel Time Aggregation

Area-based aggregation is well suited if an application demands summaries over areas. If this, however, is not the case for a specific application, it turns out that it has certain limitations. For building a dynamic, cooperative road navigation system based on VANETs, a common approach is to distribute information on the currently measured travel times along road segments. In this case, summaries over areas do not help. While certain concepts are certainly transferable, the total (or average) travel time *within an area* is not a useful concept. Therefore, further ideas are necessary in order to reduce the amount of travel time data with increasing distance, while still providing useful information to the recipients. We will now discuss another approach in detail that has already been sketched in the chapter on information dissemination. It was originally presented in Lochert et al. (2008) and tackles the aggregation problem for the specific case of travel time data supporting road navigation decisions. The key idea is to use coarser and coarser approximations of the road network with increasing distance.

4.6.1 Landmark-based aggregation

The basic idea of the aggregation scheme is based on so-called landmarks. These landmarks are points in the road network, typically junctions and intersections. Landmarks are defined on multiple levels of a hierarchy in the road network. At the highest level, these are junctions of the main roads or highways. Lower levels include all higher level landmarks plus more and more intersections of smaller streets. The lowest level is a representation of the full road network.

Cars passing a road segment can make an observation of the current travel time between two neighboring landmarks. This information is distributed within the closer surrounding. It is used by cars to calculate travel times between landmarks of the next higher level, thereby summarizing the travel times in the area. This coarser picture on the travel times is distributed in a larger area than the observations of individual cars. It is also used to calculate the travel times between landmarks of the next higher level of the hierarchy, and so on.

Figure 4.8 shows an example for a single hierarchy level. The travel time between the landmarks *Eiffel Tower* and *Arc de Triomphe* is determined. These two high-level landmarks are connected via a number of possible routes over landmarks on the next lower level, here indicated by circles. The travel times between these lower-level landmarks are known, either from direct observations or from received or previously calculated lower-level aggregates. The aggregated travel time from Eiffel Tower to Arc de Triomphe is the travel time along the minimal travel time route between these two points. Essentially, this compresses all information on all possible paths between two landmarks to a 'virtual' link connecting them.

This approach can be stated more formally as follows. The road network can be seen as a directed graph $G(E, V)$ consisting of junctions $v \in V$ and street

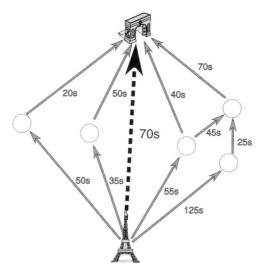

Figure 4.8 Landmark aggregation.

segments $e \in E \subseteq V^2$ connecting these junctions. The segments are rated with a weight $w(v_1, v_2)$ corresponding to the current travel time. Some junctions are distinguished as landmarks $l \in L$, $L \subseteq V$. A route $r(A, B)$ between two landmarks A and B is a sequence of junctions $(v_1, \ldots, v_n)$ such that $v_1 = A$, $v_n = B$ and all pairs of consecutive junctions are connected by a street segment, i.e., for all $i = 1, \ldots, n-1$ there exists $(v_i, v_{i+1}) \in E$. The cost of this route is

$$\|(v_1, \ldots, v_n)\| := \sum_{i=1}^{n-1} w(v_i, v_{i+1}).$$

Let $R(A, B)$ denote the set of all possible routes from A to B. We can then define the fastest route $r^*(A, B)$ as follows:

$$r^*(A, B) := \operatorname*{argmin}_{r \in R(A,B)} \|r\|.$$

$\|r^*(A, B)\|$ is used as the travel time between landmarks A and B on the next higher level, i.e., the operation performed when calculating an aggregate is to determine $\|r^*(A, B)\|$ based on lower-hierarchy travel times. Any standard routing algorithm may be used in order to calculate $\|r^*(A, B)\|$. Note that it is *not* relevant which route actually achieves this travel time. While the car is further away from both landmarks, the relevant information is that it is *possible* to travel from A to B within the given time. When it comes closer to the respective area, it will receive the locally available, more detailed information. This information can then be used for routing.

Aggregated travel times should not be calculated for each pair of landmarks. First, this would result in a number of aggregates that grows like $\Theta(n^2)$ with the number of landmarks. More importantly, however, travel time aggregates between landmarks that are very far away from each other do also not contribute much additional information: there will be many other landmarks 'in between', and a sequence of aggregates over those is likely to be a good approximation of the travel time between the distant pair of landmarks. Hence, only such pairs should be considered that are not too far apart. It is either possible to use a fixed criterion, like a maximum beeline distance of landmark pairs for which aggregates are formed on a given hierarchy level, or to mark the landmark pairs explicitly in the map data, choosing them such that a good approximation of the underlying road network is maintained. The further discussion here follows the latter approach.

4.6.2 Judging the quality of information

Before calculating the aggregated travel time between two landmarks and passing it on to other cars, a car needs to be able to judge whether its knowledge about the current traffic situation suffices for a good estimate. From a very general perspective, a very large number of road segments could lie on a possible route between two landmarks. It is therefore necessary to determine which road segments are likely to be relevant for the travel time estimate. An aggregate may be formed if information on these relevant road segments is locally available.

In order to define the set of relevant road segments, we look at what we call the *standard travel times* along the road segments. These travel times are hard-coded in the road map data and represent reasonable expectations of the travel times, as they are currently used for non-dynamic road navigation systems. One can easily calculate the *optimal standard route* $r^*_{\text{std}}(A, B)$ from A to B on the basis of this static data – this is essentially what current navigation systems do. For any route r, it is also easily possible to calculate the standard travel time $\|r\|_{\text{std}}$.

We then choose a threshold $\theta > 1$. We define the set $\mathcal{R}(A, B)$ of relevant road segments between two landmarks A and B to encompass all road segments that lie on a route for which the standard travel time is at most a factor of θ longer than the optimal standard travel time, i.e., a road segment e is in $\mathcal{R}(A, B)$ if and only if there exists a route $(v_1, \ldots, v_n)$ from A to B such that e is part of that route and

$$\|(v_1, \ldots, v_n)\|_{\text{std}} \leq \theta \cdot \|r^*_{\text{std}}(A, B)\|_{\text{std}}.$$

Since this criterion is based only on the (static) standard travel times, the set of relevant road segments does not depend on the current traffic situation or on a car's current knowledge. An aggregated travel time from A to B may be computed if information on the road segments in $\mathcal{R}(A, B)$ is available.

4.6.3 Hierarchical landmark aggregation

In order to perform hierarchical aggregation, landmarks are assigned a level in a hierarchy. Landmarks of a higher level are also members of all lower levels. More formally, for a set of landmarks L_i of an aggregation level i:

$$L_i \subset L_{i-1} \subseteq V, \quad i > 1.$$

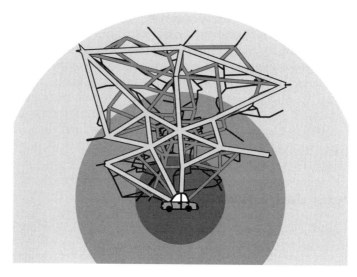

Figure 4.9 Availability of information on different levels.

To form an aggregate on hierarchy level i, the aggregates of level $i-1$ are used in the very same way as individual observations of cars are used on the first level. Thus, the landmarks on level $i-1$ are used like the junctions in the discussion above, and the aggregated travel times between them take on the role of the travel times along individual street segments.

The area in which individual observations and aggregates are distributed is limited based on their level in the hierarchy. Individual observations are distributed in a very limited range whereas the highest level aggregates are distributed in the whole network. Figure 4.9 depicts this. The circles around the car at the bottom of the scenario indicate the regions from which detailed, fine-grained level 0 information, slightly aggregated level 1 information, and more coarsely aggregated level 2 information is available to this car. As the car travels towards its destination, these regions shift accordingly.

The destination of a trip will not always be a high-level landmark position. Nevertheless, the aggregated information can of course be used for route planning. In order to do so, a navigation system would 'fill up' the missing information between the final destination and close-by landmarks by using the standard travel times hardcoded in the map data. This is reasonable, because a final decision on the last part of the route is not yet required at this stage – it is sufficient if a good choice for the immediately upcoming routing decisions can be made. While the car approaches its destination, the route can be updated and refined as more detailed information becomes available.

4.6.4 Evaluation

An evaluation that is again based on ns-2, VISSIM, and the Brunswick scenario used above is presented in Lochert et al. (2008). This reference also discusses and evaluates a scheme for finding optimal placements of roadside units (RSUs) in a city based on genetic algorithms. Here, we focus on results concerning the landmark aggregation scheme itself, but will also take the effects into account that supporting the application with a varying amount of infrastructure has. The evaluation therefore includes between 0 and 100 RSUs in the scenario. To assess the effects that result from a heavily disconnected network, the simulations are performed with a lower equipment density: 5% of the simulated vehicles participate in the VANET. For further details on the methodology we again refer the reader to the original publication.

The evaluation is application-centric and estimates the impact that the system has on the value of the application for the user. It concentrates on the travel time savings that can be achieved when a dynamic, VANET-based road navigation system is used, versus the use of a non-networked navigation system with static map data. It is based on samples of the estimated travel time savings, which are measured as the ratio between the travel time along the route suggested by the VANET-based road navigation system and the travel time along a route to the same destination that the non-networked navigation system would use.

Note that the car's current knowledge is typically not perfect. It will virtually always deviate from the current traffic situation to some extent (e.g., because the situation changes over time). The route calculated by the VANET-based system may therefore even be worse than the standard route. The travel time benefit is thus highly dependent upon the dissemination performance: it will be high if up-to-date information relevant for the route calculation is known by the car. The travel time savings are thus well suited as a metric for the application benefit.

It seems obvious that a higher number of RSUs improves the performance of the dissemination process and hence higher travel time savings can be achieved. In Figure 4.10 this expectation is confirmed. On the x-axis the number of used roadside units is shown. The y-axis shows the relative travel time, the error bars show 99.9% confidence intervals. A relative travel time of one means that the optimal route based on the car's current knowledge, denoted by r^*, is neither better nor worse than the route r^*_{std} that would be suggested based on static data, i.e., $\|r^*\| = \|r^*_{std}\|$. If RSUs are placed at 100 locations, an average car needs a relative travel time of 0.9 compared to the standard travel time. This is equivalent to a travel time saving of 10%.

Similar time reductions, however, are also possible with fewer roadside units. Even without any infrastructure support the aggregation-based dissemination scheme is able to deliver data to cars that can help to find better routes. The knee in the plot indicates that a good tradeoff between cost and utility in the considered city could be between 10 and 30 RSUs.

It should be noted that a large number of cars do not profit from the additional information, since the standard path to their destination is not congested, or despite a certain level of congestion no better alternative route exists. Those vehicles would not profit from any traffic information system at that time. However, they are included in the calculation of the average travel time savings. Cars for which better routes actually do exist often exhibit substantially larger improvements than the above average values. This can be seen by investigating the distribution of travel time savings. Figure 4.11 shows the cumulative distribution function of the individual relative travel times. The large fraction of cars with a relative travel time of one includes all those cars that would choose the same path without any dynamic information.

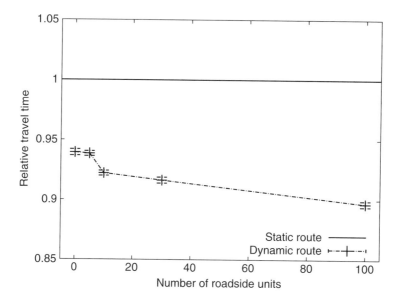

Figure 4.10 Achieved travel time savings with a varying number of roadside units.

4.7 Conclusion

In this chapter, we have considered several important aspects of VANET convenience applications. Their potentially significant role as a market introduction success factor on the one hand and the significant viability constraints imposed by limited capacity and connectivity as well as competition on the other hand define the two poles that dominate this area of research. We have illustrated how the latter constraints impact the realizability of VANET applications, and thereby also pointed out what protocol designers must keep in mind in a VANET context.

We have discussed traffic information systems in detail, in order to demonstrate that very demanding convenience applications can be realized via VANET technology. In particular, we presented two specific algorithmic approaches for designing scalable communication protocols in VANETs, and thereby demonstrated how it is possible deal with their inherent limitations.

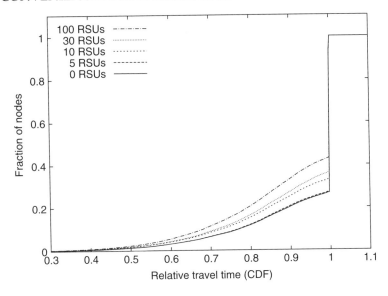

Figure 4.11 Cumulative distribution function of the relative travel times.

References

Caliskan M, Graupner D and Mauve M 2006 Decentralized discovery of free parking places. *VANET '06: Proceedings of the 3rd ACM International Workshop on Vehicular Ad Hoc Networks*, pp. 30–39.

Dornbush S and Joshi A 2007 StreetSmart Traffic: Discovering and disseminating automobile congestion using VANET's. *VTC '07-Spring: Proceedings of the 65th IEEE Vehicular Technology Conference*, pp. 11–15.

Flajolet P and Martin GN 1985 Probabilistic counting algorithms for database applications. *Journal of Computer and System Sciences* **31**(2), 182–209.

Grossglauser M and Tse DNC 2002 Mobility increases the capacity of ad hoc wireless networks. *IEEE/ACM Transactions on Networking* **10**(4), 477–486.

Gupta P and Kumar PR 2000 The capacity of wireless networks. *IEEE Transactions on Information Theory* **46**(2), 388–404.

Hull B, Bychkovsky V, Zhang Y, Chen K, Goraczko M, Miu A, Shih E, Balakrishnan H and Madden S 2006 Cartel: A distributed mobile sensor computing system. *SenSys '06: Proceedings of the 4th International Conference on Embedded Networked Sensor Systems*, pp. 125–138.

Ibrahim K and Weigle MC 2008 Optimizing CASCADE data aggregation for VANETs. *MoVeNet '08: Proceedings of the 2nd International Workshop on Mobile Vehicular Networks*.

Lee SB, Pan G, Park JS, Gerla M and Lu S 2007 Secure incentives for commercial ad dissemination in vehicular networks. *MobiHoc '07: Proceedings of the 8th ACM International Symposium on Mobile Ad Hoc Networking and Computing*, pp. 150–159.

Lochert C, Caliskan M, Scheuermann B, Barthels A, Cervantes A and Mauve M 2005 Multiple simulator interlinking environment for inter vehicle communication. *VANET '05: Proceedings of the 2nd ACM International Workshop on Vehicular Ad Hoc Networks*, pp. 87–88.

Lochert C, Scheuermann B and Mauve M 2007a Probabilistic aggregation for data dissemination in VANETs. *VANET '07: Proceedings of the 4th ACM International Workshop on Vehicular Ad Hoc Networks*, pp. 1–8.

Lochert C, Scheuermann B, Caliskan M and Mauve M 2007b The feasibility of information dissemination in vehicular ad-hoc networks. *WONS '07: Proceedings of the 4th Annual Conference on Wireless On-demand Network Systems and Services*, pp. 92–99.

Lochert C, Scheuermann B, Wewetzer C, Luebke A and Mauve M 2008 Data aggregation and roadside unit placement for a VANET traffic information system. *VANET '08: Proceedings of the 5th ACM International Workshop on VehiculAr Inter-NETworking*, pp. 58–65.

Matheus K, Morich R, Paulus I, Menig C, Lübke A, Rech B and Specks W 2005 Car-to-car communication – market introduction and success factors. *ITS '05: Proceedings of the 5th European Congress and Exhibition on Intelligent Transport Systems and Services*.

Nadeem T, Dashtinezhad S, Liao C and Iftode L 2004 TrafficView: traffic data dissemination using car-to-car communication. *ACM SIGMOBILE Mobile Computing and Communications Review* **8**(3), 6–19.

Palazzi CE 2007 *Fast Online Gaming over Wireless Networks*. PhD thesis University of California, Los Angeles.

Piorkowski M, Grossglauser M and Papaioannou A 2006 Mobile user navigation supported by WSAN: Full-fledge demo of the SmartPark system. *MobiHoc '06: Proceedings of the 7th ACM International Symposium on Mobile Ad Hoc Networking and Computing*. Technical Demonstration.

Rybicki J, Scheuermann B, Kiess W, Lochert C, Fallahi P and Mauve M 2007 Challenge: Peers on wheels – a road to new traffic information systems. *MobiCom '07: Proceedings of the 13th Annual ACM International Conference on Mobile Computing and Networking*, pp. 215–221.

Stoica I, Morris R, Liben-Nowell D, Karger DR, Kaashoek MF, Dabek F and Balakrishnan H 2003 Chord: A scalable peer-to-peer lookup protocol for Internet applications. *ACM/IEEE Transactions on Networking* **11**(1), 17–32.

Wischhof L, Ebner A, Rohling H, Lott M and Halfmann R 2003 SOTIS – a self-organizing traffic information system. *VTC '03-Spring: Proceedings of the 57th IEEE Vehicular Technology Conference*, pp. 2442–2446.

Xu H and Barth M 2006a An adaptive dissemination mechanism for intervehicle communication-based decentralized traffic information systems. *ITSC '06: Proceedings of the 9th International IEEE Conference on Intelligent Transportation Systems*, pp. 1207–1213.

Xu H and Barth M 2006b Travel time estimation techniques for traffic information systems based on inter-vehicle communications. *TRB '06: 82nd Annual Transportation Research Board Meeting*.

the huge impact that the deployment of VANET technologies could have on vehicular traffic systems, the growing effort in the development of communication protocols and mobility models specific to vehicular networks is easily understandable.

In this chapter, we are therefore going to illustrate the challenges to modeling vehicular motions and introduce the various options that may be found in literature. Figure 5.2 notably illustrates the guideline we will follow in this book, where models are considered in five categories as function of their scopes and characteristics:

- **random models** Vehicular mobility is considered random and the mobility parameters, such as speed, heading, and destination are sampled from random processes. A very limited interaction between vehicles is considered in this category.

- **flow models** Following the classification described in Hoogendoorn and Bovy (2001), single and multi-lane mobility models based on flow theory are considered from a microscopic, mesoscopic, or macroscopic point of view.

- **traffic models** Trip and path models are described in this category, where either each car has an individual trip or a path, or a flow of cars is assigned to trips or paths. Moreover, the impact of time on these models is also described.

- **behavioral models** They are not based on predefined rules but instead dynamically adapt to a particular situation by mimicking human behaviors, such as social aspects, dynamic learning, or following AI concepts.

- **trace-based models** Mobility traces may also be used in order to extract motion patterns and either create or calibrate mobility models. Another source of mobility information also comes from surveys of human behaviors.

We describe in this chapter the most popular models available in literature but address them from a networking and application point of view, notably by discussing their interactions with network simulators and emphasizing the role of VANET applications in determining their required level of precision. We introduce three major interaction classes controlling how traffic information is shared with network simulators and conversely how communication information may be used by traffic simulators to improve VANET applications. A network and a traffic simulator may be

- **isolated** Vehicular mobility traces are statically generated and parsed to a network simulator. No specific interaction is either defined or possible between the network and a traffic simulator.

- **embedded** A vehicular traffic simulator is embedded into a network simulator, or conversely a network simulator is embedded into a vehicular traffic simulator, allowing a bidirectional interaction between both simulators.

- **federated** A vehicular traffic simulator is not included into, but federated with, a network simulator through a communicating interface controlling the

OD Matrix	Home	Restaurant	Stadium	Shopping center	Park
Home	-	30%	10%	50%	10%
Restaurant	40%	-	10%	40%	10%
Stadium	30%	30%	-	10%	30%
Shopping center	40%	20%	10%	-	30%
Park	30%	30%	10%	30%	-

Origin–destination (OD) matrix

(a) OD matrix trip modeling

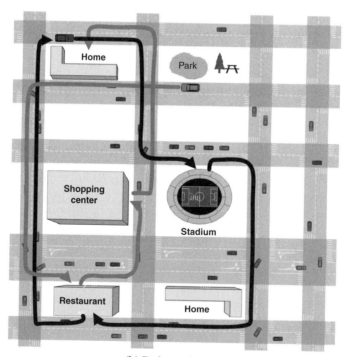

(b) Path modeling

Figure 5.1 The multilayer modeling concept of flow, path, and trip modeling as addressed in this chapter.

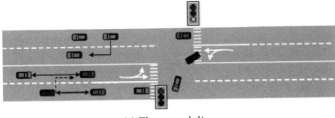

(c) Flow modeling

Figure 5.1 Continued.

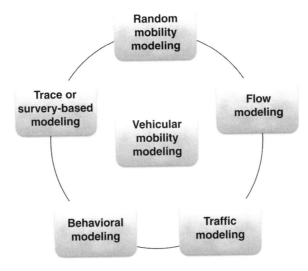

Figure 5.2 Classification of vehicular mobility modeling approaches as addressed in this chapter.

information exchanged between both simulators. Thanks to such a federation feature, other simulators, such as a VANET application simulator, may also be easily added.

We finally conclude this chapter by proposing a design framework for the generation of realistic vehicular mobility models depicting the required modeling modules, their level of precision and their interaction capabilities. Our objective is to provide readers with a clear vision of how the various options described in this chapter can be applied for the design of mobility models.

This chapter is organized as follows. We first provide in Section 5.2 a description of the general notations that we will use throughout this chapter. Section 5.3 introduces the historical random models. Then Section 5.4 provides a detailed description of how vehicular flows are modeled, while Section 5.5 illustrates how vehicular traffic is modeled. In Section 5.6 and Section 5.7 we describe the behavioral and trace-based models respectively. We finally cover the relationship between network and traffic simulators in Section 5.8 and introduce the framework describing the major required components for a vehicular mobility model in Section 5.9. Eventually, Section 5.10 discusses the current state and future directions in the area of vehicular mobility modeling, and Section 5.11 concludes this chapter.

5.2 Notation Description

The notations employed for the formal description of the vehicular mobility models discussed in this chapter are illustrated in Figure 5.3, where vehicle i will be considered as the reference vehicle.

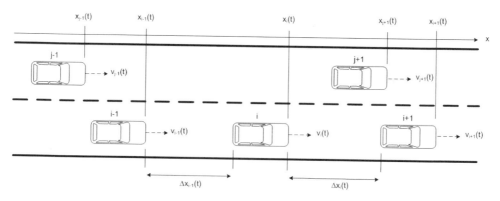

Figure 5.3 Notations employed for the vehicular mobility model descriptions in this chapter.

At time t, $x_i(t)$ and $v_i(t)$ represent respectively the position and speed of vehicle i. Indexes $i+1$ and $i-1$ represent respectively the vehicles immediately in front and behind vehicle i with positions $x_{i+1}(t)$ and $x_{i-1}(t)$, and with speed $v_{i+1}(t)$ and $v_{i-1}(t)$. The bumper-to-bumper distance between vehicle i and $i+1$ at time t is symbolized by $\Delta x_i(t)$. Indexes $j+1$ and $j-1$ represent vehicles on an immediately adjacent lane respectively in front or behind vehicle i. Their positions at time t are $x_{j+1}(t)$ and $x_{j-1}(t)$, while their speed at time t are $v_{j+1}(t)$ and $v_{j-1}(t)$. For models based on two dimensions, we also consider $\theta_i(t)$ the heading of a vehicle i at time t.

We will employ further notations throughout this chapter that are illustrated in Table 5.1. Note that these symbols may be found with indexes representing the vehicle to which they correspond. For vehicular mobility models computing the acceleration, we also employ $a_i(t)$ as the instantaneous acceleration of vehicle i at time t.

VEHICULAR MOBILITY MODELING FOR VANET

Table 5.1 Symbols employed throughout this chapter.

Symbol	Definition
Δt	time step in [s]
a	maximum acceleration in [m/s^2]
b	maximum deceleration in [m/s^2]
L	car length in [m]
v^{min}	minimum velocity in [m/s]
v^{max}	maximum velocity in [m/s]
v^{des}	desired or targeted velocity in [m/s]
θ^{max}	maximum heading in [rad]
θ^{min}	minimum heading in [rad]
T	safe time headway [s]
Δx^{safe}	safe distance headway in [m]
v^{safe}	safe velocity in [m/s]
τ	driver reaction time in [s]
μ	stochastic parameter in [0; 1]

5.3 Random Models

Random mobility models have been a long-time favorite for the modeling of random mobility patterns for computer science or telecommunication applications. Their popularity was largely due to their implementational simplicity and also to their stochastic properties that allowed one to conduct analytical studies and easily reproduce results. With the emerging of more specific applications, models have been tailored to mimic more realistic mobility patterns while keeping their stochastic nature. Considering vehicular mobility, despite the various efforts to improve their realism, these models were found inappropriate for modeling vehicular mobility for VANET applications. Nevertheless, in this section we provide brief descriptions of the major random mobility models that have been targeted at modeling vehicular mobility.

The baseline mobility models are illustrated in Figure 5.4. The most popular random model is the *Random Waypoint Model (RWM)* in which each vehicle randomly samples a destination d (a waypoint) and a speed v that will be chosen to move toward d. The major assumption is that vehicles maintain a fixed velocity between waypoints, a feature that is rarely observed in vehicular motions. A slight alternative is the *Random Walk Model (RWalk)*, which does not sample a destination, but instead randomly generates a moving azimuth θ and the journey time t. Despite their simplistic approach, these models have been heavily studied and employed in modeling mobile networks.

Observing that vehicles as well as humans tend to move in social groups, the *Reference Point Group Mobility Model (RPGM)* tried to mimic such a pattern. For that matter, nodes are separated into groups, where a group leader determines the group's general motion pattern ($v_{leader}(t)$ and $\theta_{leader}(t)$) and where the motion of any other group member ($v_{member}(t)$ and $\theta_{member}(t)$) can only slightly deviate from the group leader's movement. Formally, RPGM defines movements according to the following rules:

$$v_{leader}(t) = rand(0,1) \cdot v^{max} \tag{5.1}$$

$$\theta_{leader}(t) = rand(0,1) \cdot \theta^{max} \tag{5.2}$$

$$v_{member}(t) = v_{leader}(t) + \mu \cdot v^{max} \cdot v_{dev}^{max} \quad \text{where } 0 \le v_{dev}^{max} \le 1 \tag{5.3}$$

$$\theta_{member}(t) = \theta_{leader}(t) + \mu \cdot \theta^{max} \cdot \theta_{dev}^{max} \quad \text{where } 0 \le \theta_{dev}^{max} \le 1 \tag{5.4}$$

where v_{dev}^{max} and θ_{dev}^{max} are the maximum deviation ratio around respectively the maximum velocity and heading of the group leader and where μ is a stochastic parameter.

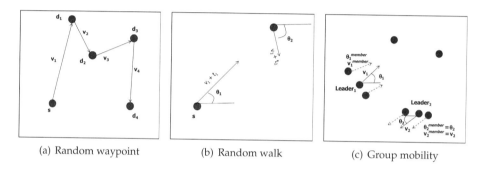

(a) Random waypoint (b) Random walk (c) Group mobility

Figure 5.4 Random mobility models.

A second step to improve the realism of these models regarding vehicular mobility has been to restrict the movement of cars on street graphs (see Figure 5.5). Vehicles therefore randomly sample destinations on graph vertices and are restricted to moving at a specific velocity on the graph edges. The graph may be adapted to a specific environment. Notably, the *Freeway* model restricts the movement on several bi-directional multi-lane freeways, while the popular *Manhattan* model restricts vehicular movements to urban grids. In both models, the movement of an individual vehicle is modeled according to the following set of rules:

$$v_i(t + \Delta t) = v(t) + \tau \cdot a \cdot \Delta t \tag{5.5}$$

$$\text{IF } \Delta x_i(t) > \Delta x^{safe} \text{ THEN } v_i(t) \le v_{i+1}(t) \, \forall i \tag{5.6}$$

where Equation (5.5) shows that the next speed is temporally dependent on the previous one, and Equation (5.6) illustrates that the velocity of a following vehicle i cannot exceed that of the preceding one $i + 1$. In the case of the Manhattan model, vehicles use a *stochastic turn* function f_{turn} that randomly chooses the next movement among all possible directions at each intersection.

A complete description of the previously described models and their impact on networking may be found in Bai et al. (2003). Extensions of these two approaches

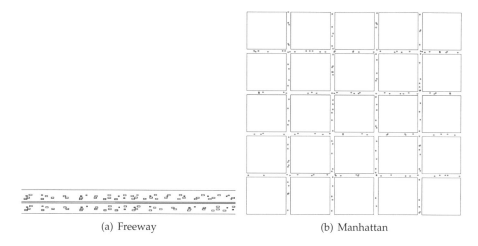

(a) Freeway   (b) Manhattan

Figure 5.5 Random mobility models on graphs.

to better mimic typical vehicular traffic could also be found. For example, Saha and Johnson (2004) extended this approach by importing topological maps from the US Census Bureau TIGER database (TIGER 2006) and by assigning speeds on a per-edge basis. The highest speed path is then computed by weighting the cost of using a specific edge considering the allowed speed. Other models falling into this category are BonnMotion (2005); Jaap et al. (2005); Zhou et al. (2004).

The *Freeway* and *Manhattan* models implement a basic interaction between cars that will be explained in detail shortly. However, such interaction is too limited to have an impact on the traffic flow. Moreover, the ability of Saha and Johnson (2004) to weight a trip based on speed constraints may also be seen as a first step toward modeling realistic vehicular trip patterns.

5.4 Flow Models

Compared to the computer scientists or the telecommunication engineers, the civil and traffic engineers were confronted very early on with a need for higher modeling details than simple random patterns. They notably had to consider the physical interactions between different vehicles and their interaction with the environment. By imitating nature, they approached the problem by modeling vehicular mobility as flows.

The literature considers the following three classes of flow models:

- **microscopic modeling** This class of traffic flow describes the mobility parameters of a specific car with respect to other cars in detail. It usually commands the car's acceleration/deceleration in order to maintain either a safe distance headway or to guarantee a safe time headway (reaction time). Its high level of precision is reflected by a similarly high computational complexity.

- **macroscopic modeling** This category does not consider the mobility parameters of a specific car but instead quantities of macroscopic meanings such as *flow, speed, or density* are modeled. Inspired by fluid theory, macroscopic flows have the advantage of a reduced computational complexity compared to the microscopic ones but they can still realistically model macroscopic quantities.

- **mesoscopic modeling** This approach proposes describing traffic flows at an intermediate level of details. Individual parameters may be modeled, yet of a macroscopic meaning. The objective is to benefit from the scalability of the macroscopic approach but still providing a detailed modeling close to microscopic models.

Typical applications under study with these flow models were, for example, the evolution of traffic on a lane merge or the study of an optimal ramp access policy. More recently, automated cruise control or even automated driving had also been investigated with these models. The common factor in each of these applications was the absolute requirement to avoid accidents. Almost all models that may be found in literature have therefore been developed for accident-free environments. Yet, the ability to actually model accidents by drivers not respecting safety or regulatory rules is also crucial to the evaluation of VANET applications such as active safety. As such features depend on an individual driver's behavior, this domain is mostly found in behavioral theory.

In the rest of this section, we are going to describe these different classes in more detail and provide examples of popular models that have been used to model driver motions.

5.4.1 Microscopic flow models

Car following models (CFM)

Car following models (CFMs) are probably the most popular class of driver model. CFMs usually represent time, position, speed, and even acceleration as continuous functions, but most have been extended to provide discrete formulations. CFMs adapt a following car's mobility according to a set of rules in order to avoid any contact with the leading vehicle. A general schema is illustrated in Figure 5.6.

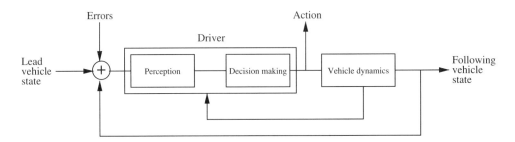

Figure 5.6 General schema for car following models.

The objective of (CFMs) is to model vehicular traffic avoiding accidents by controlling each individual car's driving dynamics to maintain a safe inter-distance between vehicles, a safe time-headway or both. The founding remark to that objective is probably the *Pipe's rule*

> A good rule for following another vehicle at a safe distance is to allow yourself at least the length of a car between you and the vehicle ahead for every 16.1 km/hours (10 mph) of speed at which you are travelling. (Pipe 1953)

Pipe's rule may be generalized by the general *Collision Avoidance (CA)* or *Safety Distance* formulation:

$$\Delta x^{\text{safe}}(v_i) = L + T \cdot v_i + \psi \cdot v_i^2 \tag{5.7}$$

where L is the vehicle length, T the safe time headway that may also be considered as reaction time, and where $\psi \cdot v_i^2$ represents the breaking distance. ψ is an adjusting parameter function on the deceleration b_i and b_{i+1} of the following and leading vehicles. When $\psi = 0$, the breaking distance between leading and following vehicles are considered equal, while ψ is maximized when the leading vehicle instantaneously comes to a stop. The safe distance headway $\Delta x^{\text{safe}}(v_i)$ is therefore the minimum distance for a driver, as a function of its speed v_i, to come to a full stop without accident, including reaction and breaking times.

The Gipps Model (Gipps 1981) is based on such CA or Safety Distance approach. The safety constraint is enhanced such that not only does vehicle i not crash into vehicle $i+1$ in case of a sudden stop but also vehicle $i-1$ is able to react to avoid an accident with vehicle i.

From a different perspective, a group of researchers at General Motors research, notably Gazis, Herman and Rothery, Chandler and Montroll, approached the problem with rational physical considerations and introduced the *Stimulus–Response* approach:

$$response = sensitivity * stimulus \tag{5.8}$$

By considering human perceptions, Chandler et al. (1958) proposed that the acceleration of a following car should be proportional to the *perceived* relative speed between leading and following vehicle, with a given reaction time between these two quantities:

$$\frac{dv_i}{dt}(t) = \gamma \cdot (v_{i+1}(t-T) - v_i(t-T)) \tag{5.9}$$

where γ represents the *sensitivity* of a driver, while the *stimulus* is defined by the relative speed difference between a leader and a follower. The response may be considered as the acceleration or braking of the following vehicle.

The sensitivity has been formulated by Gazis et al. (1961) as:

$$\gamma = c \cdot \frac{v_i^m(t)}{\Delta x_i^l(t-T)} \tag{5.10}$$

where c is an adjusting coefficient, m is a speed exponent with values typically in $[-2; 2]$ and l is a distance exponent with values typically in $[-4; 1]$. Brackstone and McDonald (1999) provide typical values for c, l, m as proposed by various studies.

By regrouping Equations (5.9) and (5.10), we obtain the formal definition of the *Gazis–Hermann–Rothery (GHR) model* also known as the *General Motor (GM)* model:

$$\frac{dv_i}{dt}(t) = c \cdot v_i^m(t) \cdot \frac{\Delta v_i(t-T)}{\Delta x_i^l(t-T)}. \tag{5.11}$$

The GHR model therefore defines the acceleration of a vehicle at time t, as a function of the speed and distance differences between two vehicles at time $t - T$.

When adapting the GHR model in a discrete formulation, we have an extra parameter Δt as the update interval during which the acceleration is considered constant. For an optimal discrete implementation, $\Delta t \leq T$ and the ratio $T/\Delta t$ should result to an integer. A discrete formulation of the GHR model is as follows:

$$x_i[t] = \frac{1}{2} \cdot a_i[t - \Delta t] \cdot \Delta t^2 + v_i[t - \Delta t] \cdot \Delta t + x_i[t - \Delta t] \tag{5.12}$$

$$v_i[t] = a_i[t - \Delta t] \cdot \Delta t + v_i[t - \Delta t] \tag{5.13}$$

$$a_i[t] = c \cdot v_i^m[t] \cdot \frac{\Delta v_i[t-T]}{\Delta x_i^l[t-T]}. \tag{5.14}$$

Different models have been later proposed in different directions. Brackstone and McDonald (1999) and Panwai and Dia (2005) classified Car Following Models in five classes: *(GHR)-like Models, Psycho-Physical Models, Linear Models, Collision Avoidance (CA)*, and *Fuzzy Logic Models*. A description of the differences between these models is out of scope of this book and refer the interested reader to these references for further details on their specificities. Next we briefly describe the most popular models that are notably used in well known traffic simulators.

Intelligent Driver Model (IDM) The IDM is also based on a *stimulus–response* approach which computes the instantaneous acceleration consisting of a free acceleration $a^{free} = a \cdot [1 - (v_i/v_i^{des})^4]$ to reach the desired speed v_i^{des}, and an interaction deceleration $a^{int} = -a \cdot (\delta/\Delta x_i(t))^2$ with respect to a leading vehicle $i+1$. The *stimulus* in IDM is the gap ratio between the current gap Δx_i and the desired gap δ, the follower trying to catch up with a pulling-away leader or slowing down with a closing-in leader. The IDM may be expressed with the following equations:

$$\frac{dv_i}{dt}(t) = a\left(1 - \left(\frac{v_i(t)}{v_i^{des}}\right)^4 - \left(\frac{\delta(v_i(t), \Delta v_i(t))}{\Delta x_i(t)}\right)^2\right) \tag{5.15}$$

$$\delta(v_i(t), \Delta v_i(t)) = \Delta x^{rest} + \left(v_i(t)T + \frac{v_i(t) \cdot \Delta v_i(t)}{2\sqrt{a \cdot b}}\right) \tag{5.16}$$

where v_i^{des} is the desired velocity distributed in $[v^{min} \cdots v^{max}]$, Δx^{rest} is the gap between two vehicles at rest and the other terms are listed in Table 5.1. As for the GHR model, a discrete formulation adds an update interval Δt between which a constant acceleration is assumed. Equations (5.12) and (5.13) may also be used for position and speed updates.

From Equations (5.15) and (5.16), we can also observe that the IDM is a pure deterministic model obtaining the instantaneous acceleration solely based on deterministic stimuli and thus cannot model irrational behaviors. This model is notably used in the traffic simulator VanetMobiSim (2009).

The Krauss Model Another approach for a pure stimulus–response approach is the Krauss model, Krauss et al. (1997). Different from IDM, the Krauss model is discrete in time as it does not compute the instantaneous acceleration but the future speed at time step $t + \Delta t$ to be reached by a vehicle i.

Also different from IDM, the Krauss model tries to model human sporadic and irrational reactions with a stochastic parameter μ, a feature that the IDM is unable to mimic. Taking as input the maximum speed $v^{\max}$, the maximum acceleration and deceleration a resp. b, and the stochastic parameter μ, Krauss defines the following set of equations:

$$v_i^{\text{safe}}(t + \Delta t) = v_{i+1}(t) + \frac{\Delta x_i(t) - v_{i+1}(t) \cdot T}{\Delta v_i(t)/(2 \cdot b + T)} \quad (5.17)$$

$$v_i^{\text{des}}(t + \Delta t) = \min[v^{\max}, v_i(t) + a \cdot \Delta t, v_i^{\text{safe}}(t + \Delta t)] \quad (5.18)$$

$$v_i(t + \Delta t) = \max[0, v_i^{\text{des}}(t + \Delta t) - \mu] \quad (5.19)$$

$$x_i(t + \Delta t) = x_i(t) + v_i(t) \cdot \Delta t. \quad (5.20)$$

Equation (5.17) computes the speed of vehicle i required to maintain a safe interdistance and avoid creating an accident with the preceding vehicle $i + 1$. Similarly to IDM, a reaction time T is also used in order to add a jitter to the response to the stimuli. Once the safe speed has been computed, Equation (5.18) computes the target speed to be reached by node i which is a simple increment from the previous speed upper-bounded by v_i^{safe} and $v^{\max}$. Equation (5.19) updates the current speed for the next time step Δt according to a stochastic variation around the target speed μ. Finally, Equation (5.20) updates the position of vehicle i for the next time step. This stochastic variation models the physical capacity to effectively provide a given speed increment and also the driver's conformance to the model.

The Krauss model is used by the traffic simulator SUMO (2009).

The Wiedemann Model This falls into the psycho-physical category and takes a different approach from the two previous models. It is possible that different drivers will show different reactions in response to a same stimulus. For example, a driver will certainly not react similarly with respect to a change of the inter-distance (the stimulus) if it is very close to, or very far from, that car. Notably, when two vehicles are close, drivers tend to apply smooth fine-grained acceleration or deceleration and not a brutal reaction. Wiedemann (1974) therefore proposed a psycho-physical model that considers the perceptual psychology for the reaction to stimuli.

The Wiedeman model identified four driving states that a driver may be in and that would control its reaction to a similar stimulus:

- **Free-driving** The driver is far from any immediate preceding vehicle and is thus not influenced by any other vehicle. It freely accelerates to reach v^{des}.

- **Approaching mode** The driver is influenced by an approaching preceding vehicle and applies a normal deceleration rate to reach a safe inter-distance.

- **Following mode** The driver is in a typical CFM with low acceleration or deceleration rate to guarantee a fine granularity of its reaction.

- **Breaking mode** The driver is critically influenced by a preceding vehicle and applies a higher deceleration rate to avoid a crash.

This model is used by the commercial traffic simulator VISSIM (2009).

Cellular automata models (CA)

CA models are a different class of driver model. Unlike CFM, CA are discrete in *space* and *time* therefore reducing the computational complexity but still being able to mimic drivers' reaction in response to their environment. CA-models describe a traffic system as a lattice of cells of equal size. Similarly to psycho-physical models, a set of rules is provided to control the movement of vehicles moving from cell to cell. The cell size is chosen such that it can host a single vehicle that can move at least to the next cell in one time step Δt. Without loss of generality, the velocity is expressed as the number of cells per time step and we consider a time step $\Delta t = 1$.

The most popular CA is the Nagel and Schreckenberg (1992) (N-SCHR) model which defines the following rules:

1. **Accelerating:** $v_i(t+1) = \min(v^{\max}, v_i(t))$

2. **Breaking:** $v_i(t+1) = \min(v_i(t), \Delta x_i(t))$

3. **Randomization:** $v_i(t+1) = RAND(0 \cdots (v^{\max} - 1))$ with probability p

4. **Moving:** $x_i(t+1) = x_i(t) + v_i(t+1)$.

Figure 5.7 illustrates a typical application of the N-SCH model to a single lane road. We used the notation $action_i : v_i(t) \to v_i(t+1)$ to represent the dynamics of the N-SCH model. For example, $Acc_i : 3 \to 4$ means vehicle i is in an acceleration phase between time t and $t+1$ and increases its speed from 3 to 4 m/s.

At time t, we have four vehicles each of them on an acceleration phase. A vehicle i that previously moved three cells accelerates to increase its speed to move four cells at the next time step as it has at least that number of free cells ahead. Yet, at time $t+1$, i is suddenly very close to vehicle $i+1$ and must break to reduce its speed to one cell as it is the number of free cells available ahead. We can observe that unless the system has a critical number of vehicles which would limit the maximum speed, sudden and unrealistic breaks may be produced by CA models. The reason comes from the synchronous movements at the next time step. Indeed, if node i had known that node $i+1$ was already breaking, it could have avoided accelerating and may even have decelerated preemptively. Yet, it can not have such information as all nodes are making a decision at the same time. Moreover, similar to a domino effect, the action of the leading vehicle itself depends on its own leading vehicle. Nevertheless, cellular automata such as the Nagel and Schreckenberg model have been shown to efficiently model vehicular traffic as they are notably used by the TRANSIMS (2009) traffic simulator from the US Los Alamos Laboratory.

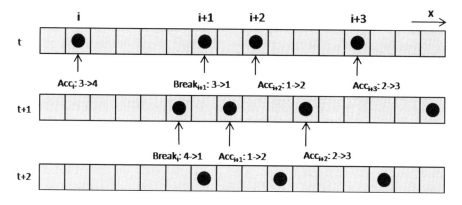

Figure 5.7 The Nagel–Schreckenberg cellular automaton model.

5.4.2 Macroscopic flow models

Unlike microscopic driver models, the macroscopic approach is inspired by hydrodynamic phenomena and do not intend to reproduce individual but rather macroscopic quantities. In order to adapt these hydrodynamic phenomena to vehicular mobility, let us consider a small road segment $[x, x + \Delta x]$, later referred to as a road segment x. Macroscopic models describe the join evolution of the vehicular *density* $\rho(x, t)$, the vehicular *velocity* $v(x, t)$ and the vehicular *flow* $m(x, t)$. The density reflects the expected number of vehicles located in x at time t. The flow is the expected number of vehicles passing by x for a time interval $[t, t + \Delta t]$, and the velocity is the expected speed of vehicles in x, and is related to density and flow following this equation:

$$v(x, t) = m(x, t)/\rho(x, t).$$

The basic equations in fluid models are the conservation of vehicles:

$$m(x, t) = \rho(x, t) \cdot v(x, t) \tag{5.21}$$

$$\frac{\partial \rho(x, t)}{\partial t} + \frac{\partial m(x, t)}{\partial x} = 0 \tag{5.22}$$

the first illustrating the relationship between flow, speed, and density, and the second describing that the vehicular density in x varies according to the incoming and outgoing flows in x. This provides a system of two equations with three unknowns. The third equation is usually the differentiating factor between the various macroscopic models available in literature.

The most well-known fluid model is the *Lighthill–Whitham–Richard (LWR) model*. It expresses the velocity as a function of the density $v(x, t) = v(\rho(x, t))$ and leads to:

$$\frac{\partial \rho(x, t)}{\partial t} + \frac{\partial \rho(x, t) \cdot v(\rho(x, t))}{\partial x} = 0 \tag{5.23}$$

$$\Longleftrightarrow$$

$$\frac{\partial \rho(x, t)}{\partial t} + \frac{\partial \rho(x, t) \cdot v(\rho(x, t))}{\partial \rho} \cdot \frac{\partial \rho(x, t)}{\partial x} = 0. \tag{5.24}$$

The simplified computational complexity of the LWR model makes it a good candidate for large-scale traffic simulations but does not reach the same level of precision as the microscopic flow models. Although the LWR model is able to model kinetic waves, it is notably unable to model cluster effects smaller than the region over which the macroscopic parameters have been measured, a serious limiting factor for urban vehicular modeling.

The assumption from kinematic waves in general and from the LWR in particular is that $m(x,t)$ depends on $\rho(x,t)$ only, which also means that $v(x,t)$ depends on $\rho(x,t)$ only. In other words, the assumption is that there are no speed adaptation delay effects when moving from a sparse area x_1 to a dense area x_2 and vice versa. When a vehicle moves from a sparse area to a dense area, the speed is immediately adapted, a behavior lacking realism especially in urban areas where density and speed could have significant variances.

Extensions of the LWR models proposed to relax this assumption by adding *inertia* to the speed. For example, considering a linear relationship between speed and density ($\partial_\rho v(\rho) = \pm C$), we can add a *diffusion* factor on the velocity by representing $v(x,t) = v(\rho(x+D,t))$ as the adaptation of the speed as a function of density at the *headway* distance D to the next vehicle. By developing and substituting back to Equation (5.22), we obtain

$$v(\rho(x+D,t)) = v(\rho(x,t)) + D \cdot \frac{\partial v(\rho(x,t))}{\partial \rho} \cdot \frac{\partial \rho(x,t)}{\partial x} \quad (5.25)$$

$$\Longleftrightarrow$$

$$\frac{\partial \rho(x,t)}{\partial t} + \frac{\partial(\rho(x,t) \cdot v(\rho(x,t)))}{\partial \rho} \cdot \frac{\partial \rho(x,t)}{\partial x} = \pm C \cdot \frac{\partial \rho(x,t)}{\partial x} \quad (5.26)$$

where $D \approx 1/\rho(x,t)$ and where Equation (5.25) is obtained from a Taylor expansion of $v(\rho(x+D,t))$. Equation (5.26) means that the velocity depends not only on the density in x but also on the gradient of the density. Further improvements have been added to this approach, notably by also modeling a reaction time to velocity changes (Payne 1973).

Numerical implementations of macroscopic models also exist. Most of them are approximations of the continuum models spatially by dividing space into cells of size Δx, temporally by considering the evolution of traffic in a time period Δt, or both.

For the case of both space and time domains, Equation (5.22) may be rewritten and the LWR equation $v(x,t) = v(\rho(x,t))$ approximated respectively as

$$\rho[\Delta x, t + \Delta t] = \rho[\Delta x, \Delta t] + m^{in}[\Delta x, \Delta t] - m^{out}[\Delta x, \Delta t] \quad (5.27)$$

$$v[\Delta x, t + \Delta t] \approx v^{max} \cdot \left(1 - \frac{\rho[\Delta x, t + \Delta t]}{\rho^{jam}}\right) \quad (5.28)$$

where $m^{in/out}(\Delta x, \Delta t)$ are respectively the number of vehicles flowing *in* and *out* of a cell during the time Δt. Equation (5.28) is the Greenshields (Greenshields 1935) linear approximation of $v(x,t) = v(\rho(x,t))$ and where ρ^{jam} is the jam density. v^{max} and ρ^{jam} are border parameters calibrated optimally for each cell as $v^{max} = v(\rho)$ for $\rho \to 0$ and $\rho^{jam} = \rho(v)$ for $v = 0$.

Equations (5.27) and (5.28) as well as border parameters ρ^{jam} and v^{max} therefore allow a representation of the LWR equation in discrete domain. Other discrete formulations of the LWR or Payne models may be found in Hoogendoorn and Bovy (2001).

The major advantage of fluid motion models is the *gross* characterization of vehicular mobility and potentially reduced computational load. We refer the reader to Hoogendoorn and Bovy (2001) and Nagel et al. (2003) for more details on this class of flow model.

5.4.3 Mesoscopic flow models

Mesoscopic models represent an intermediate modeling level of traffic flows and model gross characteristics of traffic flows at an aggregate level, typically as probability density functions, but describe the interactions between vehicles at an individual level. A specific model may describe the velocity distribution at a specific time and space or the vehicular arrival rate or time-headway at a specific time and space and at the same time control the behavior of a vehicle as a function of this information. The mesoscopic models did not initially attract as much attention as the macro- or microscopic ones but lately showed an efficient trade-off between the modeling of individual vehicles and the modeling of large quantities of vehicles. Mesoscopic models may take different forms such as the modeling of the headway distribution or the size or density of a cluster of vehicles.

A typical example of mesoscopic models is the *gas-kinetic* traffic flow model that was introduced by Prigogine and Herman (1971). This class of model is a generalization of the macroscopic models for compressible fluids and thus may also be found alternatively classified as macroscopic models. An interesting feature of gas-kinetic models is that they are able to model sparse traffic, where the interaction between vehicles is rare or simply absent, much better than LWR or Payne models. The basic concept in gas-kinetic models is the *phase-space-density* (PSD) representation $\rho(x, v, t)$ which corresponds to the distribution function of cars in an interval dx, dv, dt around x, v, t, which may be interpreted as the expected number of vehicles in that interval. The PSD is a mesoscopic generalization of the macroscopic term $\rho(x,t)$ which can be obtained from PSD as $\rho(x,t) = \int_0^\infty \rho(x,v,t)\, dv$. Similarly, the macroscopic speed may also be obtained from PSD as $v(x,t) = \int_0^\infty v \cdot \rho(x,v,t)\, dv$.

Another typical example of mesoscopic models is the queue model that was initially introduced to traffic theory by Gawron (1998) and later extended by Cetin et al. (2003). The behavior of each vehicle is modeled but its dynamics depending on the macroscopic parameters of a FIFO queue representing a road segment between two junctions. Each queue is characterized by its length l^{queue}, its capacity C, its free flow speed v_0, the number of vehicles already in the queue n_{vehicle} and the number of lanes n_{lane}. Then, similarly to data communication, each time a vehicle accesses a road, it is entered into the corresponding queue. The time required to move along the road is computed according to the following general equation:

$$t_{\text{travel}} = f(l^{\text{queue}}, v_0, n_{\text{lane}}, n_{\text{cars}}). \tag{5.29}$$

Various expressions for the travel time could be used but it has been shown by Gawron (1998) that in homogeneous traffic a simplistic expression such as l^{queue}/v_0, which neglects the impact of the vehicular density on the speed, can lead to a very good approximation of motion patterns obtained with more complex microscopic driver models. Finally, when this time expires, the vehicle i is extracted from the queue and entered to the queue corresponding to the next road that the vehicle follows.

Obviously, as vehicles cannot be 'dropped' like a packet when the size of a queue is exceeded, and as each road has a finite incoming and outgoing capacity, restrictive measures are added.

- A car cannot leave a queue if the maximum outgoing capacity is exceeded.
- A car cannot leave a queue unless it has found another queue that will accept it.
- A queue cannot accept a car if it exceeds its capacity.

The last condition particularly allows the modeling of the backward propagation of a traffic jam across an intersection. The capacity of a queue is usually computed as $l^{queue} \cdot n_{lane}/L$, where L is the length of a vehicle. Figure 5.8 illustrates the mesoscopic queue approach for driver modeling.

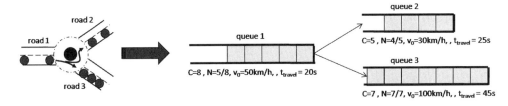

Figure 5.8 Queue-based modeling of an intersection, where C is the capacity of the queue, N is its filling ratio, v_0 is the average speed, and t_{travel} is the time to travel through the queue. A vehicle reaching road 1 is pushed into the corresponding queue 1 and is scheduled to be pulled out of it after t_{travel}. Then it is pushed into the corresponding queue 2 or 3 only if its capacity permits. Otherwise, it remains in its current queue and contributes to the congestion in the queue and to an increase of t_{travel}.

Such an approach is also able to model complex intersections and very large-scale urban topologies. For example, Cetin et al. (2003) used this approach to conduct a large-scale traffic generation of a whole county using parallel computing.

5.4.4 Lane changing models

So far, all the models we have considered are based on single-lane traffic. Multi-lane traffic also needs to be studied considering non-homogeneous vehicles and driving patterns. For instance, a vehicle approaching a slow moving truck might prefer to overtake than to slow down. Modeling lane changing behaviors is actually as complex as modeling human motion patterns because it significantly depends on human behaviors. Indeed, a realistic modeling should actually include three parts: the **need** for lane changing, the **possibility** of lane changing, and the **trajectory** used for lane changing. Each part is important for a realistic lane changing attempt. Most of the models are based on a *Gap Acceptance* threshold (Gipps 1986) or a set of rules (Chowdhury et al. 1997). But other approaches (Ahmed 1999; Treiber and Helbing 2002) also consider forced merging, behavioral aspects or game theory.

Table 5.2 Notations for the gap acceptance concept.

Symbol	Definition
$\Delta x_i^{safe}(v_i^{des})$	safe inter-distance between vehicle i and $i+1$ as a function of the desired speed of i in [m]
CF Gap	$\Delta x_i^{safe}(v_i)$ safe inter-distance between i and $i+1$ as a function of the current speed of i in [m]
lag gap	expected inter-distance between vehicle i and $j-1$ after the lane change in [m]
lead gap	expected inter-distance between vehicle i and $j+1$ after the lane change in [m]
$\Delta x_{j-1}^{safe}(v_{j-1})$	safe inter-distance between vehicle i and $j-1$ if on the same lane in [m]
$\Delta x_{j+1}^{safe}(v_{j+1})$	safe inter-distance between vehicle i and $j+1$ if on the same lane in [m]

Figure 5.9 illustrates the general concept behind lane changes. Considering the notations in Table 5.2, the **need** occurs when $CF\,gap < \Delta x_i^{safe}(v_i^{desired})$, in other words, when the current safe inter-distance with the leading vehicle would need to be increased if i would reach or keep its desired speed. The **possibility** occurs when $lag\,gap > \Delta x_{j-1}^{safe}(v_{j-1})$ AND $lead\,gap > \Delta x_{j+1}^{safe}(v_{j+1})$, in other words, when the inter-distance between i and $j-1$, resp. $j+1$, would not violate the safety distance constraints of resp. $j-1$ and $j+1$, thus avoiding accidents on the adjacent lane by the lane change. If the **possibility** is not fulfilled, i does not have any other choice than to reduce its speed v_i to respect the safety constraint. Finally, the **trajectory** is modeled by the time of the lane change and the chosen new lane. Lane changes are also modeled differently as a function of the modeling class, i.e. microscopically, mesoscopically, or macroscopically.

Microscopic lane changing models

The conceptual difference of lane changing models compared to single-lane models is that drivers not only have to consider the leading vehicle but all vehicles around them in any adjacent lane. All microscopic models previously presented were enhanced with lane-changing models.

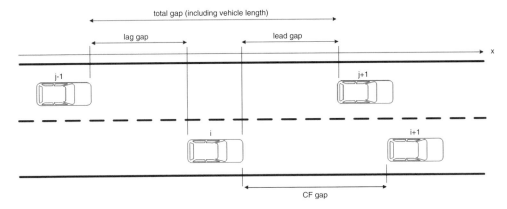

Figure 5.9 The gap acceptance approach.

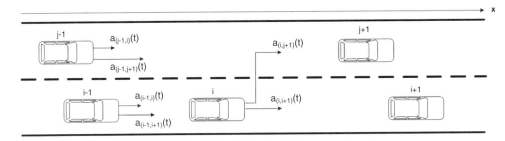

Figure 5.10 MOBIL: lane changing model for IDM.

The IDM model uses the *Minimizing Overall Braking Induced by Lane changing (MOBIL)* model, Treiber and Helbing (2002). According to Figure 5.10, the decision to change lane depends on the following equation:

$$a_{i,j+1} + p \cdot [a_{j-1,i} + a_{i-1,i+1}] > a_{i,i+1} + p \cdot [a_{i-1,i} + a_{j-1,j+1}] + \delta_{\text{thr}} \quad (5.30)$$

and $a_{j-1,i} \geq b_{j-1}^{\text{safe}}$, where $a_{\alpha,\beta} = a^{\text{IDM}}(v_\alpha, d_{\alpha-\beta}, v_\alpha - v_\beta)$ and where b_{j-1}^{safe} is the safe deceleration of $j - 1$. In practice, $a_{i,j+1}$ and $a_{i,i+1}$ represent the car's own advantage, while $[a_{j-1,i} + a_{i-1,i+1}]$ represents the required acceleration of neighboring vehicles after the lane change and $[a_{i-1,i} + a_{j-1,j+1}]$ the acceleration of vehicles before the lane change. Finally, p is a politeness factor controlling the selfishness of the car (ignoring the impact of its own decision on neighboring vehicles). To avoid frenetic lane changes, a threshold δ_{thr} is used in order to wait before a significant need to change lane occurs.

Lane changing models also exist for the Nagel–Schreckenberg model (Nagel et al. 1995 or Chowdhury et al. 1997), Krauss (1998), or Wiedemann (1991). The approaches of Krauss and of Wiedeman are conceptually similar to MOBIL. Due to its synchronous decisions, the cellular automaton approach deserves a brief

VEHICULAR MOBILITY MODELING FOR VANET

comment. As illustrated in Figure 5.11, when a vehicle i wants to change lane, it has to compute the maximum movement of following vehicles in the different lanes as it cannot know what would be their progress. It has to guarantee that its inter-distance with the following vehicle $j-1$ is at least larger than the distance moved by $j-1$ at maximum speed at the next time step. Therefore, the **need** is formulated as $v_i^{\text{desired}} > \Delta x_{i,i+1}$ and the **possibility** is expressed as $\Delta x_{i,j+1} > \Delta x_{i,i+1}$ and $\Delta x_{j-1 \to i} > v_{j-1}^{\max}$.

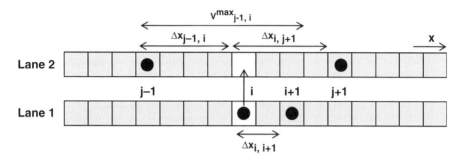

Figure 5.11 Lane changes for cellular automata models.

Mesoscopic lane changing models

It is also possible to model lane changes for mesoscopic models by controlling a specific vehicle for which the decision is made on aggregated metrics (for instance queue service time). The decision metric depends on the approach. For queue models, for example, it is when the travel time v^{travel} (the time to exit the queue) becomes too long. Considering queue models, multi-lanes are modeled by separated single-lane FIFO queues each of them having different characteristics and occupancy levels. The **need** being a too-long travel time, the **possibility** is to find an alternative queue that would provide a benefit while limiting the cost to other occupants of the queue. When the decision is made, the trajectory places the vehicle at the same position in the new queue. Figure 5.12 illustrates a typical lane change.

Macroscopic lane changing models

Modeling lane changes for macroscopic models is conceptually more problematic as individual drivers desire to change lane cannot be modeled. By observing nature, the hydrodynamic phenomenon that could represent lane changes is adjacent rivers overflowing onto each other, each lane being represented by one river. Cars cannot change lane up to a desired flow/speed/density on a specific lane. Then, when it is reached, the vehicles 'overflow' onto adjacent lanes. Note that the flow conservation is still kept but considering all available lanes instead of each ones.

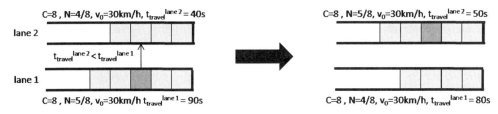

Figure 5.12 Lane changes for queue models, where C is the capacity of the queue, N is its filling ratio, v_0 is the average speed and t_{travel}^q is the time to travel through queue q. A queue represents each lane of a road and a vehicle in lane 1 estimates that its travel time on lane 2, $t_{travel}^{lane\,2}$, might be better than its current travel time on its lane, $t_{travel}^{lane\,1}$, and changes lane.

The basic flow equation is extended to also consider overflows:

$$m_l(x,t) = \rho_l(x,t) \cdot v_l(x,t) \tag{5.31}$$

$$\frac{\partial \rho_l(x,t)}{\partial t} + \frac{\partial m_l(x,t)}{\partial x} = \Phi_l = \sum_i \Phi_{l \to i} - \Phi_{l \to l} \tag{5.32}$$

where $\Phi_{l \to i}$ and $\Phi_{l \to l}$ are the flow leaving lane l to lane i and incoming to lane l from lane i respectively. Φ_l therefore represents the net flow gain as the sum of incoming flows from other lanes and the sum of outgoing flows to other lanes. Note that if multi-lanes were not supported, Φ_l would simply be 0. More details related to lane changes for traffic stream models may be found in Laval and Daganzo (2006).

5.4.5 Intersection management

A realistic flow model would not be complete without a description of its interaction with intersections or traffic signs. We mentioned in the previous sections that flow models are usually configured with a desired speed v^{des} that vehicles try to reach. In practice and to respect the law, such target speed should not be higher than a speed limitation on a specific street. Similarly, a flow model should also respect the traffic signaling at intersections. Intersection management is modeled as an obstacle to flow models. Depending on the type of signaling, such obstacle may 'appear' and 'disappear' as a function of a controlling entity (green or red light). Similarly to flow models, intersection management has different levels of modeling:

- *microscopic* An intersection with traffic lights or signs is modeled similarly to a stopped vehicle. The driver model acquires the state of the traffic sign/light and moves accordingly.

- *mesoscopic* In queuing models, an intersection is modeled as the end of a queue. Stop signs or traffic lights are modeled by varying the queue discharge capacity.

- *macroscopic* An intersection is modeled as a pipe branching. Conceptually, traffic signs or lights may be modeled by locally reducing or stopping the flow for a given time interval. Yet, as macroscopic models cannot model individual vehicles, stop or yield signs and complex traffic light patterns are difficult to model.

In order to mimic the well-known cluster effect created by intersections, flow models must therefore interact with intersections. Solely from a flow model's point of view, an intersection is just another obstacle to be considered similarly to any other car. The intersection management itself is what creates a particular traffic pattern. For example, different semaphore times for traffic lights will alter the size of the cluster effect, or traffic lights synchronization may optimize traffic flows on a particular direction or lane.

5.4.6 Impact of flow models on vehicular mobility

Before closing this section related to flow models, we would like to illustrate the impact of flow models and their configuration parameters on vehicular mobility. We compare here one random model, one macroscopic flow model, and one microscopic flow model with the help of a basic benchmark test that is used in vehicular flow theory. We refer to Fiore and Härri (2008) for a larger analysis of more vehicular mobility models with the help of a larger set of benchmark tests.

The benchmark test illustrated here is called the *flow-density* diagram and depicts the evolution of the flow of vehicles on a single lane street when the density of vehicle is increased. This test is usually also known as the *lambda shape* diagram from its recognizable shape, as the flow of vehicle typically increases by increasing the density of vehicles until the street reaches its saturation point. Then any further increase of density will decrease the vehicular flow and create congestions. The objective of a mobility model is therefore to recreate such an easily observable vehicular mobility pattern, as a failure to do so will lead to a wrong modeling of traffic density, flow, and speed.

Figure 5.13 depicts the lambda shape test for the Freeway random model, the LWR macroscopic flow model, and the IDM microscopic flow model, as previously described in this chapter. We wanted to also include a random model in order to justify the use of flow models in vehicular mobility modeling. We can first see in Figure 5.13(a) that, by missing a correct microscopic interaction between vehicles, the Freeway random model fails the test. The vehicular flow keeps on increasing as a function of an increasing vehicular density, a pattern that is never observed in real life. According to this benchmark test, the Freeway model does not appropriately model vehicular mobility as it cannot correctly recreate traffic congestion. As depicted in Figure 5.13(b) and 5.13(c), both LWR and IDM models pass the test, which shows that both flow models implement a sufficient level of realism in the interaction between vehicles. Yet, the saturation point and the lambda shape are slightly different between the LWR and IDM models: notably the vehicular flow of the LWR model drops after the saturation point significantly faster than that of the IDM. This aspect may come from granularity losses between the continuous models and their discrete implementations. It could also come from the different,

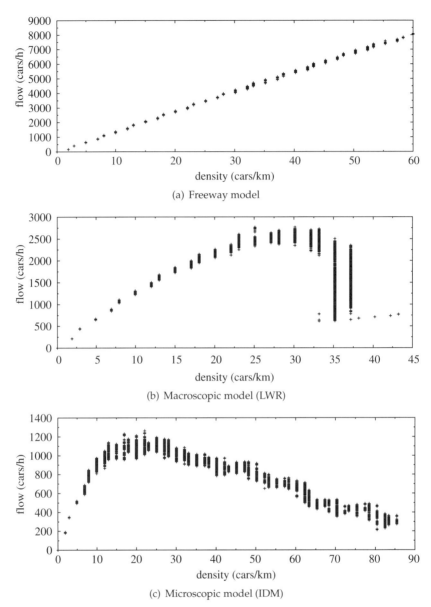

Figure 5.13 Vehicular outflow as a function of density for one random and two flow models. (Reproduced by Permission of © 2008 ACM.)

possibly inappropriate, values employed for the configuration parameters of each model and justifies a careful choice of parameter values as well as their calibration with real traces when using flow models for vehicular mobility modeling.

5.5 Traffic Models

In the previous section, we described the various approaches to modeling vehicle motion patterns as flows. While this fine-grained approach is able to efficiently model the impact of the immediate neighborhood of a vehicle on its driving dynamics, vehicular traffic is also significantly influenced by large-scale and coarse-grained traffic effects. According to the multiscale modeling introduced in Section 5.1, a traffic model class must be introduced.

In order to illustrate the differences between the flow and the traffic model classes, let us consider the motion dynamics approaching an intersection. Flow models simply consider an intersection as a potential obstacle and a vehicle either reduces its speed to adjust to the turning angle or simply decelerates to come to a full stop if the right-of-way is not granted. Yet no provision is included on what to do either at or after that intersection. On the other hand, traffic models are in charge of modeling the intersection policies (traffic light, stop sign, etc.), the turning policy (stochastic turn, precomputed turns), and the global path to be followed by a vehicle (Origin–Destination (OD) matrices).

In literature, flow and traffic models have been investigated almost independently for a very long period as each approach had its own objectives and applications. Traffic or civil engineers developed precise flow models for small-scale traffic studies, for example to understand the impact of a ramp access or a lane suppression on traffic flows. Transportation engineers developed solutions for dynamic traffic assignment (DTA) and modeled global trips and paths from origin to destination points in order to evaluate large-scale mobility, for instance to simulate city-wide evacuation plans. One of the particularities of VANET is that both approaches are necessary in order to model correct connectivity patterns at small and large scale.

Traffic models may be divided into two complementary motion patterns: a *trip* and a *path*, both influenced by a third parameter: *time*. A trip models the sequence of OD points that vehicles visit, while a path specifies the precise way followed by vehicles between an origin and a destination point. Time schedules the vehicles' departing time and thus impacts the chosen path, but also alters the OD points and the transition probabilities.

Conceptually, this approach may be considered as *macroscopic* modeling, as tendencies and coarse motion patterns rather than exact interactions between vehicles are modeled. This consideration actually created confusions between the denomination of 'macroscopic flow' models described in the previous section and the 'macroscopic' traffic models to be described in this section. We therefore opted for 'traffic models' as the denomination of this class of mobility models.

By observing vehicular traffic, we may also see that both trips and paths are far from being random. Humans do not randomly choose OD points but rather move from their residential areas to their workplaces or to the grocery store. Similarly, humans do not randomly pick a path from these OD points but tend to take the most appropriate one according to their habits (fastest path, low traffic path etc.). We will describe here the various approaches to model trips and paths for vehicular traffic.

5.5.1 Trip planning

Trip planning has the responsibility for modeling vehicles moving from an origin to a destination point. Three major approaches may be found in the literature: *random trip*, *stochastic turn*, and *OD matrix*.

The random trip is the simplest approach and which is notably used by the random models described in Section 5.3. Vehicles randomly select an origin and a destination point in the traffic environment. No correlation is either modeled between the different destinations or between vehicles. Another solution for a random trip planning is not to choose any specific destination but rather choose a new direction at each intersection or any other 'waypoint'. This approach is called *stochastic turn*, as it randomly selects the next direction according to a stochastic process. In that particular case, a path planning is not required. The calibration of the stochastic turn process is usually performed by field measurements of turning flows at intersections.

The OD matrix approach on the other hand selects *Points of Interests (PoI)* on the traffic environment and builds a transition matrix to model the correlations between various trips. For example, landmarks may be extracted from TIGER (2006) files. Figure 5.14 illustrates a typical OD matrix with transition probabilities. Although this approach is more complex as it requires an extraction of PoI and corresponding transition probabilities, it significantly contributes to mimicking the strong non-uniform distribution of OD and the correlations found in various vehicular trips. Surveys are usually the primary source of information to identify the OD points and estimate the transition probabilities.

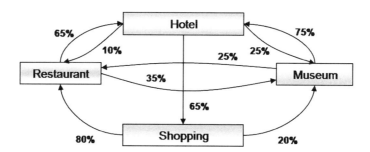

Figure 5.14 Origin–Destination (OD) matrix trip planning, where vertices are Points of Interest (PoI) and the transition probabilities are the tendency to move from one PoI to another.

Similarly to flow modeling, trip planning may be considered with different degrees of precision. Each vehicle may have a specific OD matrix with vehicle-specific transition probabilities, or all vehicles may share a common OD Matrix. Trip models are usually assigned to a flow of vehicles as it reduces the problem of modeling the large-scale mobility patterns to known physical properties. Behavioral or agent-based models, as we will discuss later, define a trip for each individual agent, who will then try to improve it in contact with other agents.

5.5.2 Path planning

Once the origin and destination points have been assigned, it is the role of the path planning to determine the sequence of directions to be followed by each vehicle to reach its destination. Path planning usually precomputes or dynamically recomputes the sequence of intersections to be followed based on a preferred optimization function (shortest path, fastest path, less crowed path, etc.).

Inspired by graph theory, path planning is a very challenging task that requires scalable and dynamic algorithms as they typically have to build paths for a very large set of vehicles over a large area. Moreover, as the best paths are usually based on dynamic weights that are altered by the paths' usage itself, paths must often be recalculated or alternative paths be precomputed. Paths are mostly based on efficient Dijkstra-type graph algorithms where the link weights depend on different parameters (distance, speed, density, habits, etc.). Efficient path planning is a very promising research field in traffic telematics as it is expected from inter-vehicular communications to provide path weights that would be more precise and more rapidly available. We describe next the major orientations that can be found in literature.

Flow-centric versus agent-centric path planning

When considering a large number of vehicles to be simulated over a large urban map consisting of thousands of vertices, a path planning algorithm may pace a scalability issue. In order to deal with this as a function of the simulation environment, path planning may be seen from a microscopic or macroscopic point of view as illustrated in Figure 5.15.

The microscopic path planning, called in this chapter *agent-centric*, creates at least one distinct path per vehicle. It controls each individual vehicle and any specific action to be conducted on the vehicle (drivers changing their mind, accident, traffic jam) is immediately applied by recomputing its optimal path. Despite its computational cost, the agent-centric approach is receiving increasing interest from the VANET community as it is able to immediately model the impact of a traffic accident not only on immediate vehicles but also on other vehicles that could have heard the information by inter-vehicular communications. The agent-centric approach is notably used by the traffic simulators MATSim (MATSim 2009), VanetMobiSim (2009), and Schroth et al. (2005).

The macroscopic path planning, called in this chapter *flow-centric*, takes a different approach. In order to increase scalability, it only builds a subset of paths, and a flow of vehicles will be following the same path. The major asset is its reduced computational complexity as the number of paths is usually significantly smaller than the number of vehicles. Yet, it has the same limitations as any macroscopic approach as it models vehicles as flows and cannot control the reaction of an individual vehicle confronted by specific traffic situations. In order to model vehicles being re-routed, a subset of alternative paths are precomputed. This proactive computation is actually also a limitation as it is either not possible or too complex to optimize these alternative paths according to the dynamic evolution of traffic. Yet, its high scalability made flow-centric models the first choice of popular traffic

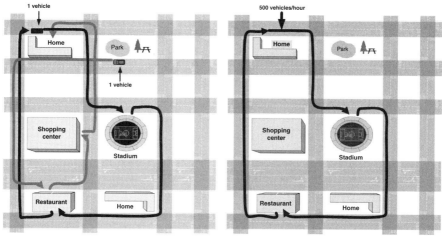

(a) Agent-centric path planning   (b) Flow-centric path planning

Figure 5.15 Agent-centric versus flow-centric path planning, where the agent-centric approach provides one path sequence for each vehicle, while the flow-centric approach provides one path sequence for each flow of vehicles (for example, a flow of 500 vehicles/hour follows a same path).

simulators, such as SUMO (2009), VISSIM (2009), Aimsun (2009), and CORSIM (2009).

5.5.3 Influence of time

By observing traffic over a single day, it is also easy to see that patterns are significantly different as a function of time. During morning rush hours, we usually observe an inbound traffic stream while the evening rush hours show the opposite. When planning trips or paths, it is therefore also crucial to include the influence of time in the equation. For example, the OD matrix of Figure 5.14 could be totally different or simply have different transition probabilities as a function of the time of day. Also, the path departure times, per agent or per flow, must also be defined as function of time. In order to estimate such time patterns and calibrate the trip or path planning, surveys or statistics from traffic traces are usually used. We will discuss this approach in Section 5.7.

5.5.4 Impact of traffic models on vehicular mobility

Before concluding this section on traffic models, we would also like to illustrate their impact on vehicular mobility. We describe here one typical aspect and refer readers to Fiore and Härri (2008) for a broader analysis of the influence of vehicular mobility models on the vehicular networking shape.

For this illustration, we used a 4 × 4-block irregular grid, each block having a 250 m length, with an average vehicular density of 20 vehicles/km/lane. Vehicles enter and leave the scenario from the borders of the grid and move according to the IDM flow model with a traffic light intersection management. Each street has the same speed limit except the first horizontal and vertical streets located at 250 m that we purposely configured to allow higher speeds. We used a random trip model between all entry/exit points but configured two different path models in order to emphasize their impacts on the mobility patterns. In Figure 5.16(a), vehicles choose the shortest path to their destinations and we can easily observe a regular and uniform accumulation of vehicles at each intersection. In Figure 5.16(b), vehicles select their paths as a function of the allowed speed limits, a pattern mimicking drivers' natural tendency to choose higher-speed streets to reduce the driving time. Vehicles therefore tend to use the two high-speed streets on their paths and we can accordingly observe a higher accumulation of vehicles at intersections located on these two streets. Despite it being a basic example, the different motion patterns depicted in Figure 5.16 show how a correct path model may be crucial in order to model observable vehicular mobility.

5.6 Behavioral Models

A major limitation of most synthetic models comes from the complexity of modeling detailed human behaviors. Drivers are far from being machines and cannot be programmed to follow a specific behavior in all cases. Instead they respond to stimuli and local perturbations that may have a global effect on traffic modeling. The stimulus–response approach of car-following models such as the Wiedeman model is an initial attempt to model human behaviors. We could also observe in the recent evolution of synthetic driver models a gradual increase of behavioral rules replacing strict physical ones. Although this is related to human mobility, Musolesi and Mascolo (2006) illustrated such an approach by developing a synthetic mobility model based on social network theory, then validated it using real traces. They showed that the model was a good approximation of human movement patterns. That was a signal showing that behavioral theory itself could also be a different methodology for modeling vehicular traffic.

In behavioral theory, actions do not only result from a stimulus–response pattern but also from social or physical influences and even artificial intelligence through a learning process. For example, Legendre et al. (2006) proposed an approach where the behavioral rules are a set of attraction or repulsion forces modeled as vectors. Considering vehicular traffic, the target destination is typically a strong attraction force. Any obstacle or any other vehicle located in between will generate repulsion forces. By summing up the influence of all attraction or repulsion forces, this approach obtains a directional movement vector. Considering that such computation must be performed at each time step, since forces change due to vehicular movements, the major drawback of this approach is its computational cost.

Balmer (2007) proposed a multi-agent-based behavioral model, from the observation that OD matrices and flow-based models were lacking enough granularities

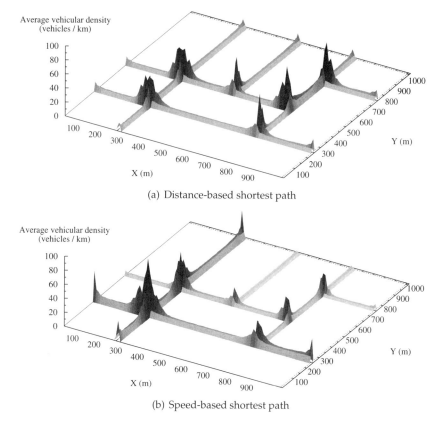

Figure 5.16 Impact of two different path models on the average vehicular density distribution. (Reproduced by Permission of © 2008 ACM.)

in adapting motion patterns to drivers' individual traffic experiences and the correlation with other drivers. The approach is similar to assigning a unique OD matrix to each agent (driver) and using an agent-based path modeling. As the behavioral model previously described, Balmer et al. proposed the use of behavioral rules and learning cycles. A driver for instance does not randomly choose to go to one point of interest but instead goes there because it has particular business to do and the path will depend on the environment and its preferred habits. Such behavior is unique to each driver and changes as a function of their own learning curve. Generalizing, each agent has their own intelligence and will react differently to similar stimuli, as each agent has their own vision of the environment, their own preferred ways of acting on the environment and their own objectives.

For simulating this multi-agent approach, Balmer et al. created the multi-agent traffic simulator MATSim (2009). It is composed of a dynamic agent-based behavioral model and a queue-based mesoscopic flow model. We briefly provide here the major steps followed by this multi-agent behavioral model. MATSim contains the following modules that are called in a cycle:

- **agent plan database** Each agent initially creates a plan, consisting of their intentions during a particular time period (trip, path, etc.). The agent database may contain several plans as agents learn and dynamically change their plans.

- **execution** The plans of each agent are executed simultaneously based on the mesoscopic queue model.

- **scoring** Plans are evaluated based on particular score functions depending on individual or global metrics of interest.

- **strategy** The agents learn during this phase. According to the scores obtained by each plan, plans are modified following a particular strategy. The process is then iterated (execution, scoring, strategy, plan change) until the objective of the strategy is fulfilled.

Behavioral theory may also be used in order to model non-normative motion patterns. Unlike approaches previously described in this chapter, Espié (2006) describes a behavioral approach that does not specifically respect traffic rules. Considering an intersection environment, Espié et al. address the relevance of drivers always respecting rules compared to behaviors observed in daily traffic and provide the agent with more autonomy with respect to violating traffic rules. Their study showed that non-normative behavioral modeling was also necessary to obtain more realistic mobility patterns. Further studies on modeling driver behaviors in automotive environment may be found in Cacciabue (2007).

Behavioral theory is currently subject to an increasing interest mainly due to the simple observation that parts of the parameters describing VANET applications in general or vehicular mobility in particular are related to human behaviors. Failing to model them correctly will significantly alter the success of VANET applications. As an example, one promising application is active traffic safety where a system will provide traffic advice to drivers in order to avoid accidents. Considering now that a critical part of the factors influencing the success of the application is the driver's conformance to the advice, how should one model a human being not following safety advice, or more generally the traffic rules? Behavioral theory might help in answering this question.

5.7 Trace or Survey-based Models

Because of the complexity of modeling vehicular mobility, only a few very complex synthetic models are able to come close to a realistic modeling of motion patterns. One could also follow a different approach. Instead of developing complex models and then calibrating them using mobility traces or surveys, crucial time could be saved by directly extracting generic mobility patterns from movement traces. Such an approach has recently become increasingly popular as mobility traces have started to be gathered through the various measurement campaigns launched by projects such as Crawdad (2009), Dieselnet (2009), MIT Reality Mining (2009), USC MobiLib (2007), or Cabspot (2006). The most difficult part of this approach is to extrapolate patterns not observed directly by traces. By using complex mathematical models, it is possible to predict mobility patterns not reported in the traces to some extent. The limitation is also often linked to the class of the measurement campaign. For instance, if motion traces have been gathered for bus systems, an extrapolated model cannot be applied to the traffic of personal vehicles.

Another limitation for the creation of trace-based vehicular mobility models is the limited availability of vehicular traces. Some research groups are currently implementing testbeds, but the outcome might not be available soon, if they are even made available to the public. We list here three existing projects providing vehicular mobility traces that may be used to extract vehicular mobility models. The first one, called Dieselnet (2009) and created by the University of Massachusetts, provides mobility traces of a bus system in the city of Amherst, MA USA. Cabspot (2006) is a second source of vehicular mobility traces. Taxis in the San Francisco Bay Area have been equipped with GPS and communication devices which periodically transmit the position of the respective taxis to a central database. A visualizer shows live taxi movements but mobility traces are also stored and made available to the public. The third one is the GATech vehicular traces (Fujimoto et al. 2006), which are GPS vehicular movement traces gathered during a network connectivity campaign on a segment of highway. Although solutions for vehicular mobility are at this time limited, a significantly larger number of solutions have been proposed and are available for human mobility on Crawdad (2009).

Surveys are also an important source of macroscopic mobility information that can be either used to calibrate a synthetic model or even to create one. One source of large-scale surveys is the US department of Labor, which gathered extensive statistics of US workers' behaviors, spanning from commuting time or lunch time, to traveling distance or preferred recreational occupancies. By including this type of statistics into a mobility model, one is able to develop a generic mobility model able to reproduce the pseudo-random or deterministic behavior observed in real urban traffic.

The University of Delaware mobility models, UDel Models (2007), typically fall into this category. The mobility simulator is based on surveys from a number of research areas including time-use studies performed by the US Department of Labor and Statistics, time-use studies by the business research communities, or pedestrians and vehicle mobility studies by the urban planning and traffic engineering communities. Based on this data, the mobility simulator models arrival times at work, lunch time, breaks/errands, pedestrian and vehicular dynamics (e.g. realistic speed-distance relationship and passing dynamics), and workday time-use such as meeting size, frequency, and duration. Vehicle traffic is derived from vehicle traffic statistics collected by state and local governments such that it is able to model vehicle dynamics and daytime street usage. We can also cite the agenda-based mobility model (Zheng et al. 2006), which combines both social activities and geographic movements. The movement of each node is based on an individual agenda, which includes all kinds of activities in a specific day. Data from the US

National Household Travel Survey has been used to obtain activity distributions, occupation distributions and dwell-time distributions.

A Multi-agent Microscopic Traffic Model (MMTS) is proposed by Cetin et al. (2003) and generates public and private vehicular traffic over real regional road maps of Switzerland with a high level of realism within a period of 24 hours. This synthetic flow model is calibrated using data from census and other local or national mobility surveys or statistics. Precomputed vehicular mobility traces are available on the MMTS (2006) website. A similar approach has been followed at the Los Alamos Research Labs, USA, but using more precise statistics from various urban traffic management systems such as sensors at traffic lights and measured traffic flows. These last two examples are an illustration of the limitation of the survey-based approach, as survey or statistical data is only able to provide a coarse-grain mobility as illustrated in the previous paragraph. If we need a more detailed and realistic mobility representation, then we still require a complex synthetic model and to calibrate it using surveys or statistics.

5.8 Integration with Network Simulators

In the previous section, we described the different methods for developing mobility models adapted to vehicular traffic. Yet, in order to be used by the communication networking community, these models need to be made available to network simulators. This simple compatibility issue has unfortunately been a conundrum for many years. Initially, the worlds of mobility models and network simulators could have been compared to a mute talking to a deaf. They had never been created to communicate, and, even worse, they had been designed to be controlled separately, with almost no interaction whatsoever. When imagining the promising traffic telematics applications that could be obtained from VANET, where communication could alter mobility, and where mobility would improve network capacity, the VANET community has worked in the past few years to define efficient communication interfaces between the two worlds.

In this section, we illustrate the need to create an interaction between a mobility model and a network simulator. We also describe the different approaches that can be followed depending on the applications' required level of interactions.

5.8.1 Network simulators

Before moving forward, we would like to shortly introduce the various network simulators that are available to the vehicular networking community, as they play a key role in the classification introduced in the rest of this chapter. As with traffic simulators, we may also classify the simulators as commercial or open source. We emphasize that commercial licenses are sometimes available free of charge for university programs.

Commercial network simulators

Probably the most widely used commercial network simulator is OPNET (2009), an efficient discrete event simulator containing a large set of network protocols, including a complete wireless suite. Although the license may be onerous, OPNET offers a university program with a free license for the fully functional OPNET Modeler. An alternative to OPNET is QualNet (2009), which has the interesting feature of managing multi-processor parallel computing. As with OPNET, it contains a large library of wireless network protocols, including a mobile ad-hoc network (MANET) suite. Unfortunately, the fully functional Qualnet, including the parallel computing feature, requires the purchase of a commercial license. Finally, the OMNeT++ (2009) simulator is another major license-based network simulator available for network research and offers various network protocols libraries. OMNeT++ is offered free of charge for academic and educational purposes, but a commercial use requires the purchase of a license.

Open source network simulators

As an alternative to commercial network simulators, open source ones are also available. However, the extent of the community behind a given simulator significantly determines the various available protocols, as only a limited set of functionalities are usually provided by default. The most widely used network simulator is NS-2 (2008), which did not earn its popularity from its simplicity or efficiency, but from its modularity and universality. Extended to wireless networks by the Monarch Project at Rice University, it de facto became the reference simulator for MANETs. It contains the major layers of the OSI stack, and because of its popularity, all major protocols for MANETs. The official support for NS-2 ended in 2008 with the version NS-2.33 and shifted to its successor NS-3 (2009). Thanks to a growing community shift from NS-2 to NS-3, a stable NS-3 version with basic wireless networks protocol stacks was released in 2009. An alternative to NS-2 is GloMoSim, which is the free version of QualNet with reduced capabilities. It has the same functionalities and objectives as NS-2/NS-3. The set of available protocols is, however, reduced due to the smaller GloMoSim community. SWANS (2004) and its extension SWANS++ (2007) are scalable network simulators coded in Java. Although having suffered from a limited popularity at the beginning, SWANS has now reached a sufficiently large community and also contains the major protocols for MANETs. Finally, The GTNetS (2008) is a fully featured network simulation environment that allows the study of the behavior of moderate to large scale networks. GTNetS's strict compliance with different protocol stack layers makes it an easier solution for the development of network protocols for MANETs.

Considering VANET, the NS-2 PHY and MAC layers have been extended and validated for the modeling of vehicular networks by Chen et al. (2008), a development which fills a major gap in previous evaluations of VANET protocols. Recently, Bingmann and Mittag (2009) enhanced and ported these modules to the official release of NS-3. NS-2/NS-3 therefore seem to be the most adapted simulators among the previously described solutions, as they contain an implementation of IEEE 802.11p PHY/MAC, including EDCA for NS-3, as described in this book.

5.8.2 Isolated mobility models

Initially, mobility was seen by network simulators as random perturbations from optimum statical configurations. Then, in order to provide a basic control on the use on the mobility patterns, network simulators became able to load mobility scenarios. However, as illustrated in Fig. 5.17, the different scenarios must be generated prior to the simulation and be parsed by the simulator according to a predefined trace format. Then, modifications to the mobility scenario are not possible, and no interaction therefore exists between these two worlds. Unfortunately, all historical models and most of the recent mobility models available to the research community fall into this category. Because of space limitation, we could not include a description of the large literature on isolated models. We refer the reader to Härri et al. (2009) for a detailed survey and taxonomy of mobility models available to vehicular networks.

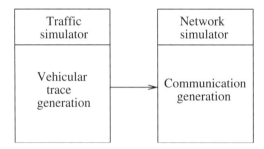

Figure 5.17 Interaction between network and traffic simulators: the isolated case.

In spite of the limitation described in the previous paragraph, the networking community furnished itself very well with isolated models, as this category has the advantage of allowing independent developments in mobility and network modeling. It is only recently that the need for a more significant interaction has appeared, as specific applications in vehicular communications, such as traffic safety or traffic management, justified a better interaction between the two worlds.

5.8.3 Embedded mobility models

If network simulators cannot fully interact with mobility simulators, another solution is to fuse the two in one single simulator. Accordingly, new simplistic network simulators were created, where the lack of elaborated protocol stacks was compensated for by a native collaboration between the networking and the mobility worlds (see Figure 5.18).

MoVes (Bononi et al. 2006) is an embedded system that generates vehicular mobility traces and also contains a basic network simulator. The major asset of this project is its ability to partition the geographical area into clusters, parallelizing and distributing the processing of each task. Although the mobility model reaches a sufficient level of detail, the project's drawback is the simplified network model,

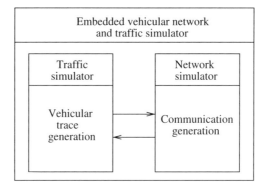

Figure 5.18 Interaction between network and traffic simulators: the embedded case.

which only includes a basic PHY and MAC layer architecture and lacks routing protocols. Gorgorin et al. (2006) developed their own embedded traffic and network simulator. Although being basic, the traffic model allows a sufficient level of detail. However, the network simulator part is its major limitation, as it is only modeled by a simplistic discrete event simulator handling a basic radio propagation and a CSMA/CA MAC protocol.

The NCTUns simulator (Wang and Chou 2008) also falls into the category of embedded models. Although being initially a network simulator, it has later been embedded with a vehicular traffic simulator providing a sufficient level of detail in the vehicular motion pattern. Moreover, the IEEE 802.11p and 1609 stacks have been implemented in its version 5.0, making NCTUns one of the first VANET simulators to support the full WAVE protocol stack.

The particular approach followed by Killat and Hartenstein (2009) is VCOM, an embedded communication module in VISSIM. In order to increase the scalability, the communication model is not based on a network simulator but on an analytical communication model providing the probability of packet reception in an IEEE 802.11p-based network given four input parameters: the distance between sender and receiver, transmission power, transmission rate, and vehicular traffic density. Killat et al. calibrated and validated their analytical model with simulation results from the network simulator NS-2 on a similar environment. By suppressing the resource demanding network simulator and by using a validated analytical communication model, the use of VCOM in VISSIM allows large-scale simulations of VANET, while still keeping the simulation's credibility.

The main asset of the embedded approach is to have both network and mobility models natively and efficiently interacting. However, their major limitation is the simplified network or mobility simulation capabilities found in many of the available solutions. This might be unfortunate as the actual direction in simulations of VANET is the use of validated vehicular mobility patterns and a strict compliance with standard protocols. The approach followed by Killat and Hartenstein (2009) might therefore give a hint on the future of embedded models.

5.8.4 Federated mobility models

The final possible approach is to federate existing network simulators and mobility models or professional traffic simulators through a set of interfaces (see Figure 5.19). This solution has been taken, for example, by Prof. Fujimoto and his group in Georgia Tech (Wu et al. 2005). They generated a simulation infrastructure composed of two independent commercial simulation packages running in a distributed fashion over multiple networked computers. They federated a validated traffic simulator CORSIM with a state-of-the-art network simulator QualNet using a distributed simulation software package called the Federated Simulations Development Kit (FDK), which provides services to exchange data and synchronize computation. In order to allow a direct interaction between the two simulators, a common message format has been defined between CORSIM and QualNet for vehicle status and position information. During the initialization, the road network topology is transmitted to QualNet. Once the distributed simulation begins, position updates are sent to QualNet and are mapped to mobile nodes in the wireless simulation. Accordingly, both simulators work in parallel and thus may dynamically interact with each other by altering for example mobility patterns based on network flows, and vice versa. As mentioned by the authors, a major limitation comes, however, from the complex calibration of CORSIM and from its large number of configuration parameters that must be tweaked in order to fit to the modeled urban area. A similar solution has been taken by a team from UC Davis (VGrid 2009) with a simulation tool federating the network simulator SWANS and a synthetic traffic model.

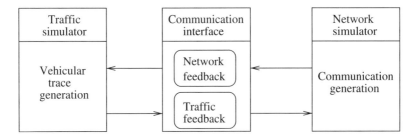

Figure 5.19 Interaction between network and traffic simulators: the federated case.

Another promising approach is called TraNS and federates the traffic simulator SUMO with either the network simulator NS-2 or OMNeT++. Using an interface called Interpreter[1], traces extracted from SUMO are transmitted to NS-2 or OMNeT++, and conversely, instructions from NS-2/OMNeT++ are sent to SUMO for traffic tuning. The MSIE (2008) project is an alternative approach federating NS-2 to VISSIM instead of SUMO. This project is also more complete, as it proposes interlinking different simulators for traffic, network, and application simulations. The major actual limitation is the communication latency between the different

---
[1]On the latest version of TraNS, the Interpreter has been replaced by the TraCI interface.

simulators. Unfortunately, the interlinking interface itself is also not freely available at the time of writing. Schroth et al. chose to replace VISSIM by a complete tool called the CARISMA traffic simulator. While not being as complete as VISSIM or SUMO, it helps to accurately evaluate the effects of car-to-car messaging systems in the presence of urban impediments by benefiting from the federated approach and a 'real-time' trip (re)configuration.

Recently, Queck et al. (2008) proposed a generalization of the federated approach by developing an interface called V2X Simulation Runtime Infrastructure (VSimRTI) which uses concepts defined in the IEEE Standards for Simulation and Modeling (M&S) High Level Architecture (HLA) to provide a federation capability and required interfaces for arbitrary but specific simulators (see Figure 5.20). The current version of VSimRTI federates the traffic simulator SUMO, the network simulator SWANS, a configuration environment eWorld (2009) and a basic vehicular application container.

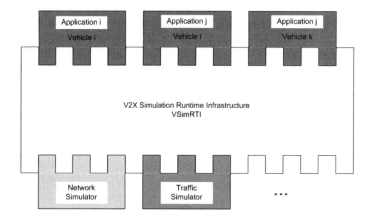

Figure 5.20 VSimRTI: A federation model for simulators based on High Level Architecture (HLA).

By interlinking independent and validated simulators, the federated approach is able to benefit from the best of both worlds, as state-of-the-art mobility models or traffic simulators may be adapted to work with modern and efficient network simulators. However, it is computationally demanding as both simulators need to be run simultaneously, and the development of the interlinking interface may not be a straightforward task depending on the targeted network and traffic simulators. Moreover, configuration complexity may also be seen as a limitation, as the calibration of traffic simulators usually requires the tweaking of a large set of parameters. But more importantly, the level of detail required for vehicular network simulators may not be as demanding as that for traffic analysis. Nevertheless, the networking and traffic modeling communities have a mutual interest in working together. At this time of promising benefits obtained from the various cross-layer approaches in network research, the ability to proactively or reactively act on mobility patterns in order to improve network efficiency and radio propagation, or the ability to alter mobility patterns based on dynamic traffic events received by means of communication protocols, or even more promising, the ability to evaluate the impact of such bi-directional interaction on the efficiency of VANET applications, will probably be a central approach in future networking research projects.

5.8.5 Application-centric versus network-centric simulations

Initiated by an increased interaction between a traffic and a network simulator, we observe the emergence of the simulation of VANET applications, for which one is not only interested in providing efficient VANET capabilities but also in the evaluation of the impact of information provided by a VANET. One particular aspect of VANET applications is that they do not only relate to a networking aspect but also to a mobility aspect, typically by altering mobility based on information provided by a VANET. Initially located on the network side, application could also be located at the mobility side with the appearance of VANET applications, or more generally as an independent simulator that would be federated to network and traffic simulators as illustrated by Queck et al. (2008).

With more interactions between simulators, new applications therefore become possible. Yet, with new applications come also new challenges, for instance with the design of a proper interaction methodology between simulators with various capabilities. For example, a possible traffic safety application, such as a *collision avoidance system*, provides driving advice in order to avoid an imminent impact. From a simulative point of view, the application's requirements for the traffic simulator are to alter cars' trajectories and to evaluate such behavior on other vehicles and on impact avoidance. That means that the interaction interface should have provision for controlling the trajectory of individual vehicles, a feature that is not common to all traffic simulators. For example, flow-centric traffic models might not be appropriate as they can only alter the trajectory of a flow of vehicles, while macroscopic or mesoscopic flow models might be too coarse to be able to evaluate the influence of a *single* course change on surrounding traffic.

Other very important requirements are the latency and precision of the interactions, as wrong or late information is no better than no information at all. For that matter, Killat et al. (2007) studied the latency and synchronization requirements between the network and mobility modules. But then, if the objective is a perfect synchronization between node positions in the mobility model and in the network simulator, then a significant computational load is only used by the interface module and alters the simulation performance. On the other hand, if the synchronization task is relaxed, the positions in the mobility model and the network simulator will be different, thus creating inconsistent decisions by both modules. Killat et al. showed that a trade-off should and could be found depending on the application needs.

Also, an application acts as a function of information provided by a VANET. Its efficiency therefore significantly depends on the availability and freshness of the required information. If bad channel or networking conditions simulated by a network simulator are responsible for the delivery of false information to the application, the latter may not act optimally. For example, considering a safe overtaking application which would monitor the immediate neighborhood of a vehicle to see if it is safe to change lane and overtake, Killat et al. (2008) showed the critical relationship between the application efficiency in avoiding accidents and the freshness of the information regarding potential traffic pitfalls from neighboring vehicles. The interaction interface should therefore have provision for the evaluation of the various levels of quality of the information provided by the network and traffic simulators to the application simulator.

These are actually just brief examples of a new paradigm called *application centric* simulation, where an application takes the central role to control and alter mobility based on information from other sources. Opposite to this view is traditional *network centric* simulation, where the network plays the central role because the objective is to test the robustness to mobility of a particular network protocol or a communication solution.

The isolated category is, unfortunately, unable to model application-centric approaches as motion patterns cannot be altered according to dynamic traffic information. The embedded category may also model application-centric approaches to the condition that the level of realism of the traffic or network modules is sufficient to the applications. The federated category perfectly fits to the requirements of an application-centric approach, first because a total mutual interaction is possible, and second because it regroups independent solutions, potentially allowing the choice of the optimal traffic or network simulator as a function of the application requirements. Yet the challenge is in finding an efficient design for the interaction interface between simulators.

5.8.6 Discussion

In the community, the three described integration approaches of mobility and network simulator (i.e. the isolated, embedded, and federated ones) exist and are still currently developed in parallel, despite a slight tendency toward federated interactions. The success of the isolated case, and thus its survivability, comes from its simplicity and universality. Once a mobility model is developed for a network simulator, it could, for instance, be easily extended to support another one. The choice between the three different approaches therefore clearly depends on the application requirements. For example, if only a limited interaction between the mobility and network simulators is necessary (the network centric approach), the isolated case should be chosen. Instead, the federated or embedded cases should find favor with the community working in safety and traffic management applications. The federation-based interaction between simulators proposed by Queck et al. (2008), which might be seen as a generalization of the federated case as described in this chapter, is also expected to attract an increasing attention from the VANET community due to its modularity with respect to a variety of simulators and to its support to application-centric approaches.

5.9 A Design Framework for Realistic Vehicular Mobility Models

In the previous sections we described the various challenges and options for modeling vehicular motions at various level of detail and to integrate or interact with network simulators. It is now important to see how these different approaches can be connected to create a realistic vehicular traffic simulator adapted to a targeted application. We therefore propose a concept map for a comprehensible representation of the functionalities of a realistic vehicular mobility model. As can be seen in Figure 5.21, the concept map is organized around two major module: the *motion constraints* and the *traffic generator*. Additional modules such as *time* and *external influences* are also required for a fine tuning of the mobility patterns.

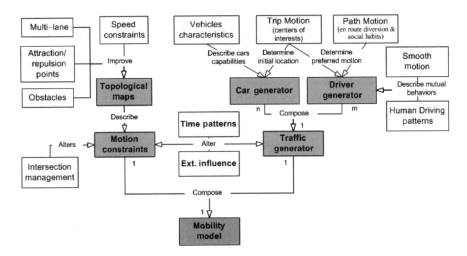

Figure 5.21 Proposed concept map for the generation of realistic vehicular mobility models.

5.9.1 Motion constraints

Motion constraints describe the relative degree of freedom available to each vehicle. Macroscopically, motion constraints are streets or buildings, but microscopically, constraints are modeled by neighboring cars, pedestrians, or by modelization's diversities either due to the type of car or to the driver's habits.

Modules that belong to the motion constraints functional block are

- *Accurate and realistic topological maps* Street topologies should manage different densities of intersections, contain multiple lanes, different categories of streets and their associated speed limitations.

- *Obstacles* Obstacles should be understood in a wide sense, as both constraints to cars' mobility and hurdles to wireless communications.

- *Attraction/repulsion points* Initial and final destinations of road trips are not random. Most of the time, drivers are moving to similar final destinations, called attraction points (e.g. office), or from similar initial locations, called repulsion points (e.g. home), a feature that creates bottlenecks.

- *Intersection management* This corresponds to the process of controlling an intersection, and may either be modeled as a static obstacle (stop signs), a conditional obstacle (yield sign), or a time-dependent obstacle (traffic lights).

5.9.2 Traffic generator

It defines different kinds of vehicles, and deals with their interactions according to the environment under study. Macroscopically, it models traffic densities, speeds and flows, while microscopically it deals with properties such as the inter-distance between cars, acceleration, braking, and overtaking.

Modules that belong to the traffic generator functional block are:

- *Vehicles characteristics* Each category of vehicle has its own characteristics, which have an impact on a set of traffic parameters. For example, macroscopically speaking, some urban streets and highways are forbidden to trucks depending on the time of the day. Microscopically speaking, acceleration, deceleration, and speed capabilities of cars or trucks are different. Accounting for these characteristics alters the traffic generator engine when modeling realistic vehicular motions.

- *Trip motion* A trip is macroscopically seen as a set of source and destination points in the urban area. Different drivers may have diverse interests that affect its trip selection.

- *Path motion* A path is macroscopically seen as the set of road segments taken by a car on its trip between an initial and a destination point. As may also be observed in real life, drivers do not randomly choose the next heading when reaching an intersection, as is currently the case in most vehicular networking traffic simulations. Instead, they choose their paths according to a set of constraints such as speed limitations, time of the day, road congestion, distance, and even drivers' personal habits.

- *Deceleration and acceleration models* Vehicles do not abruptly break and accelerate. Models for decelerations and accelerations should consequently be considered.

- *Human driving patterns* Drivers interact with their environments, not only with respect to static obstacles, but also to dynamic obstacles, such as neighboring cars and pedestrians. Accordingly, the mobility model should control the mutual interactions between vehicles, such as overtaking, traffic jam, or preferred paths.

Time It can be seen as the third functional block that describes different mobility configurations for a specific time of the day or day of the week. Traffic density is indeed not identical during the day. A heterogeneous traffic density is always observed at peak times, such as rush hours or during special events. This block influences the motion constraints and the traffic generator functional blocks, as it may alter the trip or path computation, and also the attraction/repulsion points.

External influences These model the impact of a communication protocol or any other source of information on the motion patterns. Some motion patterns cannot be proactively configured by vehicular mobility models as they are externally influenced. This category models the impact of accidents, temporary road works, or real-time knowledge of the traffic status on the motion constraints and the traffic generator blocks.

5.9.3 Application-based level of realism

We previously described the various building blocks to be included into the design of a realistic vehicular mobility model. The content itself of each building block may be configured to a specific application. For example, the *trip motion* block could include complex OD matrices enhanced with a *path motion* or simply a stochastic turn approach. It is up to the designer to use the required complexity of each module.

The particularity of vehicular mobility models compared to commercial traffic simulators is that their complexity and requirements for realism depend on the target application and are usually less demanding as traffic simulators. The Car2Car Communication Consortium (C2CCC) listed three potential key applications. We briefly describe them here and discuss the level of realism required for each of them.

Applications of vehicular communications

In its manifesto (C2CCC 2009), the C2CCC proposed the following key applications for inter-vehicular communications.

Traffic safety This provides active safety solutions based on inter-vehicular communication such as cooperative forward collision warning, pre-crash sensing or warning, or hazardous location notification. Although requiring a significant level of precision in the modeling of vehicular motions, traffic safety applications usually have a small scale and might only require a precise microscopic flow model and generally a simplistic traffic model as the small-scale nature would boil trips and paths down to arrival rate and turning probabilities. As radio communications are precisely used to alter vehicular traffic, a total bidirectional interaction is required between the network and the traffic simulators. Furthermore, as the reception or the non-reception of a safety message may have a significant impact on the overall efficiency of a safety application, the level of realism for the network simulator must also be high. Consequently, the federated approach may be recommended for more flexibility in the choices of the appropriate required features.

Traffic efficiency Its objective is to improve the traffic throughput or fuel consumption and reduce traffic delay by providing the required traffic information faster and more precisely. Typical applications could be green light optimal speed advisory or lane merging assistance. Traffic efficiency applications usually have a large scale but require less precision in the modeling of vehicular motions. For example, a mesoscopic or macroscopic flow model and an agent-centric trip model could create an acceptable traffic generator. As static or dynamic obstacles, or more precisely drivers' unawareness of them, are the major reasons of traffic inefficiency, the motion constraints functional block should contain as many modules as possible. Considering network simulators, a full bidirectional interaction with the mobility simulator is also necessary. Yet, unlike traffic safety, the required granularity of the network simulator is reduced as a coarse grain packet reception rate and not a precise packet-based reception probability is sufficient. For its flexibility, the federated approach may also be recommended but embedded models may also be considered.

Infotainment This represents all other applications that could be provided to drivers, from context-specific advertisement to opportunistic music download and Internet connectivity. Fiore and Härri (2008) described the minimum requirements for such applications. As the average information distribution is more important than the exact one and that global vehicular flows are sufficient, the requirements are lower than the two previous application types. Moreover, no specific interaction with network simulators is required. The isolated approach may therefore be considered for this category, but the embedded or federated approaches might still be needed should mobility be altered.

5.10 Discussion and Outlook

For the past 70 years, vehicular mobility has been analyzed in order to improve traffic infrastructures or to design coherent traffic management systems. It recently benefited from a sudden increase of interest for vehicular networks due to the lack of vehicular mobility models natively available in network simulators. The objective was first to reach a higher degree of realism in the representation of vehicular movements and second to be able to influence these movements with respect of the applications under study.

In this chapter, we described various approaches to modeling vehicular motions at various modeling granularities. It is important to understand that these approaches, notably the flow and traffic models, are not mutually exclusive and can be individually or jointly used as a function of the level of precision required by a specific application. For example, some applications might only need a macroscopic flow model, while some safety protocols could need to obtain a microscopic flow model in order to realistically model accidents and their impacts on urban traffic. Identifying the minimum requirements to realistically model vehicular motions under study by a particular application is a critical task. An initial attempt has been conducted by Fiore and Härri (2008). Yet, the question is still open: what are the *sufficient* requirements?

Another domain that we have not covered in this chapter is the concept of *imperfect driving*. In all models described in this chapter, decisions are based on perfect information related to traffic such that an accident can always be avoided. Yet, if we are unable to model or create accidents, how can we evaluate the impact of an accident prevention application? Killat et al. (2008) approached the problem by altering traffic information related to overtaking such that drivers create accidents simply by not seeing some vehicles on adjacent lanes. The safe overtaking application would therefore compensate such reduced vision by obtaining the required information by vehicular communication. Killat et al. showed that the success of such an application would come in part from the quality of the traffic information provided through vehicular communication and in part by the driver's faith in the advice provided by the application.

We discussed in this chapter the modeling of vehicular motion patterns. The straightforward objective has been to evaluate how mobility influences network and application protocols. The answer is not clear yet, as vehicular networks are also a wireless environment and it is expected from the radio link to have an impact at least as important as mobility. As a matter of fact, the quality of the radio channel between two vehicles significantly depends on their mutual mobility. A realistic radio propagation and fading model should therefore be implemented, which should also be able to consider radio obstacles possibly blocking the signal. A radio module could do this task but its precise location should be discussed. At this time, a very simplified version of such a module is almost exclusively contained in network simulators. The radio module needs to have access to mobility patterns and topographic information to model the radio channel. It should therefore be included in traffic simulators. Yet, the radio module also requires access to the data traffic model from each of the potential transmitters, and justifies its inclusion in network simulators. Similarly to the federated or embedded approaches, a *trium vira* could be created, where three modules control each part of a realistic simulation of vehicular networks, and interact with each others in order for each parameter to impact other modules (see Figure 5.22). This might be a motivating future direction for the community working on vehicular mobility modeling, and as a matter of fact, most of the key players in that field have already envisioned this next step.

We finally described in this chapter the challenges, options, and design guidelines of vehicular mobility modeling for VANET. We covered the most popular models available in literature but addressed them from an architecture, networking and application point of view. For a survey and a larger description of mobility models available to VANET, we refer readers to Härri et al. (2009).

5.11 Conclusion

As a prospective technology, vehicular ad-hoc networks (VANET) have recently been attracting increasing attention from both research and industry communities.

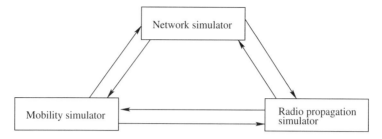

Figure 5.22 *Trium vira* in VANET simulations, where a network, a traffic, and a radio propagation simulator are federated for full mutual interaction.

One of the fastest-growing domains of interest in VANET is safety, where communications are exchanged in order to improve the driver's responsiveness and safety in case of road incidents. VANET's characteristics are a higher mobility and a limited degree of freedom in the mobility patterns. Such particular features make standard networking protocols inefficient or unusable in VANET. Accordingly, one of the critical aspects when testing VANET protocols is the use of mobility models that reflect as closely as possible the real behavior of vehicular traffic. In this chapter, we covered the different approaches in vehicular mobility modeling, their granularities and their complexities. Unlike MANETs, the major objective of VANET protocols is a direct alteration of the traffic patterns for safety or trip optimizations. Accordingly, we also described the new trend to interlink traffic and network simulators in order to create a cross-layer collaboration between routing and mobility schemes. We finally presented a framework that should be followed for the generation of realistic vehicular mobility patterns and provided a brief discussion on the relationship between the needs of the applications and the selected mobility modules. We hope that this chapter will be a good guideline for people interested in understanding the unique relationship between traffic models and network protocols in vehicular networks.

References

Ahmed K 1999 *Modeling Drivers' Acceleration and Lane Changing Behavior*. PhD thesis, Massachusetts Institute of Technology (MIT).

Aimsun 2009 TSS Aimsun – The Integrated Transport Modeling Software, http://www.aimsun.com.

Bai F, Sadagopan N and Helmy A 2003 The IMPORTANT Framework for Analyzing the Impact of Mobility on Performance of Routing for Ad Hoc Networks. *AdHoc Networks Journal – Elsevier* **1**, 383–403.

Balmer M 2007 *Travel demand modeling for multi-agent traffic simulations: Algorithms and systems*, PhD thesis, ETH Zurich, Switzerland.

Bingmann T and Mittag J 2009 An overview of PHY-layer models in NS-3, Marseille, France.

BonnMotion 2005 http://web.informatik.uni-bonn.de/IV/BonnMotion.

Bononi L, Di Felice M, Bertini M and Croci E 2006 Parallel and Distributed Simulation of Wireless Vehicular Ad Hoc Networks. *Proc. of the ACM/IEEE International Symposium on Modeling, Analysis and Simulation of Wireless and Mobile Systems (MSWiM '06)*, pp. 28–35, Torremolinos, Spain.

Brackstone M and McDonald M 1999 Car-following: A Historical Review. *Transportation Research Part F: Traffic Psychology and Behaviour* **2**, 181–196.

C2CCC 2009 CAR 2 CAR Communication Consortium, (C2C-CC), http://www.car-to-car.org/.

Cabspot 2006 The Cabspotting Project, http://cabspotting.org.

Cacciabue P 2007 *Modelling Driver Behaviour in Automotive Environments Critical Issues in Driver Interactions with Intelligent Transport Systems*. Springer.

Cetin N, Burri A and Nagel K 2003 A large-scale Multi-agent Traffic Microsimulation based on Queue Model. *Swiss Transport Research Conference (STRC)*.

Chandler RE, Herman R and Montroll EW 1958 Traffic Dynamics: Studies in Car Following. *Operations Research* **6**, 165–184.

Chen Q, Schmidt-Eisenlohr F, Jiang D, Torrent-Moreno M, Delgrossi L and Hartenstein H 2008 Overhaul of IEEE 802.11 Modeling and Simulation in NS-2. *Proc. of the ACM/IEEE International Symposium on Modeling, Analysis and Simulation of Wireless and Mobile Systems (MSWIM '08)*, Chania, Greece. Code part of the official NS-2.33 release.

Chowdhury D, Wolf DE and Schreckenberg M 1997 Particle Hopping Models for Two-lane Traffic with Two Kinds of Vehicles: Effects of Lane-changing Rules. *Physica A Statistical Mechanics and its Applications* **235**, 417–439.

CORSIM 2009 CORSIM 6.1 – Microscopic Traffic Simulation Model, http://mctrans.ce.ufl.edu/featured/TSIS.

Crawdad 2009 Community Resource for Archiving Wireless Data At Dartmouth, http://crawdad.cs.dartmouth.edu.

Dieselnet 2009 UMass Diverse Outdoor Mobile Environment (DOME), http://prisms.cs.umass.edu/dome.

Espié S 2006 Non-normative Behaviour in Multi-agent System: Some Experiments in Traffic Simulation, pp. 30–36, Hong Kong, SAR China.

eWorld 2009 The eWorld Framework http://eworld.sourceforge.net/.

Fiore M and Härri J 2008 The Networking Shape of Vehicular Mobility. *Proc. of the 9th ACM International Symposium on Mobile Ad Hoc Networking and Computing (MobiHoc '08)*, pp. 261–272, Hong-Kong, SAR China.

Fujimoto R, Guensler R, Hunter M, Wu H, Palekar M, Lee J and Ko J 2006 CRAWDAD Data Set Gatech/vehicular (v. 2006-03-15). Downloaded from http://crawdad.cs.dartmouth.edu/gatech/vehicular.

Gawron C 1998 An Iterative Algorithm to Determine the Dynamic User Equilibrium in a Traffic Simulation Model. *International Journal of Modern Physics C* **9**(3), 393–407.

Gazis DC, Herman R and Rothery RW 1961 Nonlinear Follow-the-Leader Models of Traffic Flow. *Operations Research* **9**, 545–567.

Gipps PG 1981 A Behavioral Car Following Model for Computer Simulation. *Transportation Research Board* **15**, 105–111.

Gipps PG 1986 A Model for the Structure of Lane Changing Decisions. *Transportation Research Board* **20**, 107–120.

Gorgorin C, Gradinescu V, Diaconescu R, Cristea V and Iftode L 2006 An Integrated Vehicular and Network Simulator for Vehicular Ad-Hoc Networks. *Proc. of the 20th European Simulation and Modeling Conference (ESM)*.

Greenshields B 1935 A study of highway capacity. *Proceedings Highway Research Record* **14**, 448–477.

GTNetS 2008 The Georgia Tech Network Simulator (GTNetS), http://www.ece.gatech.edu/research/labs/MANIACS/GTNetS/.

Härri J, Filali F and Bonnet C 2009 Mobility Models for Vehicular Ad Hoc Networks: A Survey and Taxonomy. *IEEE Communications Surveys & Tutorials*.

Hoogendoorn SP and Bovy PHL 2001 State-of-the-art of Vehicular Traffic Flow Modelling. *Journal of Systems and Control Engineering – Special Issue on Road Traffic Modeling and Control*, 283–303.

Jaap S, Bechler M and Wolf L 2005 Evaluation of Routing Protocols for Vehicular Ad Hoc Networks in City Traffic Scenarios. *Proc. of the 5th International Conference on Intelligent Transportation Systems Telecommunications (ITST '05)*, Brest, France.

Killat M and Hartenstein H 2009 An Empirical Model for Probability of Packet Reception in Vehicular Ad Hoc Networks. *EURASIP Journal on Wireless Communications and Networking*.

Killat M, Gaugel T and Hartenstein H 2008 Enabling traffic safety assessment of VANETs by means of accident simulations. *Proc. of the 19th IEEE Symposium on Personal, Indoor and Mobile Radio Communications (PIMRC '08)*, pp. 1–6, Cannes, France.

Killat M, Schmidt-Eisenlohr F, Göbel G, Kosch T and Hartenstein H 2007 On the Accuracy of Coupling a Mobility and a Communication Simulator for VANETs. *4th International Workshop on Intelligent Transportation (WIT)*, Hamburg, Germany.

Krauss S 1998 *Microscopic Modeling of Traffic Flow: Investigation of Collision Free Vehicle Dynamics*. PhD thesis, Universität zu Köln.

Krauss S, Wagner P and Gawron C 1997 Metastable States in a Microscopic Model of Traffic Flow *Physical Review E* **55**, 55–97.

Laval JA and Daganzo CF 2006 Lane-changing in Traffic Streams. *Transportation Research Part B: Methodological* **40**, 251–264.

Legendre F, Borrel V, Dias de Amorim M and Fdida S 2006 Reconsidering Microscopic Mobility Modeling for Self-Organizing Networks. *IEEE Network Magazine* **20**, 4–12.

MATSim 2009 The MATSim Framework, http://www.matsim.org.

MIT Reality Mining 2009 MIT Media Lab: Reality Mining, http://reality.media.mit.edu.

MMTS 2006 Realistic Vehicular Traces from the Multi-agent Microscopic Traffic Simulator (MMTS), http://lst.inf.ethz.ch/ad-hoc/car-traces/.

MobiLib 2007 Community-wide Library of Mobility and Wireless Networks Measurements, http://nile.cise.ufl.edu/MobiLib/.

MSIE 2008 Multiple Simulator Interlinking Environment for C2CC in VANETs, http://www.cn.uni-duesseldorf.de/projects/MSIE.

Musolesi M and Mascolo C 2006 A Community Based Mobility Model for Ad Hoc Network Research. *2nd ACM/SIGMOBILE International Workshop on Multi-hop Ad Hoc Networks: From Theory to Reality (REALMAN '06)*, pp. 31–38.

Nagel K and Schreckenberg M 1992 A Cellular Automaton Model for Freeway Traffic. *Journal de Physique I* **12**, 2221–2229.

Nagel K, Schreckenberg M, Latour A, and Rickert M 1995 Two Lane Traffic Simulations. *Physica A* **231**.

Nagel K, Wagner P and Woesler R 2003 Still Flowing: Approaches to Traffic Flow and Traffic Jam Modeling. *Operations Research* **51**, 681–710.

NS-2 2008 The network simulator NS-2, http://www.isi.edu/nsnam/ns.

NS-3 2009 The network simulator NS-3, http://www.nsnam.org/.

OMNeT++ 2009 OMNeT++ – Discrete Event Simulation System, http://www.omnetpp.org/.

OPNET 2009 the OPNET Modeler, http://www.opnet.com.

Panwai S and Dia H 2005 Comparative evaluation of microscopic car-following behavior. *IEEE Transactions on Intelligent Transportation Systems* **6**(3), 314–325.

Payne H 1973 Freeway Traffic Control and Surveillance Model. *Journal of the Transportation Engineering Division* **99**, 767–783.

Pipe L 1953 An Operational Analysis of Traffic Dynamics. *Journal of Applied Physics* **24**, 274–281.

Prigogine I and Herman R 1971 *Kinetic Theory of Vehicular Traffic*. Elsevier, New York.

QualNet 2009 QualNet Developer http://www.qualnet.com.

Queck T, Schüenemann B and Radusch I 2008 Runtime infrastructure for simulating vehicle-2-x communication scenarios. *Proc. of the 5th ACM International Workshop on Vehicular Ad Hoc Networks (VANET '08)*, pp. 78–78, San Fransisco, CA, USA.

Saha AK and Johnson DB 2004 Modeling Mobility for Vehicular Ad Hoc Networks. *Proc. of the 1st ACM International Workshop on Vehicular Ad Hoc Networks (VANET'04)*, pp. 91–92, Philadelphia, PA, USA.

Schroth C, Dötzer F, Kosch T, Ostermaier B and Strassberger M 2005 Simulating the Traffic Effects of Vehicle-to-vehicle Messaging Systems. *5th International Conference on ITS Telecommunications (ITST '05)*.

SUMO 2009 SUMO – Simulation of Urban Mobility, http://sumo.sourceforge.net.

SWANS 2004 Scalable Wireless Ad hoc Network Simulator (SWANS), http://jist.ece.cornell.edu/.

SWANS++ 2007 Extensions to the Scalable Wireless Ad-hoc Network Simulator (SWANS++), http://www.aqualab.cs.northwestern.edu/projects/swans++/.

TIGER 2006 U.S. Census Bureau – Topologically Integrated Geographic Encoding and Referencing, http://www.census.gov/geo/www/tiger.

TRANSIMS 2009 TRANSIMS – Transportation Analysis and Simulation System – Open Source, http://transims-opensource.org/.

Treiber M and Helbing D 2002 Realistische Mikrosimulation von Strassenverkehr mit einem einfachen Modell. *16th Symposium Simulationstechnik (ASIM '02)*, pp. 514–520, Rostock, Germany.

UDel Models 2007 UDel Models For Simulation of Urban Mobile Wireless Networks, http://udelmodels.eecis.udel.edu/.

VanetMobiSim 2009 The Vanet Mobility Simulator (VanetMobiSim) 2.0, http://vanet.eurecom.fr.

VGrid 2009 UCD Vehicular Grid Computing (VGrid), http://wwwcsif.cs.ucdavis.edu/~VGrid.

VISSIM 2009 PTV simulation – VISSIM, http://www.ptvag.com/traffic/software-system-solutions/vissim.

Wang S and Chou C 2008 NCTUns Simulator for Wireless Vehicular Ad Hoc Network Research. *Ad Hoc Networks: New Research* Nova Science Publishers. NCTUns website: http://nsl.csie.nctu.edu.tw/nctuns.html.

Wiedemann R 1974 Simulation des Straenverkehrsflusses *Schriftenreihe des Instituts fur Verkehrswesen der Universitat Karlruhe* **8**.

Wiedemann R 1991 Modeling of RTI-Elements on Multi-lane Roads. *Advanced Telematics in Road Transport* vol. DG XIII Commission of the European Community.

Wu H, Lee J, Hunter M, Fujimoto RM, Guensler RL and Ko J 2005 Simulated Vehicle-to-Vehicle Message Propagation Efficiency on Atlanta's I-75 Corridor. *Transportation Research Board Conference*, pp. 82–89.

Zheng Q, Hongm X and Liu J 2006 An Agenda-based Mobility Model. *39th IEEE Annual Simulation Symposium (ANSS-39-2006)*, Huntsville, AL, USA.

Zhou B, Xu K and Gerla M 2004 Group and Swarm Mobility Models for Ad Hoc Network Scenarios using Virtual Tracks. *Proc. of the IEEE Military Communications Conference (MILCOM '04)*, pp. 289–294.

6

Physical Layer Considerations for Vehicular Communications

Ian Tan and Ahmad Bahai

University of California at Berkeley

The Federal Communications Commission (FCC) has allocated the frequency spectrum between 5.850 and 5.925 GHz for dedicated short-range communications (DSRC) in vehicular environments. With its nationwide availability, this allocation provides an ideal opportunity for automakers, government agencies, commercial entities, and motorists to work in concert to increase highway safety and provide transportation-related services; however, realizing this vision requires reliable, low-latency, wireless communication methods.

The vehicular environment presents a number of challenges that must be understood and appropriately managed in order to enable reliable wireless communications. One core issue is understanding the nature of the wireless channel encountered by vehicular radios. DSRC, once fully deployed, will be used in urban, suburban, and highway environments at a variety of speeds. This implies that any communication method employed in the DSRC band will face channels with different delay and Doppler spreads. Knowledge of how these statistics vary with location and speed is essential to designing DSRC band communications.

To obtain this knowledge, we constructed a complete RF channel sounding system using high-quality test equipment and Matlab® postprocessing code. Enabled with differential GPS units to accurately record location data, it measured both delay and Doppler characteristics simultaneously while correlating these results to the position and velocity of the mobile nodes. After conducting a measurement campaign encompassing urban, highway, and rural environments in Michigan, we derived a comprehensive set of statistics summarizing channel behavior in terms of various Doppler and delay metrics. In general, these measurements indicate that delay spreads are in the hundreds of nanoseconds, while Doppler spreads are on the order of 1–3 kHz, on average. The implications for the DSRC standard of these measurements center around the need to compensate for time-varying channels over the course of a packet transmission, as well as lesser issues of equalization and power control.

The chapter is organized as follows: we begin with an overview of the proposed DSRC standard and the specific parameters of the orthogonal frequency division multiplexing (OFDM) architecture it employs. A development of wireless communications channel theory follows in Section 6.3, along with an examination of common metrics used to quantify the performance of wireless channels in Section 6.4. After establishing these fundamentals, Section 6.5 elaborates upon our measurement methods. Finally, measurements and accompanying analysis lie within Section 6.6. For the impatient, an aggregate summary of most results may be found in Section 6.6.4 with analysis closely following in Section 6.6.5.

6.1 Standards Overview

6.1.1 A brief history

The standard for DSRC began within the American Society for Testing and Materials (ASTM) subcommittee E17.51, which was tasked with examining issues concerning vehicle roadside communications. In July 2003, the ASTM published standard E2213-03 (ASTM 2003), the last revision of their standard dealing with DSRC communications. It leverages the IEEE 802.11a standard heavily, adopting minimal changes to the physical (PHY) layer specified in IEEE (1999) and the media access control (MAC) layer specified in IEEE (2003).

Development of the standard then passed primarily into the hands of the IEEE, specifically to two working groups (WG), P1609 and 802.11p. The P1609 WG began with the MAC layer and moved upwards, concerning itself mainly with systemic issues such as multi-channel operation, cross-layer interfaces, security, and overall architecture. They have produced a number of standards which are (as of December 2008) either in trial usage or approaching the end of their initial trial use period. These standards are numbered as P1609.1 through P1609.4 (IEEE 2006a,b,c, 2007c). However, since these standards do not affect the fundamentals of the PHY layer's operation, we do not cover them further here.

The 802.11p WG currently holds ownership of any PHY and MAC changes for DSRC, with some of the members of the E17.51 also participating in the 11p WG. Due to the political difficulties of making extensive alterations to a widely used standard like 802.11, the initial PHY changes put in place by E2213-03 have remained. In fact, the current version of the 802.11 standard (IEEE 2007b) has absorbed into its main body the PHY layer modifications of ASTM E2213-03. Consequently, the current 802.11p amendment draft (IEEE 2007a) concerns itself with MAC and layer interface issues. Because little technical content has changed in the PHY between the publication of E2213-03 and the current version of IEEE 802.11, we refer to both of them subsequently as the 'DSRC' or '802.11p' standards. To our knowledge, the 802.11p amendment (IEEE 2007a) still holds draft status, though the current revision of IEEE 802.11 (IEEE 2007b) containing the DSRC-relevant PHY is in force.

6.1.2 Technical alterations and operation

Given this history, the number of changes made to transform the 802.11a PHY into 802.11p are relatively small. Like 802.11a, 802.11p leverages OFDM to compensate for both time- and frequency-selective fading since OFDM copes exceptionally well with the dispersive linear channels found in mobile environments. Unlike 802.11a, however, 802.11p will be used in drastically different environments. While the IEEE intended 802.11a for short range, low mobility, indoor use, 802.11p will encounter medium ranges (up to 1 km), extremely high mobility, and rapidly changing channel conditions.

To compensate for these differences, only two primary changes have been put in place. Firstly, DSRC will operate at a slightly higher frequency band than 802.11a. Its allocated bandwidth of 75 MHz spans 5.850 GHz to 5.925 GHZ, while 802.11a operates in the 5.170–5.230 GHz and 5.735–5.835 GHz bands. The DSRC band is largely free of interference from competing wireless devices, as it is not designated for ISM (industrial, scientific, medical) usage. Because the primary motivator behind DSRC is safety-related communications, regulators wished to minimize extraneous interference as much as possible.

Secondly, the DSRC standard reduces channel widths to 10 MHz from the 20 MHz width of 802.11a. This has a number of cascading side effects, some of which aid in compensating for vehicular wireless channels. Table 6.1 shows a side-by-side comparison of the PHY layer parameters for 20 MHz and 10 MHz operation. Note that, by moving to a narrower channelization while keeping the number of occupied subcarriers constant at 52, the spacing between subcarriers has been halved to 156.25 kHz. This new spacing, in turn, results in a longer fast Fourier transform (FFT) interval (6.4 µs) and guard interval (1.6 µs). Consequently, the length of OFDM symbols doubles from 4 µs at 20 MHz to 8 µs at 10 MHz.

An OFDM primer

In order to understand the implications of the changes transforming 802.11a to DSRC, we give a basic overview of how a modern OFDM system operates here. Interested readers may consult any number of references on OFDM such as Bahai et al. (2004). The expansion of the acronym OFDM largely summarizes its core operational concept – orthogonal frequency division multiplexing. OFDM systems transmit streams of time-domain digital data simultaneously by multiplexing them onto orthogonal frequency bands. Conceptually, begin with ζ streams of digital data, where $X_s[m]$ represents the data subsymbol on stream s at time m. For simplicity, let $X_s[m]$ take values in $\{\pm 1\}$ (1-bit data), and allocate B Hz of bandwidth to contain all ζ streams. Note that, in general, the number of bits representable by $X_s[m]$ could be greater than one. A way to send all streams simultaneously is through ζ amplitude modulated subcarriers with $\Delta f = (B/\zeta)$ Hz separation between each

Table 6.1 Comparison of PHY layer parameters between 20 MHz in 802.11a and 10 MHz operation used in 802.11p (IEEE 2007b).

Parameter	20 MHz channels	10 MHz channels
N_{SD}: Number of data subcarriers	48	48
N_{SP}: Number of pilot subcarriers	4	4
N_{ST}: Number of total subcarriers	52	52
Δ_f: Subcarrier frequency spacing	$\frac{20 \text{ MHz}}{64} = 312.5$ kHz	$\frac{10 \text{ MHz}}{64} = 156.25$ kHz
T_{FFT}: FFT and inverse FFT (IFFT) periods	3.2 μs ($\frac{1}{\Delta_f}$)	6.4 μs
$T_{PREAMBLE}$: PLCP preamble duration	16 μs	32 μs
T_{SIGNAL}: Duration of the SIGNAL BPSK-OFDM symbol	4.0 μs ($T_{GI} + T_{FFT}$)	8.0 μs
T_{GI}: GI duration	0.8 μs ($\frac{T_{FFT}}{4}$)	1.6 μs
T_{GI2}: Training symbol GI duration	1.6 μs ($\frac{T_{FFT}}{2}$)	3.2 μs
T_{SYM}: Symbol interval	4 μs ($T_{GI} + T_{FFT}$)	8 μs
T_{SHORT}: Short training sequence duration	8 μs ($10 \times \frac{T_{FFT}}{4}$)	16 μs
T_{LONG}: Long training sequence duration	8 μs ($T_{GI2} + 2 \times T_{FFT}$)	16 μs

subcarrier. Figure 6.1 illustrates this method conceptually, omitting details regarding filtering, sampling, and detection. For convenience, assume that ξ is odd and that streams are numbered from $(-\xi + 1)/2$ to $(\xi - 1)/2$.

Observe that the sth data stream modulates a cosinusoidal subcarrier which is offset from the center frequency by $s\Delta f$. Provided certain conditions are met (e.g. stream data rate less than B/ξ and filtering in place), the streams are orthogonal to each other in frequency. Now, consider the shape of the overall spectrum of width B at some fixed time m_0.[1] It can be completely described with the set of values $\{X_s[m_0]\}$ because for each s, $X_s[m_0]$ represents the value of the sth subcarrier. For one-bit values per subsymbol, only the polarity of $X_s[m_0]$ is significant, while for higher bit values per subsymbol, both the amplitude and sign deserve attention. Regardless of the complexity of $X_s[m_0]$, the collection of ξ subcarriers spans the allocated bandwidth. Based on this, conclude that the data subsymbols $X_s[m_0]$ may be interpreted as *frequency domain samples* which are spaced (B/ξ) Hz apart. The discrete Fourier transform (DFT) is a well-known method of producing spectral samples, and so for a given m, the subsymbols $X_s[m]$ may be viewed as outputs of this transform. Figure 6.1 may then be restructured as shown in Figure 6.2.

Recall that the FFT is merely an algorithm to implement the DFT. Figure 6.2 also adds blocks for serial-to-parallel and parallel-to-serial conversion, since serial, high rate data streams are usually fed as input to OFDM systems. We refer to a

---

[1] The theoretically rigid reader may object to the concept of examining a spectrum at a fixed point in time. As an alternative, imagine running a spectrum analyzer whose input is the transmitted signal and freezing the display at time m_0.

PHYSICAL LAYER CONSIDERATIONS

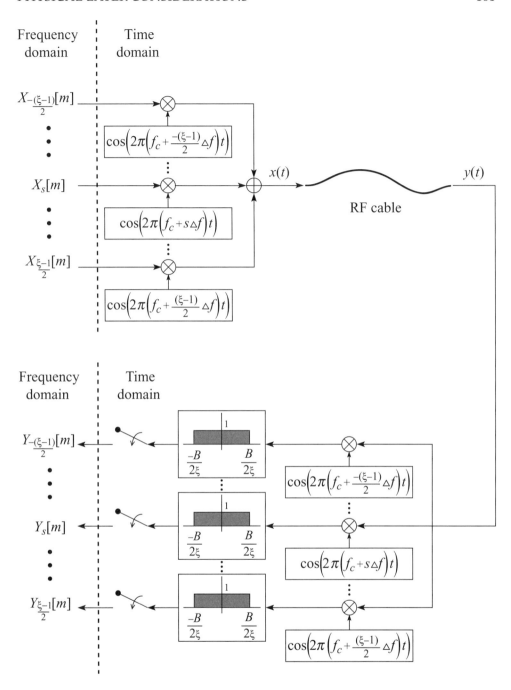

Figure 6.1 A multicarrier communications system with ξ simultaneous streams of amplitude modulated digital data being transmitted over an ideal channel (RF cable) requiring no additional equalization.

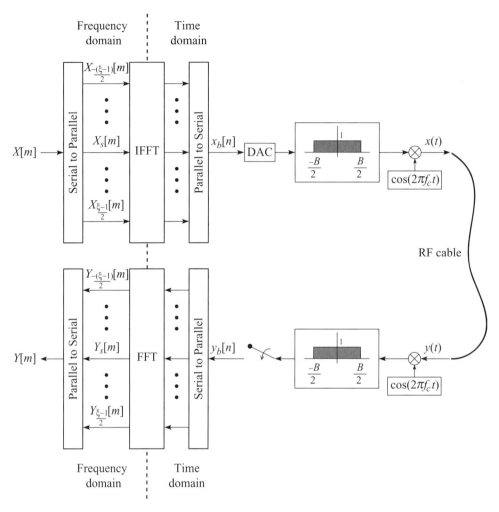

Figure 6.2 A representative OFDM system utilizing inverse and forward FFT blocks. As with Figure 6.1, an RF cable represents an ideal channel requiring no equalization.

set of frequency-domain subsymbols, taken across subcarriers at a fixed time m, as an OFDM frequency symbol. For every OFDM frequency symbol, the inverse FFT (IFFT) produces ξ time-domain subsymbols $x_b[n]$ with spacing $1/B$ between successive subsymbols after parallel-to-serial conversion; we refer to the entire set of ξ time-domain subsymbols as an OFDM time symbol.

To summarize developments to this point: modern OFDM transmitters take digital data as input, convert it into several parallel data streams, and perform an IFFT to transform it into time-domain data. This time-domain data is then filtered, up-converted, and sent over the communications channel. The OFDM receiver then

PHYSICAL LAYER CONSIDERATIONS

takes the time-domain data, down-converts, filters, and performs an FFT to recover the digital data streams. Finally, the parallel streams are converted back into a serial output stream.

What happens, though, when transmission occurs over a wireless channel? The presence of multipath in these channels corrupts the desired signal with time-delayed and distorted copies of itself. This 'echoing' or 'ringing' phenomena, known as *delay spread*, hampers the receiver's ability to determine the transmitted symbol, especially if the receiver is simultaneously receiving the most recent symbol and echoes from previous symbols. Intuitively, the receiver would have a much simpler job if the transmitter were to wait to send the current symbol until the echoes from previous symbols had abated. However, this would reduce communications efficiency due to the additional time required for each transmission. Delay spread quantifies the length of time that 'echoes' are present in the channel and will be covered in more detail in Section 6.4.1. To combat it, a guard interval is prepended to the beginning of each time-domain symbol. This interval prevents intersymbol interference (ISI) between consecutive OFDM symbols, provided that the typical delay spread of the channel is less than the guard interval.[2]

To understand how this operates, suppose that the wireless channel is modeled at baseband by a discrete function $h_b[n]$, where

$$h_b[n] = \begin{cases} \neq 0 & \text{for } 0 \leq n \leq N_D \\ 0 & \text{otherwise.} \end{cases} \quad (6.1)$$

Passband, baseband, and sampled baseband channels will be discussed more extensively in Section 6.3. For now, let $h_b[n]$ represent the nature of the passband wireless channel in an equivalent baseband form. The quantity N_D indicates the length of delay spread in the channel. The baseband output $y_b[n]$ of the channel may be written with the convolution operator $*$ as

$$y_b[n] = x_b[n] * h_b[n] \quad (6.2)$$

$$= \sum_{i=-\infty}^{+\infty} x_b[i] h_b[n-i]. \quad (6.3)$$

Two problems must be dealt with. The first concerns ISI, both between subsymbols and OFDM time-domain symbols. Assuming that OFDM symbols are sent with no intervening gaps, as they might be for a data packet, any $N_D \neq 0$ will lead to subsymbol and symbol ISI. The second problem concerns the fact that the receiver FFT utilizes circular convolution to process $y_b[n]$. Representing the DFT of $y_b[n]$ as $Y_s[m]$, we would like to assign the interpretation

$$Y_s[m] = X_s[m] H_s[m], \quad (6.4)$$

where $H_s[m]$ is the ζ-point DFT of the $h_b[n]$. However, for this to hold true, $x_b[n]$ and $h_b[n]$ must undergo a circular convolution, rather than the linear one represented by

---
[2]The use of actual data instead of merely silence for the prefix will become clear shortly.

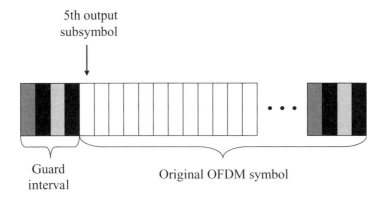

Figure 6.3 Relation of guard interval to original OFDM time-domain symbol with $L = 4$. Note that the 5th subsymbol in the extended symbol is actually the first subsymbol in the original symbol.

Equation (6.2). Recall that circular convolution is defined as

$$y_b[n] = x_b[n] \otimes h_b[n], \qquad (6.5)$$

$$= \sum_{i=-\infty}^{+\infty} x_b[i] h_b[(n-i) \% N], \qquad (6.6)$$

where N is the common length of $x_b[n]$ and $h_b[n]$ (possibly requiring zero-padding) and % is the modulus operator.

Solving the second problem first, observe that appropriate modification of the OFDM time-domain symbol will cause the linear convolution to behave like a circular convolution, at least for the ζ subsymbols concerning the receiver FFT. The modification, illustrated in Figure 6.3, requires duplicating a portion of the time-domain symbol's tail and affixing it to the start of the symbol. This duplicated data is known as the cyclic prefix. The amount to duplicate, L, depends on the expected value of the delay spread, N_D. Choosing $L \geq N_D$ permits subsymbols after the Lth in the output symbol to appear circularly convolved. Figure 6.4 redraws the OFDM system with the addition of this guard interval (GI) processing. Observe that the receiver removes the GI prior to passing ζ subsymbols to the FFT block. Since the presence of the GI was to ensure that the $(L+1)$th and subsequent subsymbols appear circularly convolved by the channel, the GI subsymbols are not needed during the FFT.

Through insertion of the GI, we have also solved the problem of symbol ISI. Provided that the GI length is chosen appropriately, any 'echoes' from a prior OFDM symbol will be contained entirely within the GI and affect only subsymbols 1 through N_D (which are discarded by the receiver) of the current symbol. ISI at the subsymbol level is also resolved, though in a different fashion. Since the use of a GI has ensured circular convolution of the channel and data, the interpretation of $Y_s[m]$ given in Equation (6.4) holds true – each frequency-domain subsymbol is scaled by

PHYSICAL LAYER CONSIDERATIONS

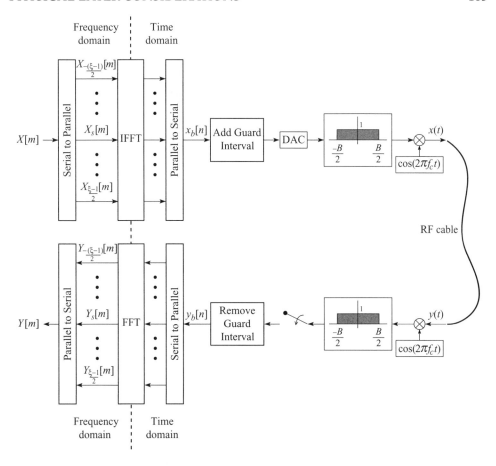

Figure 6.4 A representative OFDM system utilizing inverse and forward FFT blocks with guard interval insertion.

some quantity determined by the channel. Subcarriers thus remain orthogonal to one another, though they may be amplified or attenuated by the channel.

From the above, we see that the choice of the GI is extremely important in determining an OFDM system's tolerance to increased delay spreads. For 802.11p, the GI is chosen as 1.6 μs, which is twice that of 802.11a. However, by keeping the number of occupied subcarriers identical at 52, DSRC has potentially lower tolerance to Doppler spread given the smaller channel bandwidths. It is against these design decisions that we evaluate our measurement results.

6.2 Previous Work

Previous measurement-based research has included packet-level tests that were used to examine the performance of 802.11a in outdoor and mobile situations

(Bradaric et al. 2003; Sibecas et al. 2002). However, because fundamental characteristics of the physical channel were not explored nor were differences between the 802.11a and DSRC physical layers, it is difficult to extend their conclusions directly to DSRC performance. Other research included a thorough examination of path loss by Schwengler and Gilbert (2000) and Durgin et al. (1998) but neither addressed multipath and Doppler effects. Extensive measurement and modeling for channel emulators at 5.9 GHz has been shown by Acosta and Ingram (2006). While providing a modeling technique and an in-depth analysis of particular expressway locations, the study does not span a diverse range of possible environments. The measurements by Matolak et al. (2005) and Zhao et al. (2002, 2003) give time and frequency information in urban and suburban scenarios, but they either neglect extreme Doppler in highway situations (Zhao et al. 2002, 2003) or do not investigate non-line-of-sight (NLOS) situations with buildings or trucks as blockers (Matolak et al. 2005). In addition, the previous works do not interpret channel characteristics and their effect upon the DSRC standard.

6.3 Wireless Propagation Theory

6.3.1 Deterministic multipath models

One of the distinguishing characteristics of wireless channels is the presence of multipath. Multipath refers to the ability of the wireless channel to generate, through reflections and scattering, many time-delayed and distorted copies of a single transmitted signal. These reflections originate as the wavefront of a transmitted signal impinges upon objects in the environment such as buildings, vehicles, and even people. Furthermore, because of movements and changes in the environment, the nature of these reflections evolves over time.

We include this section to provide background material for interested readers and to establish notations that will recur later. Readers may wish to skip ahead to Section 6.4 and refer back to this section for clarification as needed. Those desiring deeper background should refer to Tse and Viswanath (2005), upon which portions of Section 6.3 are based.

To establish a deterministic model for the wireless multipath channel, first note that it can be written as a linear time-varying (LTV) system connecting a passband input $x(t)$ and passband output $y(t)$ (Tse and Viswanath 2005):

$$y(t) = \sum_i a_i(t) x(t - \tau_i(t)). \tag{6.7}$$

Equation (6.7) states that the output of the wireless channel, $y(t)$, is the sum of copies of the input $x(t)$ which are generated by reflections. Each copy of $x(t)$ is delayed by a time-varying quantity $\tau_i(t)$ and scaled by a real value $a_i(t)$. Note that both the delays and the scaling values may change over time. Since the input–output relationship is linear in t, it may also be described through a convolution of the input with the time-varying channel impulse response $h(\tau, t)$:

$$y(t) = h(\tau, t) * x(t) = \int_{-\infty}^{+\infty} h(\tau, t) x(t - \tau) \, d\tau. \tag{6.8}$$

PHYSICAL LAYER CONSIDERATIONS

To obtain the impulse response $h(\tau, t)$ and its Fourier transform $H(f; t)$, compare Equations (6.7) and (6.8). The presence of delayed copies of $x(t)$ suggests the use of the Dirac delta function $\delta(t)$, yielding

$$h(\tau, t) = \sum_i a_i(t)\delta(\tau - \tau_i(t)) \quad \text{and} \tag{6.9}$$

$$H(f; t) = \sum_i a_i(t)e^{-j2\pi f \tau_i(t)}. \tag{6.10}$$

In Equation (6.10), the frequency variable f is the dual of τ, not t. As a LTV channel, $H(f; t)$ can evolve as time t changes.

Since most processing is done at the baseband, it is instructive to translate this passband model into an equivalent, complex-valued baseband form. The complex-baseband representation combines information about the in-phase, $x_I(t)$, and quadrature, $x_Q(t)$, channels, shown in Figure 6.5, into a single compact form represented by:

$$x_b(t) = x_I(t) + jx_Q(t). \tag{6.11}$$

Both quadrature and in-phase data streams are sent simultaneously across the channel; however, because they are mixed onto two carriers offset by 90°, the receiver may separate them after some processing. The complex-baseband representation $x_b(t)$ serves merely as a compact way to represent both data streams simultaneously. Figure 6.5 represents the end-to-end communication system, beginning with $x_b(t)$ and ending with $y_b(t)$, the complex baseband representation of $y(t)$.

The relationship between $x_b(t)$ and $x(t)$ may be derived by first observing that in the frequency domain, a reasonable complex-passband representation of $X_b(f)$ is

$$X_c(f) = X_b(f - f_c), \tag{6.12}$$

where f_c is the carrier frequency. In time domain, this becomes

$$x_c(t) = x_b(t)e^{j2\pi f_c t}. \tag{6.13}$$

However, actual communications systems cannot generate complex-valued mixing signals; they always use real-valued voltages and currents. Hence, we take the real portion of Equation (6.13) and obtain

$$\Re[x_c(t)] = \Re[x_b(t)e^{j2\pi f_c t}], \tag{6.14}$$

$$= \Re[(x_I(t) + jx_Q(t))(\cos(2\pi f_c t) + j\sin(2\pi f_c t)], \tag{6.15}$$

$$= x_I(t)\cos(2\pi f_c t) - x_Q(t)\sin(2\pi f_c t), \tag{6.16}$$

$$= x(t). \tag{6.17}$$

Observe that $x(t)$ still contains both in-phase and quadrature data streams, and they are both projected onto two offset carriers in Equation (6.16). A similar derivation may be made for the relation between $y_b(t)$ and $y(t)$. Summarizing these relations,

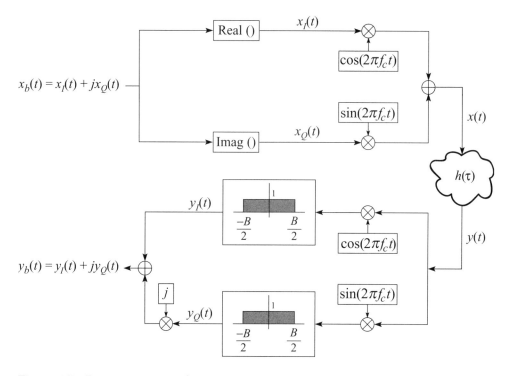

Figure 6.5 Representation of a passband communication system with complex-baseband I and Q channels.

observe that

$$x(t) = \Re[x_b(t)e^{j2\pi f_c t}] \quad \text{and} \quad (6.18)$$

$$y(t) = \Re[y_b(t)e^{j2\pi f_c t}]. \quad (6.19)$$

The next task requires determining the complex-baseband representation of the passband channel $h(\tau, t)$. Appendix 6.9.1 contains a detailed derivation of $h(\tau, t)$ based on Equations (6.8), (6.18), and (6.19). However, intuition should suggest the correct relationship between $y_b(t)$, $h_b(\tau, t)$, and $x_b(t)$ as

$$y_b(t) = h_b(\tau, t) * x_b(t). \quad (6.20)$$

The derivation in Appendix 6.9.1 shows further that

$$h_b(\tau, t) = \sum_i a_i(t)\delta(\tau - \tau_i(t))e^{-j2\pi f_c \tau_i(t)} \quad (6.21)$$

$$= \sum_i a_{i,b}(t)\delta(\tau - \tau_i(t)) \quad (6.22)$$

with

$$a_{i,b}(t) = a_i(t)e^{-j2\pi f_c \tau_i(t)}. \quad (6.23)$$

PHYSICAL LAYER CONSIDERATIONS

The complex-baseband channel is effectively the passband channel after a frequency shift, by the carrier frequency, down to the baseband.

Figure 6.5 illustrates an analog communication system. Since many modern systems are digital, Figure 6.6 shows a modified version of Figure 6.5. The sequence $x_b[n]$ is a sequence of data-bearing, possibly complex-valued, discrete symbols. The transmitter's filters pulse-shape the signal and limit the bandwidth occupied.

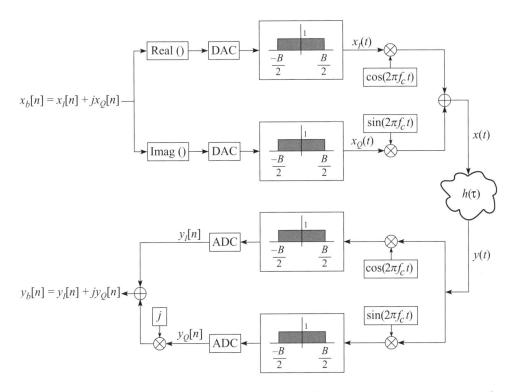

Figure 6.6 Representation of a digital passband communication system with complex-baseband I and Q data streams.

Furthermore, because most systems are sampled, the next step in establishing a deterministic channel model requires converting the baseband channel response into discrete time. Appendix 6.9.2 conducts a derivation in detail, and relevant results appear below. For a LTV channel with an allocated bandwidth of $B = 1/T_B$ Hz, the sampled impulse response is

$$h_b[p, m] = \sum_i a_i(mT_B) e^{-j2\pi f_c \tau_i(mT_B)} \operatorname{sinc}\left(\pi\left(p - \frac{\tau_i(mT_B)}{T_B}\right)\right). \quad (6.24)$$

The quantity $h_b[p, m]$ should be interpreted as the pth tap of the channel at the time instant m. If the channel was LTI, then the quantities $\tau_i(t)$ and $a_i(t)$ would no longer be time-varying, which changes Equation (6.24) into

$$h_b[p] = \sum_i a_i e^{-j2\pi f_c \tau_i} \operatorname{sinc}\left(\pi\left(p - \frac{\tau_i}{T_B}\right)\right). \tag{6.25}$$

Such LTI assumptions may be made when the rate at which the channel varies is significantly lower than the rate at which the communication system estimates the channel. The rate of channel variation is quantified by a metric known as *coherence time*, which is developed in Section 6.4.4. In fact, the degree to which these assumptions are or are not valid will play a crucial role in the assessment of measurement results.

LTI assumptions notwithstanding, much interest in wireless channels lies in how to capture and compensate for their time-varying nature. The introduction of Doppler shifts and spreads into a LTV channel model helps to accomplish this goal. We limit ourselves to an examination of a single tap LTV channel experiencing a Doppler shift, choosing to leave an explication of Doppler spread to Section 6.4.3. However, the discussion here should provide a foothold for understanding how Doppler spreads arise.

Consider the situation shown in Figure 6.7, with a mobile receiver (RX) station moving directly towards a stationary transmitter (TX) station with velocity v. Three possible signal paths are indicated, and we will consider each path separately and ignore any attenuation due to reflection or path loss. Assume that only the LOS path (path 1) contributes to the wireless channel and that both vehicles utilize omnidirectional antennas. If the transmitter sends a simple sine wave at a frequency f_0, $\sin(2\pi f_0 t)$, the receiver will receive a sine wave but at a shifted frequency, $\sin(2\pi (f_0 + (f_0 v)/c)t)$. The amount of the frequency shift, $(f_0 v)/c$, is known as the *Doppler shift*. The goal is to capture the notion of a changing channel (and by proxy, velocity) in the channel model. Now, the use of a LTI channel model will not suffice because sinusoids are eigenfunctions of any LTI system – a sinusoid of frequency f_0 as input should yield a sinusoid of identical frequency as output, which does not occur here. Thus, we must use the more general LTV model described in Equation (6.9). Let the general one-tap channel model be represented as

$$h(\tau, t) = a_0 \delta(\tau - \tau_0(t)). \tag{6.26}$$

Setting $\tau_0(t)$ appropriately to account for Doppler shift gives

$$\tau_0(t) = -\frac{vt}{c} \quad \text{and} \tag{6.27}$$

$$h(\tau, t) = a_0 \delta\left(\tau + \frac{vt}{c}\right), \tag{6.28}$$

where c is the speed of light. Consequently, a Doppler shift manifests itself in the LTV channel as a time-varying delay which is dependent upon the relative velocity between the TX and RX. If the receiver was traveling away from the transmitter, v would be negative and the received frequency would be lower than the transmitted frequency.

PHYSICAL LAYER CONSIDERATIONS   171

Next, consider the case in Figure 6.7 with only signal path 2. Although the RX will still see a positive Doppler shift, the magnitude of the shift will be less than in the previous case. Specifically, the Doppler shift will be scaled by $\cos(\theta)$, where θ is the angle formed by the receiver's velocity vector and the incident path's direction vector. If only path 3 is considered, the receiver will see no Doppler shift since the path's arrival angle is orthogonal to the RX velocity vector.

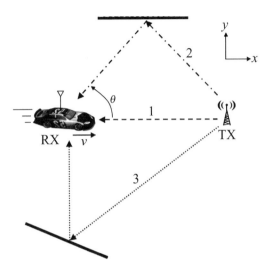

Figure 6.7 RX–TX setup for single-tap Doppler shift. Receiver travels towards transmitter and encounters LOS and reflected paths.

As the situations above illustrate, the Doppler shift encountered by a single tap in the wireless channel is affected by both the angle of arrival (AoA) of the path and the relative velocity between the transmitter and receiver. Doppler spreads, which will be covered further in Section 6.4.3, occur when several reflected signals arrive at the receiver with approximately the same delay but different AoAs or relative velocities. This would be the case in Figure 6.7 if all three paths had equal lengths. Additionally, it is possible for different taps in the same multipath profile to possess different Doppler spreads.

6.3.2 Statistical multipath models

To this point, the wireless channel has been modeled as a deterministic quantity. Although this could be true, given complete knowledge of the multipath environment (i.e. locations and types of scatterers, velocities of the transmitter, receiver, and scatterers, etc.), such detailed information is usually not available. Consequently, a statistical model of the channel is required in order to connect the rapidly varying multipath environment with communication system performance.

The first assumption that is usually made regards the nature of scatterers. Specifically, if the scattering environment is rich enough (i.e. many scatters) so that a large number of statistically independent paths contribute to each channel tap, then each tap $h_b[p,m]$ can be modeled as a complex Gaussian random variable. To see this consequence more clearly, recall from Equation (6.24) the contribution that one path makes to the pth tap:

$$a_i(mT_B)e^{-j2\pi f_c \tau_i(mT_B)} \operatorname{sinc}\left(\pi\left(p - \frac{\tau_i(mT_B)}{T_B}\right)\right). \tag{6.29}$$

Since the delay $\tau_i(mT_B)$ is on the order of microseconds while the carrier frequency f_c is of the order on megahertz or gigahertz, it is reasonable to assume that the contribution of each path has a uniformly distributed phase in $[0, 2\pi]$ and is hence circularly symmetric. This phase is encapsulated in the argument to the exponential, $-j2\pi f_c \tau_i(mT_B)$. Furthermore, given the randomness of scatterers, it is also safe to assume that all paths are independent of one another. The pth channel tap is the aggregation of many such paths. Thus, by appealing to the central limit theorem for both the real and complex portions of the pth channel tap, we may conclude that it can be modeled as a complex Gaussian random variable $CN(0, \sigma_p^2)$. The magnitude $|h_b[p,m]|$ and squared-magnitude of the pth channel tap have Rayleigh (Equation (6.30)) and exponential (Equation (6.31)) distributions, respectively (Tse and Viswanath 2005):

$$f(|h_b[m,p]|) = \frac{|h_b[m,p]|}{\sigma_p^2} \exp\left(\frac{-x^2}{2\sigma_p^2}\right) \tag{6.30}$$

$$f(|h_p[m,p]|^2) = \frac{1}{\sigma_p^2} \exp\left(\frac{-|h_b[m,p]|^2}{\sigma_p^2}\right). \tag{6.31}$$

To understand the rate at which the channel taps are changing, a metric that is often examined is the tap autocorrelation function $R_p[k]$:

$$R_p[k] = E[h_b^*[p,m]h_b[p,m+k]]. \tag{6.32}$$

When writing the autocorrelation function in this form, an implicit assumption of wide-sense stationarity is made on the underlying process $h_b[p,m]$. Intuitively, the autocorrelation function relates the current state of the channel to its future state. By noting that $\sum_p R_p[0]$ is proportional to the total amount of received energy across all taps, one definition of delay spread involves determining how many channel taps P are needed to collect a certain percentage of the transmitted signal energy. This may be represented as

$$T_d = T_B \sum_{p=0}^{P-1} R_p[0] \tag{6.33}$$

such that

$$P_d = 100 \frac{\sum_{p=0}^{P-1} R_p[0]}{\sum_p R_p[0]}. \tag{6.34}$$

PHYSICAL LAYER CONSIDERATIONS

P_d is the percentage of signal energy that should be recovered with P total taps. This allows an alternative method to calculate a delay spread value T_d given some predetermined value for P_d in a statistical model.

In the frequency domain, the Fourier transform of the autocorrelation function specifies the power spectral density (PSD) of the channel tap process. Under the assumption of a Rayleigh fading environment with a large number of uniformly distributed scatterers around the receiver, the Rayleigh PSD takes on a 'bathtub' shape as indicated in Figure 6.8.

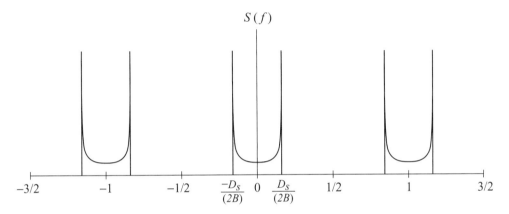

Figure 6.8 Plot of three periods of the periodic PSD of a Rayleigh-faded channel tap with delay spread D_s and system bandwidth B.

The Rayleigh fading channel assumes that there is no dominant or line of sight (LOS) path present. However, in real channels this is typically not the case, so a Ricean channel model may be more appropriate. This model assumes the presence of a dominant path, often to be taken as an LOS path, along with a number of diffuse, Rayleigh-like scatterers. For such a model, the pth channel tap may be modeled as:

$$h_b[p,m] = \sqrt{\frac{\kappa}{\kappa+1}} \sigma_p e^{j\theta} + \sqrt{\frac{1}{\kappa+1}} CN(0, \sigma_p^2) \qquad (6.35)$$

where κ is the ratio of energy in the dominant path to that contained (on average) in the scattered paths. The dominant path is assumed to arrive with some uniform phase θ. The shape of the PSD is similar to that of a Rayleigh-faded tap, except that a significant amount of energy will be present around D_{shift} Hz, where D_{shift} is the Doppler shift encountered by the dominant path.

6.3.3 Path loss modeling

For a given separation between a receiver and transmitter, it is well known that signal power decays inversely with distance. For free space, this decay rate is proportional to $1/(d^2)$, where d is the RX–TX separation and its exponent is known as the path loss exponent. For a specific, static environment, path loss is defined as the ratio of received power at some reference distance d_0 (P_{R,d_0}) to the received power at the desired distance d, $P_{R,d}$:

$$PL(d) = \frac{P_{R,d_0}}{P_{R,d}} = \left(\frac{d}{d_0}\right)^\alpha. \tag{6.36}$$

The path loss exponent in Equation (6.36) is the variable α. Path loss is often stated in decibels with the path loss at the reference distance ($PL(d_0)$) implicitly included (Rappaport 2002). Observe that $PL(d_0)$ must be measured, and d_0 must be large enough that neither antenna suffers from near-field effects. Path loss is calculated as

$$PL(d) = PL(d_0) + 10\alpha \log\left(\frac{d}{d_0}\right), \tag{6.37}$$

and it is a positive, relative quantity that must be subtracted from the power (in dBm) of the transmitted signal.

The formula in Equation (6.37), however, will not hold exactly for every situation where the RX and TX are separated by a distance d because it does not account for shadowing. Shadowing is the effect that large-scale blockers and scatterers have on the transmitted signal. Whether an object provides shadowing depends on its relative environment and position. Outdoors, hills and large buildings provide shadowing, while indoors, walls and doors may be shadowers. Given a certain environment, different instantiations of that environment will have different arrangements of shadowers. Therefore, even at an identical RX–TX separation, measurements of path loss will not be identical between, say, two different outdoor, urban settings. In the model, this discrepancy is accounted for with a log–normal random variable appended to Equation (6.37) (Rappaport 2002):

$$PL(d) = PL(d_0) + 10\alpha \log\left(\frac{d}{d_0}\right) + \chi_\sigma \tag{6.38}$$

where χ_σ is an $N(0, \sigma^2)$ random variable in decibel units.

6.4 Channel Metrics

Given that the wireless channel may vary wildly over different locations and timescales, several metrics are used to model the behavior of the channel at a given time and place. Four metrics – delay spread, coherence bandwidth, Doppler spread, and coherence time – summarize small-scale multipath effects, while the path loss exponent mentioned in Section 6.3.3 illustrates a large-scale effect. After examining each impairment in this section, their effects specifically upon OFDM are noted in Section 6.4.5.

6.4.1 Delay spread

We now elaborate further on *delay spread*, first introduced in Section 6.1.2, which quantifies the number and severity of 'echoes' in the wireless channel. It gives a

PHYSICAL LAYER CONSIDERATIONS

concrete measurement of how many copies of the transmitted signal are present at the receiver and how badly they were delayed (amount of channel memory). It is usually parameterized with three quantities. The *mean excess delay* indicates the average echo's delay in the channel. Delays are weighted by the squared magnitude of the channel tap generating the delay:

$$\bar{\tau} = \frac{\sum_p |h_b[p,m]|^2 \tau_p}{\sum_p |h_b[p,m]|^2}. \quad (6.39)$$

The quantity $\bar{\tau}$ is also known as the first moment of the power–delay profile, or PDP. The PDP shows the relative power content among the channel taps and is defined in discrete time as:

$$h_{\text{PDP}}[p,m] = \sum_k |h_b[k,m]|^2 \delta[p-k], \quad (6.40)$$

where $\delta[k]$ is the discrete-time delta function. Naturally, delays vary about this mean in reality. The standard deviation of the delays, or *RMS delay spread*, is the second central moment of the PDP (Rappaport 2002). It is calculated as:

$$\tau_{\text{RMS}} = \sqrt{\overline{\tau^2} - (\bar{\tau})^2}, \quad (6.41)$$

where

$$\overline{\tau^2} = \frac{\sum_p |h_b[p,m]|^2 \tau_p^2}{\sum_p |h_b[p,m]|^2}. \quad (6.42)$$

A third quantity, known as *maximum excess delay*, denotes the time between the first and last detectable echoes, where the threshold of detectability is referenced on the strongest echo. This threshold is typically specified in dB below the strongest echo and should be included when referencing maximum excess delay measurements. The relationship between the three delay spread parameters is indicated in Figure 6.9. In traditional communications systems, the delay spread tends to cause intersymbol interference (ISI). To combat ISI, a wealth of equalization algorithms, which attempt to reverse the delaying effects of the channel, are often used. Applying equalization requires timely channel estimates via training data, which is typically provided by the preamble of a packet. However, note that OFDM does not typically resort to time-domain equalization for ISI mitigation. As mentioned in Section 6.1.2, OFDM systems use a guard interval containing a cyclic prefix to account for ISI.

6.4.2 Coherence bandwidth

The frequency-domain counterpart to delay spread is known as *coherence bandwidth*. Conceptually, coherence bandwidth captures the severity of frequency-selective fading in the wireless channel. The magnitude of a channel with a small coherence bandwidth will vary greatly across the band of interest, whereas the magnitude of a channel with a large coherence bandwidth will be relatively flat. Naturally, the question of what constitutes 'large' and 'small' coherence bandwidths must be addressed. Coherence bandwidth size is defined relative to the bandwidth occupied

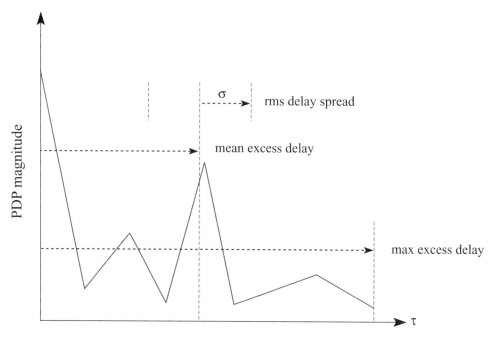

Figure 6.9 Relationship between delay spread parameters.

by the signal of interest. For example, a channel centered at 5.8 GHz may have a coherence bandwidth of 500 kHz. If a narrowband signal, such as an AM audio signal, is sent at 5.8 GHz, it will see a relatively flat channel and the coherence bandwidth may be considered 'large'. However, if a 802.11a signal, centered at 5.8 GHz, is sent over the same channel, the coherence bandwidth will be considered 'small', since the magnitude of the channel will vary over the occupied bandwidth of the signal. Figure 6.10 illustrates how, depending on the communication scheme used, a channel could have both large and small coherence bandwidth. A channel is said to have *frequency-selective fading* when its coherence bandwidth is small and *flat fading* when its coherence bandwidth is large.

Mathematically, a relationship may be derived between delay spread and coherence bandwidth. Start by taking the Fourier transform of $h_b(\tau, t)$ with respect to τ at a fixed time t. Recall that t is parameterizing $h_b(\tau, t)$ in order to make the channel LTV, so by fixing t we have a LTI channel. Drawing from Equation (6.22) and suppressing dependence on t:

$$H(f) = \int_{-\infty}^{+\infty} \sum_i a_{i,b} \delta(\tau - \tau_i) e^{-j2\pi f \tau} \, d\tau \tag{6.43}$$

$$= \int_{-\infty}^{+\infty} \sum_i a_i \delta(\tau - \tau_i) e^{-j2\pi f_c \tau_i} e^{-j2\pi f \tau} \, d\tau \tag{6.44}$$

PHYSICAL LAYER CONSIDERATIONS

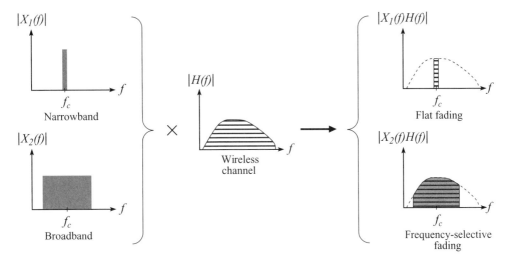

Figure 6.10 Classification of a channel as flat or frequency-selective depends on transmitted signal bandwidth.

$$= \sum_i \int_{-\infty}^{+\infty} a_i \delta(\tau - \tau_i) e^{-j2\pi(f_c \tau_i + f\tau)} \, d\tau \tag{6.45}$$

$$= \sum_i a_i \int_{-\infty}^{+\infty} \delta(\tau - \tau_i) e^{-j2\pi(f_c \tau_i + f\tau)} \, d\tau \tag{6.46}$$

$$= \sum_i a_i e^{-j2\pi(f_c \tau_i + f\tau_i)} \tag{6.47}$$

$$= \sum_i a_i e^{-j2\pi(f_c + f)\tau_i}. \tag{6.48}$$

From (6.48), notice that the wireless channel can be represented as the sum of vectors in the complex plane, where the angle to the real axis is dictated by the argument to the exponential. As f varies around f_c, each of the vectors rotates at a different rate governed by its delay τ_i. This is illustrated in Figure 6.11 with two different three-tap channels. Channel A has a delay spread of 450 ns, and channel B has a spread of 1150 ns. When all the τ_i's are close to one another, the vectors rotate at roughly the same rate, and $|H(f)|$ is relatively constant over f, as shown in Figure 6.11(a). In contrast, if the τ_i's differ drastically, then the vectors rotate at different rates, leading to a variation of $|H(f)|$ as the relative angles between the vectors change. Observe that over the indicated 500 kHz frequency span, the black vector in Figure 6.11(a) remains at a relatively constant magnitude, while its counterpart in Figure 6.11(b) changes noticeably.

To approximate the coherence bandwidth, say that the channel has changed significantly when the angle between the fastest rotating vector (largest τ_i) and slowest rotating vector (smallest τ_i) has grown by $2\pi\theta$. If the initial angle between

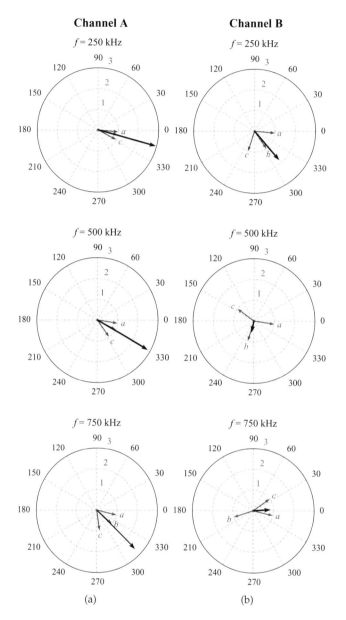

Figure 6.11 Subfigure (a) (left-hand column) shows a three-tap channel with taps a, b, and c at 50, 250, and 500 ns, respectively, evolving as the frequency f is changed. Tap b is obscured by the sum vector. Subfigure (b) (right-hand column) shows a channel with taps a, b, and c at 50, 600, and 1200 ns, respectively, evolving across the same three frequencies. Lighter vectors indicate the individual taps, while the darker vector is the sum of the channel tap vectors and represents the channel response at the indicated frequency.

PHYSICAL LAYER CONSIDERATIONS

the two vectors, occurring at frequency f_1, is

$$2\pi(f_c + f_1)\tau_{max} - 2\pi(f_c + f_1)\tau_{min} = 2\pi\theta_{initial}, \qquad (6.49)$$

then growth by θ between the vectors, occurring at frequency f_2, is

$$2\pi(f_c + f_2)\tau_{max} - 2\pi(f_c + f_2)\tau_{min} = 2\pi(\theta_{initial} + \theta). \qquad (6.50)$$

Setting $\Omega_1 = f_c + f_1$ and $\Omega_2 = f_c + f_2$, use Equations (6.49) and (6.50) to write:

$$\Omega_2(\tau_{max} - \tau_{min}) - \Omega_1(\tau_{max} - \tau_{min}) = \theta, \qquad (6.51)$$

$$\Omega_2 - \Omega_1 = \frac{\theta}{\tau_{max} - \tau_{min}}. \qquad (6.52)$$

Note that $\tau_{max} - \tau_{min}$ is an estimate of the delay spread and that $\Omega_2 - \Omega_1$ is the coherence bandwidth. Hence, Equation (6.52) shows that coherence bandwidth, B_{coh}, is inversely proportional to delay spread, τ_d:

$$B_{coh} \propto \frac{1}{\tau_d}. \qquad (6.53)$$

6.4.3 Doppler spread

Doppler spread, as previously mentioned in Sections 6.3.1 and 6.3.2, indicates how the signal energy conveyed by a given channel tap is spread in frequency. To illustrate this, consider again a single-tap passband channel $h_1(\tau, t) = a_1(t)\delta(\tau - \tau_1)$, where the amplitude of the tap varies over time, with input $x_1(t)$ and output $y_1(t)$. This is a LTV system and may be represented as:

$$y_1(t) = \int_{-\infty}^{+\infty} h_1(\tau, t) x_1(t - \tau) \, d\tau, \qquad (6.54)$$

$$= \int_{-\infty}^{+\infty} a_1(t)\delta(\tau - \tau_1) x_1(t - \tau) \, d\tau, \qquad (6.55)$$

$$= a_1(t) x_1(t - \tau_1). \qquad (6.56)$$

Equation (6.56) shows that $y_1(t)$ is the time-delayed input modulated by the amplitude of the channel tap. Based on the discussion in Appendix 6.10, which details the manifestation of LTV effects on the channel's output, the Fourier transform of $y_1(t)$ first requires the transform of $h_1(\tau, t)$:

$$H(f_\tau, f_t) = \int_{-\infty}^{+\infty} \int_{-\infty}^{+\infty} a_1(t)\delta(\tau - \tau_1) e^{-j2\pi f_\tau \tau} e^{-j2\pi f_t t} \, d\tau \, dt, \qquad (6.57)$$

$$= \int_{-\infty}^{+\infty} a_1(t) e^{-j2\pi f_t t} \, dt \int_{-\infty}^{+\infty} \delta(\tau - \tau_1) e^{-j2\pi f_\tau \tau} \, d\tau, \qquad (6.58)$$

$$= A_1(f_t) e^{-j2\pi f_\tau \tau_1}. \qquad (6.59)$$

Using Equation (6.113) to determine $Y_1(f_t)$ yields

$$Y_1(f_t) = \int_{-\infty}^{+\infty} X_1(f_\tau) A_1(f_t - f_\tau) e^{-j2\pi f_\tau \tau_1} \, df_\tau. \tag{6.60}$$

Figure 6.12 shows the consequences of this frequency 'smearing' effect on a pure tone transmission. This is in contrast to a Doppler shift, which shifts the signal's frequency content by a quantity proportional to the relative velocity between the transmitter and receiver. Doppler shifts can, however, be unified into this framework by taking the channel as $h_b(\tau, t) = \delta(\tau + (vt)/c)$, where v is the relative speed between the transmitter and receiver.

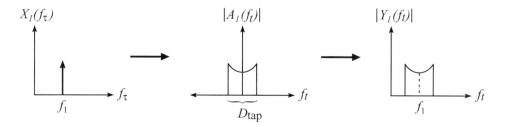

Figure 6.12 Effect of a single-tap channel's Doppler spread, D_{tap}, on a pure tone transmission at frequency f_1. The channel output yields a collection of frequencies instead of a single tone. Note that the frequency variable's dual changes between delay and time domains.

Doppler spreads occur because multiple signal reflections arrive at the receiver with different incident angles, which endows them with different Doppler shifts (see Figure 6.13). If the reflections arrive at roughly the same time, they contribute to the same channel tap. Consequently, that channel tap experiences the sum effect of several Doppler shifts, leading to a Doppler spread on the tap (D_{tap}). To quantify how closely reflections must arrive to contribute to the same tap, examine the sampled baseband model from Equation (6.24). After fixing p, notice that, due to the behavior of the sinc() term, only paths with delays falling within $[(p - 1/2)T_B, (p + 1/2)T_B]$ will contribute to the pth tap in the sampled baseband representation.

As the per-tap Doppler spread D_{tap} increases, so does the frequency span over which the channel tap's energy is distributed. If the Doppler spread is sufficiently high, then energy originally transmitted at a single frequency may be spread across a band. Doppler spreads are typically measured on a per-tap basis after estimating the spectra of each tap via an FFT. This requires several estimates of the channel in order to determine each tap's variation over time, and hence, its frequency content.

For the channel as a whole, the Doppler spread, D_S, accounts for all taps in the estimated channel and is calculated as the support of the Doppler spectrum as given in Tranter et al. (2004). To obtain the channel's overall Doppler spectra, recall that the baseband equivalent LTV channel is $h_b(\tau, t)$. For each value of τ that yields non-zero

PHYSICAL LAYER CONSIDERATIONS

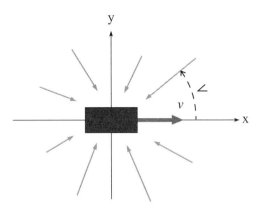

Figure 6.13 A vehicle, traveling at velocity v, will encounter reflections arriving from a variety of angles.

tap values for some t, taking a Fourier transform with respect to t produces the per-tap Doppler spectra, D_{tap}, for that particular τ. Denoting this function as $H_b(\tau, f_t)$, we may write

$$H_b(\tau, f_t) = \mathfrak{F}_t\{h_b(\tau, t)\}. \tag{6.61}$$

Next, Equation (6.62) shows that integrating over all values of τ for each f_t results in the channel's Doppler spectra, $H_{b,\text{Dop}}(f_t)$:

$$H_{b,\text{Dop}}(f_t) = \int_{-\infty}^{\infty} H_b(\tau, f_t)\, d\tau. \tag{6.62}$$

The support of $H_{b,\text{Dop}}(f_t)$ is taken as the Doppler spread of the channel.

6.4.4 Coherence time

Finally, because the wireless vehicular environment changes rapidly with respect to speed, location, and signal scatterers, channel characteristics are not static. To quantify this, *coherence time* is defined as the period of time that a channel may be approximated as time-invariant, and it is inversely related to a channel's Doppler spread. To justify this mathematically, appeal to an argument similar to that used for examining coherence bandwidth. Returning to Equation (6.21), reproduced below, observe that the expression for the channel may be viewed as the sum of several vectors in the complex plane:

$$h_b(\tau, t) = \sum_i a_i(t)\delta(\tau - \tau_i(t))e^{-j2\pi f_c \tau_i(t)}. \tag{6.63}$$

Fixing $\tau = \tilde{p}T_B$, the goal is to observe how quickly $h_b(\tilde{p}T_B, t)$ changes as a function of t. First, note that we only need to examine paths contributing to the $\tilde{p}$th tap. Define the set of such paths as $\tilde{P}$ and rewrite Equation (6.63) as

$$h_b(\tilde{p}T_B, t) = \sum_{i \in \tilde{P}} a_i(t)\delta(\tilde{p}T_B - \tau_i(t))e^{-j2\pi f_c \tau_i(t)}. \tag{6.64}$$

Next, note that the timescales over which the terms in Equation (6.64) vary are quite different. The scaling applied to the ith path, $a_i(t)$, is largely a property of the physical obstructions that the signal encounters. Its variations occur on the order of milliseconds or seconds. The second term indicates which paths contribute to the $\tilde{p}$th tap. As discussed earlier in relation to Doppler spread, paths in $\tilde{P}$ arrive within T_B of each other. For a path to contribute to another tap, it would need to lengthen by at least $(T_B/2)c$ meters. Depending on relative velocity, this path migration usually occurs on the order of hundreds of milliseconds. Coming to the third term, notice that the argument to the exponential is quite sensitive to variations in $\tau_i(t)$ due to multiplication by f_c, which is usually on the order of 10^6 or 10^9. Thus, it is again the phase of the vectors that plays the principal role in coherence time.

All the vectors contributing to the $\tilde{p}$th tap hold some relative phase to one another. However, the rate at which each vector rotates in the complex plane varies with each $\tau_i(t)$, so that over time, the relative phases between the vectors changes. This leads to a different value for $h_b(\tilde{p}T_B, t)$. The time required for such a significant phase change to occur between any two vectors can be thought of as the tap's coherence time. Define $2\pi\theta$ as a value, in radians, as the angular threshold at which a new channel is declared to exist. In other words, a new channel occurs when the angle between any two vectors in $\tilde{P}$ grows by $2\pi\theta$. Examining the argument of the exponential, write

$$2\pi f_c(\tau_{\max}(t_1) - \tau_{\min}(t_1)) + 2\pi\theta = 2\pi f_c(\tau_{\max}(t_2) - \tau_{\min}(t_2)) \qquad (6.65)$$

where $\tau_{\max}$ and $\tau_{\min}$ are the delays associated with the paths that have the maximum and minimum *derivatives* with respect to t. The derivative of delay with respect to t is important here as the coherence time measures how long it takes the channel (in t, not τ) to change the relative phase between two of its vectors by $2\pi\theta$ radians. The pair of vectors that will meet this threshold first contain $\tau_{\max}$ and $\tau_{\min}$. The relationship between the coherence time, T_c, delay-time derivatives, and the angular threshold may be written as:

$$T_c = \frac{2\pi\theta}{2\pi f_c\left(\frac{d\tau_{\max}}{dt} - \frac{d\tau_{\min}}{dt}\right)}. \qquad (6.66)$$

With the further observation that $f_c\left(\frac{d\tau_i}{dt}\right)$ is the Doppler shift on the ith path, conclude that

$$T_c \propto \frac{1}{D_s} \quad \text{where } D_s = \frac{d\tau_{\max}}{dt} - \frac{d\tau_{\min}}{dt}. \qquad (6.67)$$

The effect of coherence time upon a system depends on the relation between the channel fluctuation rate and some system-defined parameter. This parameter could be a symbol time, the length of a typical packet, or how often the channel is equalized, and it could be affected by layers above the PHY, such as application-level timing requirements. If the coherence time is less than this parameter, the wireless channel is said to be *fast fading*; otherwise, it is a *slow fading* channel.

6.4.5 Impact on OFDM systems

As mentioned in Section 6.1, wideband DSRC communications will almost certainly involve the use of OFDM. It is thus instructive to comment on how the metrics in

Sections 6.4.1 to 6.4.4 should be viewed in the light of an OFDM-based communication system.

Traditional time-domain communication systems require equalization to correct for the ISI-inducing effects of delay spread. In contrast, OFDM-based communication systems are largely immune to ISI, provided that they are designed correctly for their intended environment. As mentioned in Section 6.1.2, as long as the delay spread for a channel is less than the OFDM system's GI, the system avoids ISI. Consequently, as long as upper bounds on delay spread are reasonably certain, OFDM can be implemented with immunity to this impairment. However, immunity to ISI does not imply that OFDM is exempt from channel effects. Even if the GI is properly set, it only prevents interference from adjacent symbols. Received data is still subject to multiplication by the channel gain on each subcarrier. If the channel's gain for a subcarrier is extremely small, then deep fading occurs and that subcarrier's data could be lost.

The frequency-domain counterpart to delay spread, coherence bandwidth, causes broadband signals to suffer from variations across the band in the form of frequency-selective fading. As the name implies, OFDM systems modulate data on multiple subcarriers which are mutually orthogonal. Usually, a subset of these subcarriers are reserved as pilot tones to compensate for frequency offsets and to aid in coherent detection. Specifically, their continual presence assists in tracking variations in the channel. If the coherence bandwidth is equal to or greater than the spacing between pilot tones, then the system may implement simple, flat-fading corrections to subcarriers around each pilot. These flat-fading corrections simply consist of amplitude scaling. On the other hand, if the coherence bandwidth is smaller than the pilot spacing, then the channel varies significantly between each pilot tone, and more complex methods, such as interpolation between pilots, may be employed.

While delay spread is the primary concern for traditional communication systems, Doppler spread holds similar significance for OFDM. In a non-mobile environment, subcarriers are naturally orthogonal as each occupies different frequency bands. However, mobility induces Doppler spread, which may cause one carrier to 'bleed' information into another. This phenomenon is known as inter-carrier interference (ICI), and it is the frequency domain analog to ISI. Ideally, the subcarrier spacing is large enough such that any expected Doppler spread is much smaller than this spacing. ICI begins to impact communications when the Doppler spread is of the order of 5 to 10% of the subcarrier spacing.

Despite accounting for the previous three impairments, the fourth, coherence time, may still adversely affect an OFDM (indeed any) communication system. Most systems (including 802.11a and 802.11p) divide their data transmissions into frames or packets and place a series of training symbols at the start of each packet. These training symbols aid in clock timing recovery and channel estimation for various receiver-side equalization algorithms. They typically occur at the start of each packet, implying that channel estimation is done only once per packet. Therein lies the problem – short coherence times (compared to packet length) can induce errors since the channel may change significantly over the course of a packet's reception. Consequently, if not updated, the initial equalization will be quickly rendered incorrect and fail to account for variations in the channel.

Please note that nothing inherent in general OFDM architectures precludes the use of dynamic equalization or channel tracking to account for short coherence times. In fact, both 802.11a and 802.11p allocate four subcarriers as pilot tones, but do not mandate how or if they should be used. Since 802.11a systems do not expect to encounter short coherence times (with respect to average packet lengths), most 802.11a chipsets do not implement dynamic equalization due to the additional complexity that such schemes would entail. Consequently, without intentional modifications, DSRC chipsets based heavily upon 802.11a products may not perform well under short coherence times.

Recalling the evolution of the DSRC standard, note that the signal integrity structures for 802.11a have been optimized for use in indoor situations with slow-moving radios. However, vehicular environments involve severe multipath, higher velocities, and a wider dynamic range of signal strengths. Other than the channel bandwidth, allocated spectrum, and transmission power limits, 802.11p shares the same structure, modulation, and training sequences of the 802.11a PHY. Only DSRC's channel bandwidth of 10 MHz (compared to 20 MHz in 802.11a), alters parameters directly affecting its ability to cope with multipath, such as symbol guard intervals and subcarrier spacing, while highway velocities may change a channel's coherence time.

6.5 Measurement Theory

At first glance, one might expect that measuring the channel would simply require the generation of an 'impulse' at the transmitter and recording the signal echoes at the receiver. While intuitively pleasing, such a method is impractical in reality for a variety of reasons. One of the primary reasons lies in the difficulty of generating such a pulse, which would require a large amount of power over an extremely short duration – a task that is challenging for most power amplifiers and signal generators. The receiver would also require digitizing hardware that sampled at an extremely high rate and with high dynamic range in order to avoid missing any of the channel's reflections.

For these reasons, many channel measurements employ what are known as pulse-compression techniques. The measurement method detailed here is similar to the matched filter method detailed by Acosta-Marum (2007), which is a variation of the swept time-delay cross-correlator (STDCC) first used by Cox (1972). It relies on the impulse-like time-domain cross-correlation properties of maximal-length, pseudorandom noise sequences (MLS sequences); namely, that for some periodic MLS sequence $x_{\text{MLS}}[n]$ of length and period N,

$$R_{x_{\text{MLS}}}[k] = \frac{1}{N} \sum_{n=-\infty}^{+\infty} x_{\text{MLS}}[n] x_{\text{MLS}}[n-k] \qquad (6.68)$$

is a N-periodic signal which resembles an impulse train, as illustrated in Figure 6.14. Such signals are binary in nature, and the contents of one particular sounding sequence may be found in Appendix 6.11.1. In addition to their impulsive correlation properties, these sequences also provide a wide dynamic range of measurement.

PHYSICAL LAYER CONSIDERATIONS

For example, the autocorrelation of a sequence 511 symbols in length contains a peak that is approximately 54 dB higher than the minimum value in the autocorrelation. MLS sequences may also be referred to as PN sequences throughout the remainder of this chapter.

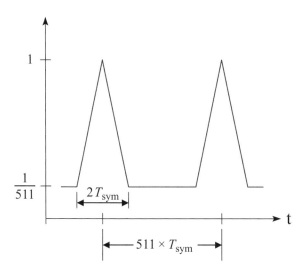

Figure 6.14 Plot of the periodic autocorrelation function $R_{x_{\text{MLS}}}$ with a period of 511 symbols and a symbol time of T_{sym}.

Measuring the LTV wireless channel may be accomplished by obtaining successive estimates of $h_b[p, m]$, the sampled complex baseband response. Each individual estimate of $h_b[p, m]$ assumes that the channel is LTI over the estimate's duration. This assumption is valid, provided that the length of the measuring signal is less than the maximum coherence time expected. To obtain these LTI channel snapshots, a PN sequence of appropriate length is passed through the communication system of Figure 6.15. At the receiver, this sequence is processed via a cross-correlation algorithm to produce a PDP and other LTI statistics. By taking many *consecutive* estimates of the channel, LTV statistics, such as Doppler spread and coherence time, may be determined. Appendix 6.11.2 details how individual LTI estimates are generated.

To enable rapid, consecutive channel estimates, the transmitter continually sends copies of the PN sequence interspersed with null periods of equivalent length, as shown in Figure 6.16. The null period prevents bleeding of multipath from one channel estimate into adjacent estimates. After down-conversion and digitization, the receiver stores copies of the received signal for offline postprocessing.

The aggregation of a number of consecutive snapshots will be referred to as a *superframe*, while the term *frame* will describe any individual snapshot. The number of frames contained in one superframe is dependent on the memory depth of the receiving equipment. By repeatedly applying the operations described

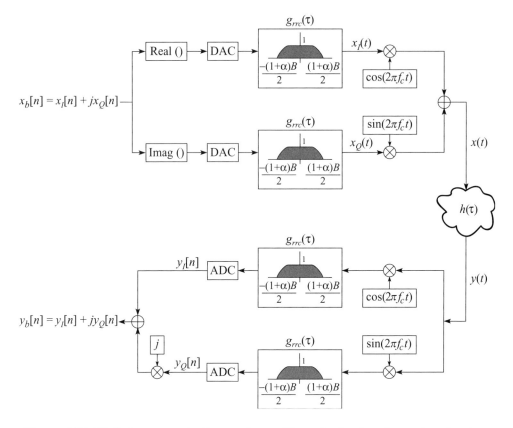

Figure 6.15 Digital communication system with root raised-cosine pulse shaping.

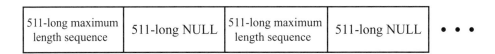

Figure 6.16 Probe waveform.

in Appendix 6.11.2 (specifically Equation (6.128)) on the frames in a superframe, many consecutive LTI estimates of the channel are obtained. They are represented by $\widehat{h_b[k,n]}$, where n is the time instant at which the estimate was taken and k is the index for the discrete channel taps.

After processing each frame in a superframe, the channel estimates are compiled into a matrix with the delay parameter k spanning the columns and each estimate occupying a single row. This matrix, known as the channel impulse response matrix (CIR matrix) contains all the estimation information for a single superframe.

PHYSICAL LAYER CONSIDERATIONS

Given a superframe, individual PDPs are first generated from each frame (row in the CIR matrix) via

$$P[p,n] = \sum_{k=0}^{510} |\widehat{h_b[k,n]}|^2 \delta[p-k], \tag{6.69}$$

where n is the row index, p is a discrete delay index, and $\widehat{h_b[k,n]}$ indicates the element in the kth column and nth row of the CIR matrix. Each individual PDP is also normalized to the strength of the strongest tap. An overall PDP for the superframe is obtained by averaging the individual PDPs. Also, a 30 dB threshold, relative to the strongest tap, is applied to the overall PDP in order to remove weak taps and spurious responses due to system noise.

Similarly, delay spread statistics (mean excess, max excess, and RMS) are obtained by applying the formulas in Equations (6.39) and (6.41) to each row of the CIR matrix and then averaging the results. Coherence bandwidths follow as the inverse of the average mean excess delay.

To calculate the effects of vehicular speed on the wireless channel, the Doppler spectra for each estimated channel tap (per-tap Doppler spectra) is extracted by taking a Hamming-windowed FFT down the columns of the CIR matrix,

$$D[n] = \frac{1}{M} \sum_{k=0}^{M-1} \widehat{h_b[k,n]} W[k] e^{-2\pi k n / M}. \tag{6.70}$$

The Hamming window function $W[k]$ is

$$W[k] = 1 - \left(\frac{k - \frac{M-1}{2}}{\frac{M+1}{2}} \right). \tag{6.71}$$

The repetition rate of the frame dictates the range of Doppler spectra (D_{range}) that may be measured, while the length of the superframe, in seconds, dictates the resolution (D_{res}) of the generated spectra. Taking T_{frame} as the length of a single frame (PN sequence plus subsequent null period) and T_{sframe} as the length of a superframe, notice that

$$D_{\text{range}} = \frac{1}{T_{\text{frame}}} \tag{6.72}$$

while

$$D_{\text{res}} = \frac{D_{\text{range}}}{\frac{T_{\text{sframe}}}{T_{\text{frame}}}}, \tag{6.73}$$

$$= \frac{1}{T_{\text{sframe}}}. \tag{6.74}$$

As a result, Doppler resolution depends mainly on the receiver's storage capabilities. Doppler spread is calculated as the support of the Doppler spectra across all taps, as described in Equation (6.62).

In situations where a line-of-sight (LOS) path is present along with other diffuse paths, knowing the relative strengths of the LOS path to other paths assists in quantifying the degree of determinism in the channel. We calculate this statistic, known as the Ricean K-factor, for each tap via the moment-method detailed by Greenstein et al. (1999). Appendix 6.11.3 summarizes pertinent details of this method. Both Greenstein et al. (1999) and Abdi et al. (2001) verified the accuracy of this method with measured data, and Abdi et al. (2001) additionally noted a slight degradation of estimation accuracy with correlated samples. Even with this degradation, however, the method still performs quite well.

6.6 Empirical Channel Characterization at 5.9 GHz

Using the methods detailed in Section 6.5, we conducted an extensive channel measurement campaign in and around Detroit, Michigan. Our campaign examined the channel responses present in an open-field test track, urban streets and canyons, rural roads, and highways. Wireless channels in these situations experienced different scatterer densities, speeds, and LOS conditions. By examining many channels with varying realizations of these parameters, we obtained a broad cross section of the environments under which DSRC will be primarily utilized and, in so doing, obtain a basis on which to evaluate the proposed DSRC standard.

6.6.1 Highway environments

The need for DSRC robustness to Doppler spread is most evident in high-speed situations. To examine this, tests were conducted on divided multi-lane highways. Speeds typically exceeded 50 mph, and separations ranged from 100 m to 1000 m. The surrounding environment varied greatly due to the presence of hills, bridges, and other concrete structures. Traffic was also uncontrolled, with a mix of passenger vehicles and large tractor–trailer trucks. Non-line-of-sight (NLOS) conditions were created either by terrain or by the imposition of blocking vehicles (e.g. large trucks or towed trailers) between our transmitter and receiver. Transmitter power was set at the maximum (33 dBm at the TX antenna) during all highway tests.

LOS test locations

When LOS is present, highway scenarios are characterized by short delay spreads but potentially large Doppler spreads. Highway speeds for our tests were about 30 m/s, and the waveform carrier frequency was 5860 MHz. Thus, Doppler shifts of 1.1 kHz or higher are possible when reflections off obstructions directly ahead of the transmitter and receiver are considered. For example, a reflection off a bridge passing over the road (i.e. an 'overpass') gives an effective path closure rate of 60 m/s and a Doppler shift of 1172 Hz. Reflections off approaching vehicles (also traveling at 30 m/s) lead to a closure rate of 120 m/s and a Doppler shift of 2.3 kHz. Since the Doppler spread accounts for a range of reflections from many angles (and hence different closure rates), spreads in excess of 2 kHz are possible and indeed likely.

Table 6.2 summarizes the average behavior of highway LOS channels. Figures 6.17, 6.18, and 6.19 present the empirical CDFs of mean excess delay, RMS delay,

PHYSICAL LAYER CONSIDERATIONS

Table 6.2 Average highway LOS channel characteristics.

Separation (m)	Speed (m/s) RX	Speed (m/s) TX	RSS (dBm)	Delay metrics (ns) $\bar{\tau}$	τ_{RMS}	τ_{max} (30 dB)	D_S (Hz)	K-factor[a]
0–50	25.56	26.12	−8.54	69.18	21.09	349.43	567.88	400.32
50–200	28.10	30.23	−25.50	208.21	95.02	1227.27	2714.29	31.77
200–500	30.33	30.71	−27.63	162.63	112.33	1527.27	1520.39	176.36
>500	35.02	33.11	−33.19	119.39	50.88	924.24	1248.61	409.67
Overall	27.81	28.41	−17.58	117.58	54.06	778.79	1223.99	290.22

[a]The K-factors indicated are associated with the strongest tap in each separation bin. This convention holds in subsequent tables.

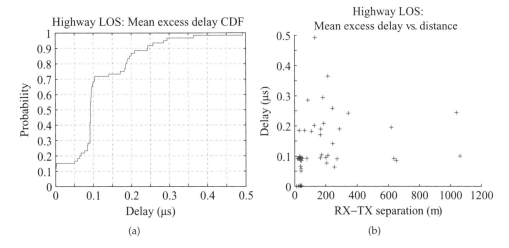

Figure 6.17 Empirical CDF for mean excess delay and scatter plot against distance for all highway LOS tests.

and Doppler spread, along with their scatter plots versus distance. These scatter plots suggested segmentation into the RX–TX separation ranges in Table 6.2. For each separation range in Table 6.2 and subsequent tables, several superframes were analyzed. The indicated quantities were calculated by averaging across all obtained measurements within each distance bin. The 'overall' row is obtained by averaging across all superframes in the highway LOS category.

The results show that, in general, delay spread is not a major concern for highway LOS channels. On average, all RX–TX separation ranges have mean excess and RMS delay spreads below the 1.6 μs maximum anticipated by DSRC. This is an unexpected result, as higher degrees of reflectivity were initially anticipated due to the increased number of reflectors (e.g. buildings and vehicles) present outdoors. On the other hand, Doppler spreads are noticeable at an average of 1223.99 Hz, while 90% of the Doppler spreads are below 2384 Hz. Though Doppler spreads must

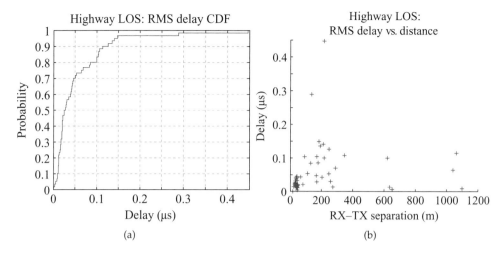

Figure 6.18 Empirical CDF for RMS delay and scatter plot against distance for all highway LOS tests.

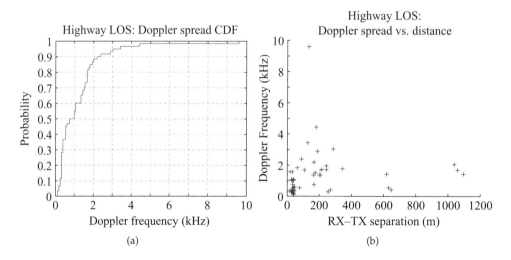

Figure 6.19 Empirical CDF for Doppler spread and scatter plot against distance for all highway LOS tests.

typically be of the order of 5% to 10% of the subcarrier spacing to cause significant ICI, issues regarding channel coherence times may be more cause for concern well before the 5% threshold is reached. Section 6.6.5 elaborates upon this further after all test categories have been covered.

PHYSICAL LAYER CONSIDERATIONS

Table 6.3 Average highway NLOS channel characteristics.

Separation (m)	Speed (m/s) RX	TX	RSS (dBm)	Delay metrics (ns) $\bar{\tau}$	τ_{RMS}	τ_{max} (30 dB)	D_S (Hz)	K-factor
0–150	28.43	27.79	−22.48	266.69	267.59	5939.39	3382.01	515.29
150–300	28.43	27.59	−30.85	255.66	136.67	1575.76	3750.94	44.03
>300	31.82	27.97	−36.48	293.01	379.58	6109.09	3289.93	74.79
Overall	29.22	27.74	−29.24	268.19	238.83	4152.22	3515.03	215.58

Figure 6.18(b) shows slight positive correlation between RMS delay and distance up to 400 m, which is corroborated by Table 6.2. A similar correlation appears in Figure 6.19(b) between 0 and 400 m.

NLOS test locations

In contrast to other channel sounding campaigns, our measurements examined high-speed NLOS scenarios along interstate highways. These situations require attention for two primary reasons. From an application standpoint, the safety-oriented nature of DSRC seeks to use wireless communications to warn drivers about hazardous conditions not directly observable to them – in essence, to act as a 'radio LOS' where visual LOS is unavailable (e.g. electronic brake light warnings). It seems reasonable to assume that many of the safety benefits of DSRC will be realized during situations where a driver cannot visually observe a hazard until it may be too late to react. Additionally, NLOS scenarios typically will introduce heavier signal attenuation and greater multipath, both of which contribute to a harsher wireless environment.

On the highway, NLOS situations were induced either by terrain (e.g. hills, curves, overpasses) or by the imposition of blocking vehicles inline or abreast of the transmitter and receiver.

Table 6.3 summarize the average behavior of highway NLOS channels. Figures 6.20, 6.21, and 6.22 present the empirical CDFs of mean excess delay, RMS delay, and Doppler spread, along with their scatter plots versus distance. The ranges suggested by these scatter plots are slightly different than those for the Highway LOS case.

Compared to the highway LOS cases, notice immediately that substantially higher Doppler spreads are present over all distances, which suggests higher angular spread and richer scattering. Such behavior is expected in high-speed, NLOS cases. Examining the Doppler CDF in Figure 6.22(a) and comparing it to its highway LOS counterpart in Figure 6.19(a), it is clear that highway NLOS channels will experience more severe Doppler effects more often than their LOS counterparts. For this case, the CDF crosses the 90% threshold at 10 220 Hz, and the 50% threshold is reached at 2180 Hz. In Figure 6.19(a), Doppler spreads of 2 kHz are not reached until almost the 90% threshold. Again, though the average Doppler spread is only about 2.25% of the 156.25 kHz subcarrier spacing, its implications for channel coherence time are more immediate.

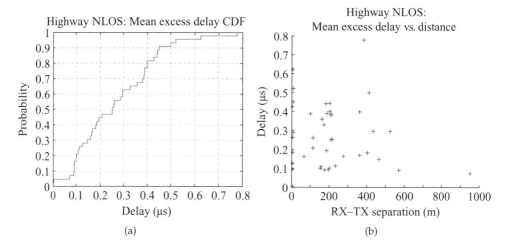

Figure 6.20 Empirical CDF for mean excess delay and scatter plot against distance for all highway NLOS tests.

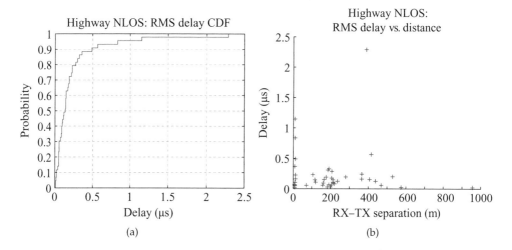

Figure 6.21 Empirical CDF for RMS delay and scatter plot against distance for all highway NLOS tests.

In terms of delay spread, the highway NLOS channels are not much more severe than their LOS counterparts. Although max excess delays are quite large and sometimes exceed DSRC's 1.6 μs guard interval, these distant taps are usually very weak compared to the strongest channel tap; hence, their effect is negligible. On the other hand, mean excess and RMS delays exceed the LOS case by only 151 ns and

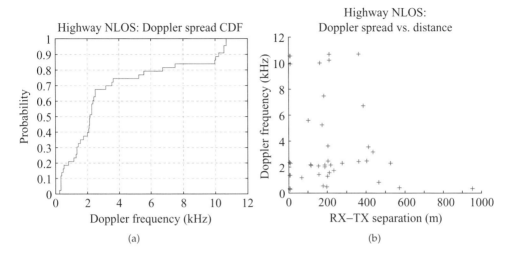

Figure 6.22 Empirical CDF for Doppler spread and scatter plot against distance for all highway NLOS tests.

184 ns on average, respectively. Both remain individually under the 1.6 μs limit, as does their sum.

No correlations with distance were noticed, though at particular distances, wide ranges in values could be observed (under 50 m for all three scatterplots, and at 200 m for Figure 6.22(b)).

6.6.2 Urban environments

To characterize urban vehicular channels, we obtained measurements on several downtown streets in metropolitan Detroit. We tested on a number of roads that were flanked by large, multistory buildings, thereby forming 'urban canyons'. LOS tests involved both same direction and opposite direction driving at low to medium speeds. NLOS tests used buildings and intersections to block LOS between the testing vehicles. Traffic was again uncontrolled throughout all tests, and RX–TX separations ranged from 12 m to over 1200 m, depending on the scenario. Transmitter power was again set to 33 dBm at all times.

Due to the location of the tests, GPS positioning data was not always available. Hence, some environmental parameters, such as speed and separation, are approximated when needed or removed from averaging calculations where appropriate.

LOS test locations

LOS tests fall into three categories: inline same-direction tests, opposite-direction tests, and urban path loss tests. The first situation examines cases where both testing vehicles travel in the same direction, usually with separations of the order of 100–200 m. Next, in order to increase relative speeds and potentially Doppler spreads,

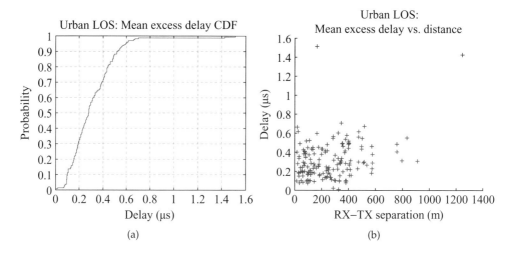

Figure 6.23 Empirical CDF for mean excess delay and scatter plot against distance for all urban LOS tests.

opposite direction tests place each vehicle at either end of an urban canyon about 800 m in length. The vehicles then approach each other at approximately 30 mph. Finally, path loss tests use a stretch of road extending over 1 km out from downtown Detroit to determine the effect that buildings have on signal attenuation.

Table 6.4 summarizes the urban LOS results. Empirical CDFs and scatterplots of mean excess delay, RMS delay, and Doppler spread may be found in Figures 6.23, 6.24, and 6.25.

Table 6.4 Average urban LOS channel characteristics.

Separation (m)	Speed (m/s)		RSS (dBm)	Delay metrics (ns)			D_S (Hz)	K-factor
	RX	TX		$\bar{\tau}$	τ_{RMS}	τ_{max} (30 dB)		
0–75	3.31	3.77	−10.45	281.15	774.58	10870.13	1270.35	84.18
75–150	4.41	2.31	−20.91	268.27	247.66	3347.11	2309.40	372.36
150–250	5.81	2.17	−30.60	330.32	430.25	5232.32	2236.47	118.37
250–350	5.18	2.89	−31.56	259.68	232.42	3188.55	1700.61	250.74
350–450	5.39	3.25	−36.06	330.00	281.06	3396.36	1918.13	57.22
>450	9.42	1.80	−40.81	439.38	726.00	10933.88	2533.66	57.16
Overall	5.41	2.71	−27.53	312.17	436.74	5973.06	1986.26	168.51

From Figure 6.23(a), observe that all mean excess delays lie well under the 1.6 μs, but Figure 6.24(a) shows RMS delay spreads could reach 6 μs. Typically, though, 90% of RMS delay spreads are less than 940 ns. Despite relatively low speeds, high Doppler spreads are still apparent, though 90% are less than 5852 Hz and 80% are

PHYSICAL LAYER CONSIDERATIONS

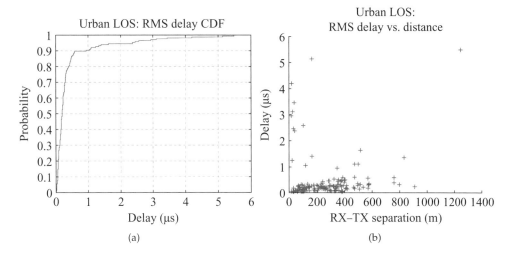

Figure 6.24 Empirical CDF for RMS delay and scatter plot against distance for all urban LOS tests.

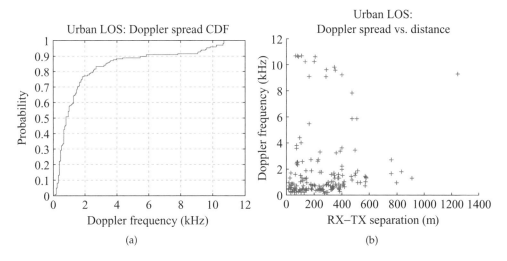

Figure 6.25 Empirical CDF for Doppler spread and scatter plot against distance for all urban LOS tests.

less than 2520 Hz. This may be attributed to the high degree of scattering present in urban environments, leading to multiple angles-of-arrival at the receiver.

Based on the Figures 6.23(b) and 6.24(b), both mean excess delay and RMS delay appear relatively flat over distance, though the 0–75 m bin shows a number of high RMS delay spreads. Otherwise, neither delay statistic appears to show any strong correlation with vehicle separation. For Doppler spreads, Figure 6.25(b) seems to show a similar insensitivity to distance, though it also indicates that while most Doppler spreads remain below 4 kHz, cases do exist where scattering richness leads to Doppler spreads between 9 and 11 kHz.

Path loss A subset of our urban LOS test collection focused on measuring path loss. The receiver was not fixed for the measurements, and normal traffic was present. Figure 6.26 presents the results with a piecewise linear fit overlaid. The path loss exponent from an RX–TX separation of 39 m up to approximately 120 m is 2.0763. Beyond 120 m, it increases to 3.9369, indicating increased attenuation over distance. This is likely due to a lost LOS path that was present at short distances, but was lost at longer distances.

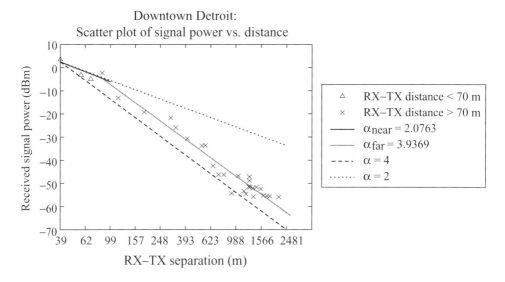

Figure 6.26 Received signal powers versus logarithmic distance in downtown Detroit.

NLOS test locations

To collect NLOS data, we primarily used the four-way intersection at the corner of E. Larned St and Brush St in downtown Detroit. A stationary vehicle parked about 50 m from the intersection on Brush St, while the second vehicle approached the intersection from E. Larned St. The intersection contained tall buildings populating three of the four corners, and the moving vehicle would pass through a tunnel 100 m long just prior to entering the intersection. Consequently, this area provided an ideal environment for maintaining NLOS as long as possible.

Other tests included here impose NLOS conditions in a number of other manners. One test utilized buildings in downtown Detroit, but with a less structured driving pattern compared to those done at the corner of E. Larned and Brush. Some data is culled from LOS tests when a reasonable determination, based on GPS data, received signal strength, and test notes indicated that NLOS conditions predominated for specific superframes. Blockages in these cases could be due to buildings, intervening vehicles, or terrain.

As with other tests in urban areas, GPS information could be unreliable. Although the NLOS tests tried to minimize this effect by keeping one vehicle fixed, distances and speeds, where listed, should be treated only as general guidelines of test outcomes. Furthermore, due to low signal strengths, superframes often required thresholds of 20 dB, instead of the 30 dB range used in previous scenarios. The data presented here uses a mix of 30 dB and 20 dB thresholded data. While this is suboptimal, the use of a lower threshold typically presents a more conservative estimate on channel properties, as extremely weak taps are not factored into calculations. It also permits a broader geographical sampling of channel characteristics.

Unlike previous scenarios, the CDF plots and scatterplots are drawn from slightly different data sets due to the absence of GPS data for some superframes. To increase data coverage, the CDF plots are calculated from all data captures within the urban NLOS tests, whereas the scatterplots are drawn from only those captures with valid RX and TX GPS data. The summary figures in Table 6.5 are based on the latter data set.

Table 6.5 Average urban NLOS channel characteristics.

Separation (m)	Speed (m/s)		RSS (dBm)	Delay metrics (ns)			D_S (Hz)	K-factor
	RX	TX		$\bar{\tau}$	τ_{RMS}	τ_{max} (X^a dB)		
0–150	4.71	1.50	−32.25	378.60	227.26	1739.39	2374.21	1301.30
150–250	3.04	3.42	−42.97	451.02	368.84	2174.24	950.23	90.77
250–500	6.30	5.04	−46.03	490.80	406.63	2310.61	1146.29	9.37
500–800	9.90	2.82	−49.66	366.82	402.33	2370.63	491.92	15.67
800–1000	6.72	1.19	−50.71	319.47	343.87	2227.27	478.04	18.63
>1000	13.75	0.46	−51.30	236.99	249.25	1765.15	439.24	24.53
Overall	7.35	2.41	−44.87	375.52	329.38	2083.54	1043.48	289.27

[a] Data thresholds are mixed between 20 and 30 dB.

From the empirical CDFs, notice that both mean excess and RMS delays are well under 1.2 µs, and that 90% and 95% of these delays, respectively, are below 60 ns. Interestingly, the RMS delay CDF in Figure 6.28(a) appears to follow an almost linear path between 10 ns and 60 ns, suggesting a uniform distribution of RMS delays over this range. From Figure 6.29(a), Doppler spreads are subdued compared to other test scenarios, with 90% of spreads below 1907 Hz and 80% below 1009 Hz. This may be accounted for in some tests by the fact that one vehicle was held stationary throughout the test duration.

Slight negative correlation versus distance is present in Figures 6.27(b) and 6.29(b), though the behavior is more short-lived in the latter. In contrast to the

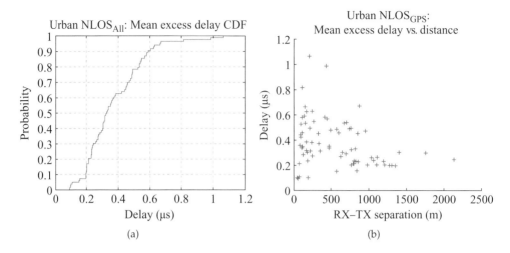

Figure 6.27 Empirical CDF for mean excess delay and scatter plot against distance for all urban NLOS tests. Subfigure (a) includes all available measurements, while subfigure (b) only includes those with valid GPS data.

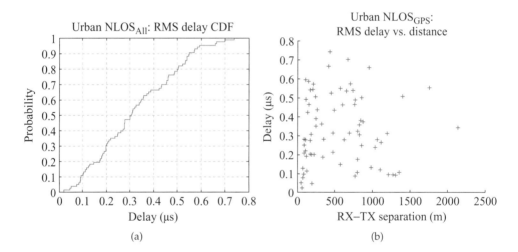

Figure 6.28 Empirical CDF for RMS delay and scatter plot against distance for all urban NLOS tests. Subfigure (a) includes all available measurements, while subfigure (b) only includes those with valid GPS data.

positive correlation with distance seen previously, negative correlation may be due to the high degree of attenuation experienced by reflections in urban situations, as reflectors are concrete or stone buildings instead of metallic vehicles, which are the case in highway situations. Hence, reflected paths may be received at a power

PHYSICAL LAYER CONSIDERATIONS

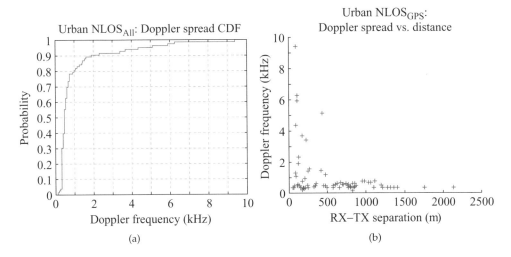

Figure 6.29 Empirical CDF for Doppler spread and scatter plot against distance for all urban NLOS tests. Subfigure (a) includes all available measurements, while subfigure (b) only includes those with valid GPS data.

level outside the dynamic range of the receiver and escape detection. Focusing on Figure 6.29(b), Doppler spreads are highest while at close range (<150 m), suggesting a high degree of angular spread in the arriving paths. As RX–TX separation increases, the effects of signal attenuation appear to reduce Doppler spreads significantly and reduce any correlation with inter-vehicle distance. An examination of these situations with an antenna array or directional antenna would contribute to substantiating these observations further.

Additionally, these known NLOS tests have markedly lower K-factors across most distance intervals as shown in Table 6.5. Except for the closest distances, average K-factors for the strongest channel taps were under 100, and often less than 50. Contrast this to the highway LOS cases, where K-factors are often over 100. This further indicates the more diffuse scattering induced by NLOS situations.

6.6.3 Rural LOS environments

A select number of tests were conducted as the RX and TX vehicles passed through a number of small towns and rural locales. LOS conditions were maintained, and vehicle speeds ranged from 19 mph to 60 mph. Though the total number of superframes captured was less than 15, the data is presented here for completeness. Figures 6.30, 6.31, and 6.32 illustrate rural delay and Doppler characteristics.

Delay spreads are all below 1.6 μs, and all Doppler spreads lie below 1400 Hz. Both factors point towards a low number of environmental scatterers, although it is not immediately apparent why the overall K-factor is only 78.63.

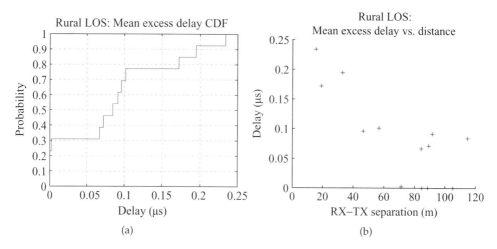

Figure 6.30 Empirical CDF for mean excess delay and scatter plot against distance for all rural LOS tests.

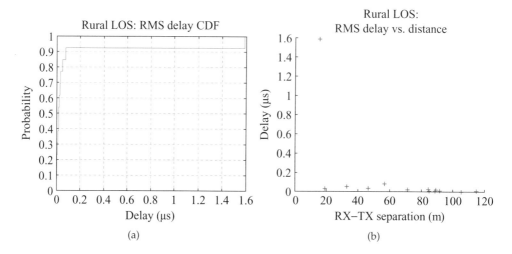

Figure 6.31 Empirical CDF for RMS delay and scatter plot against distance for all rural LOS tests.

6.6.4 Results summary

For convenience, Table 6.7 condenses the distanced-partitioned data presented in Tables 6.2 through 6.6. Recall that all reported measurements are averaged among all superframes falling into each distance bin.

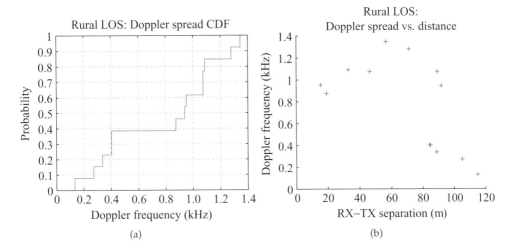

Figure 6.32 Empirical CDF for Doppler spread and scatter plot against distance for all rural LOS tests.

Table 6.6 Average rural LOS channel characteristics.

Separation (m)	Speed (m/s) RX	Speed (m/s) TX	RSS (dBm)	Delay metrics (ns) $\bar{\tau}$	τ_{RMS}	τ_{max} (30 dB)	D_S (Hz)	K-factor
0–120	22.10	22.62	−11.93	85.84	145.29	3132.87	783.59	78.63

6.6.5 Analysis

Whereas the previous section examined each environment individually and the mechanisms contributing to the results, here we compare the environments and highlight what we believe is the main impact on the DSRC standard – short coherence times.

Environment comparisons

Among the high speed scenarios (highway LOS/NLOS and rural), it is apparent that the rural environment is the least hostile among the three due to its small delay spreads resulting from a lack of scatterers. It is followed closely by the highway LOS environment, which tended to have higher mean excess delays (32 ns greater) but lower RMS delays (91.23 ns less) overall. At close ranges (0–50 m), highway LOS channels have the most favorable performance compared to the others when delay and Doppler characteristic are both considered. On the other hand, highway NLOS channels perform noticeably worse. Although average mean and RMS delays exceed the highway LOS equivalents by only 151 ns and 184 ns, respectively, Doppler spread is much higher at 3515.03 Hz, a gain of 2291 Hz over the highway LOS equivalent.

Table 6.7 Condensed table of distance-partitioned data.

Environment	Separation (m)	RSS (dBm)	$\bar{\tau}$	τ_{RMS}	τ_{max} (30 dB[a])	D_S (Hz)	K-factor	Sframe qty[b]
Highway LOS	0–50	−8.54	69.18	21.09	349.43	567.88	400.32	32
	50–200	−25.50	208.21	95.02	1227.27	2714.29	31.77	12
	200–500	−27.63	162.63	112.33	1527.27	1520.39	176.36	10
	>500	−33.19	119.39	50.88	924.24	1248.61	409.67	6
	Overall	27.81	117.58	54.06	778.79	1223.99	290.22	60
Highway NLOS	0–150	−22.48	266.69	267.59	5939.39	3382.01	515.29	15
	150–300	−30.85	255.66	136.67	1575.76	3750.94	44.03	18
	>300	−36.48	293.01	379.58	6109.09	3289.93	74.79	10
	Overall	29.22	268.19	238.83	4152.22	3515.03	215.58	43
Urban LOS	0–75	−10.45	281.15	774.58	10870.13	1270.35	84.18	28
	75–150	−20.91	268.27	247.66	3347.11	2309.40	372.36	33
	150–250	−30.60	330.32	430.25	5232.32	2236.47	118.37	27
	250–350	−31.56	259.68	232.42	3188.55	1700.61	250.74	27
	350–450	−36.06	330.00	281.06	3396.36	1918.13	57.22	25
	>450	−40.81	439.38	726.00	10933.88	2533.66	57.16	22
	Overall	−27.53	312.17	436.74	5973.06	1986.26	168.51	162
Urban NLOS	0–150	−32.25	378.60	227.26	1739.39	2374.21	1301.30	15
	150–250	−42.97	451.02	368.84	2174.24	950.23	90.77	12
	250–500	−46.03	490.80	406.63	2310.61	1146.29	9.37	12
	500–800	−49.66	366.82	402.33	2370.63	491.92	15.67	13
	800–1000	−50.71	319.47	343.87	2227.27	478.04	18.63	10
	>1000	−51.30	236.99	249.25	1765.15	439.24	24.53	12
	Overall	−44.87	375.52	329.38	2083.54	1043.48	289.27	74
Rural LOS	0–120	−11.93	85.84	145.29	3132.87	783.59	78.63	13

[a] 20 dB threshold for urban NLOS.
[b] Superframe quantity.

In fact, highway NLOS tests with RX–TX separations of 150–300 m have the highest average Doppler spread of all the cases at 3750.94 Hz. Since large trucks were used to create NLOS conditions, metal trailers appear to be responsible for creating a rich multipath environment. As stated earlier, the combination of increased multipath and high speed led to this Doppler performance. We also note that, from a safety perspective, the highway NLOS environment is one that would benefit significantly from DSRC assistance due to the combination of high speeds and reduced driver reaction time.

Looking next at the urban LOS/NLOS environments, average mean excess delays appear on the same order of magnitude and differ by 100 ns or less. However, Doppler spread is higher by 943 Hz in the LOS environment. This may be attributed to two factors: signal attenuation and the measurement threshold used. Addressing the former, observe that RSS is significantly lower across all RX–TX separation ranges in the NLOS cases. For separations of 0–150 m, average urban NLOS RSS is -32.25 dBm, while the equivalent for the LOS case is calculated to be -16.11 dBm, a difference of over 16 dBm. Received signal strength differences of over 10 dB continue throughout greater RX–TX separations: 12 dB for both 150–250 m and 250–500 m. Because of the dynamic range limitations of our receiver, these lower RSS values prompted the use of a 20 dB PDP and Doppler threshold in most NLOS cases. As we stated in Section 6.6.2, a lower threshold provides a more conservative estimate of both delay and Doppler statistics.

Comparisons between all five environments give results that match well with intuition. The urban environments have the highest delays overall due to the close proximity of buildings which enable rich scattering, while the highway NLOS environment has the highest Doppler spread from a combination of scattering and high vehicle speeds. A differentiation may also be made between LOS and NLOS environments based upon K-factors. Note that in Table 6.7, for highway NLOS and urban NLOS situations, most K-factors lie below 100. In fact, for most of the urban NLOS cases, the average K-factor value is below 50. Only when vehicles are at close range (less than 150 m separation) does the K-factor rise markedly. Taken in conjunction with lowered RSS values and additional data from Tan (2008), these factors suggest that reflection and scattering, not penetration, is the dominating mechanism in NLOS channels at 5.9 GHz.

DSRC impact

With regards to the proposed DSRC standard, examining Table 6.7 suggests that the proposed design parameters of 802.11p should account for many, but not all, of the channel impairments. We start by comparing the guard interval with the delay spreads to determine the degree to which multipath will introduce ISI. For 802.11p, this guard interval (GI) is 1.6 µs, which is twice as long as 802.11a's GI of 0.8 µs. To quantify delay spread, we examine the sum of mean excess delay and RMS delay, as opposed to examining each individually. This is a reasonable metric as RMS delay measures the spread of delays about the mean excess delay, so their sum gives a measure of the delay spread starting from the beginning of the PDP. We refer to this metric as the *sum delay spread*.

Although many max delay spreads are over the 1.6 µs interval, the sum delay spreads across all scenarios lie below the 1.6 µs guard interval for any case. This implies that ISI should not be a significant problem. Taps contributing to the max delay will always cause a slight amount of interference, but their magnitude (usually around 30 dB below the strongest tap) are not great enough to impact performance significantly.

Avoiding ISI, however, does not necessarily ensure that DSRC structures are perfectly optimized. Recall from Section 6.4.5 that an OFDM transmitter typically inserts a small number of pilot tones into each data symbol. For DSRC, each symbol contains four pilots spaced 2.1875 MHz (14 subcarriers) apart. Based on the inverse relationship between coherence time and delay spread from Equation (6.53), estimate that the delay spread should be no more than approximately 457.14 ns. Taking the figure of merit momentarily as mean excess delay, an examination of Table 6.7 shows that all environments should have coherence bandwidths larger than the spacing between pilot tones. If the figure of merit is changed to sum delay spread, however, only the highway LOS and rural LOS environments consistently have a sufficiently wide coherence bandwidth. Highway NLOS has an average sum delay spread of 507.02 ns (1.972 MHz), while urban LOS/NLOS environments are at 748.91 ns (1.335 MHz) and 704.9 ns (1.419 MHz), respectively. Consequently, more complex equalization algorithms, such as pilot interpolation schemes, should be examined as alternatives to flat-fading corrections in these environments.

Similarly, the standard accounts for some Doppler impairments but not for others. As stated in Section 6.4, high Doppler spreads may cause ICI between OFDM subcarriers. However, from the table we see that the highest average Doppler spread is 3750.94 Hz for the highway NLOS 150–300 m case. This is a small fraction (2.40%) of the 156.25 kHz inter-carrier spacing for 802.11p. In fact, it was rare to see a channel tap with a Doppler spread over 2 kHz, which is only 1.28% of the intercarrier spacing. Hence, even under the most hazardous Doppler conditions, the amount of ICI should be minimal.

However, after switching to an alternative domain, we again find that DSRC equalization structures may be suboptimal. Instead of the bandwidth between pilot tones, the problem now lies in how often the receiver estimates the channel. As noted in Section 6.4.5, specially formed data symbols are prepended to the start of every packet expressly for the purpose of estimating the channel. Despite the presence of pilot tones throughout the packet, often channel estimation is performed only at the beginning of each packet. This is the case in the large majority of commercially available 802.11 chipsets, as it lowers costs and is appropriate for the relatively static environments in which 802.11 systems often operate. Consequently, the rate of channel evolution, or coherence time, becomes a concern for DSRC. Recall the relation of coherence time to Doppler spread in Equation (6.67) as

$$T_c = \frac{M}{D_s}, \qquad (6.75)$$

where the constant M varies anywhere from 0.25 to 1 (Rappaport 2002; Tse and Viswanath 2005). Correspondingly, a Doppler spread of 3750.94 Hz gives a coherence time anywhere between 66.650 µs and 0.267 ms. For short packets, the channel

PHYSICAL LAYER CONSIDERATIONS

may remain relatively invariant over the course of the packet. However, longer packets may experience more fluctuations in the channel which are not compensated for by initial equalization settings.

Figure 6.33 illustrates the relationship among data transmission time, PHY payload size, and rate. Assuming an upper estimate of $T_{coh} = 0.267$ ms and a transmission rate of 3 Mbps, it is likely that any packets with PHY payloads over 85 bytes will experience multiple channel fades. Moving to higher-order modulations will increase the probability of having an invariant channel but this risks increased sensitivity to Doppler and ICI. Furthermore, if the lower estimate of $T_{coh} = 66.65$ μs applies, then even at 24 Mbps packets over 79 bytes will likely experience multiple fades. These calculations assume fixed overhead from the PHY layer header of 40 μs.

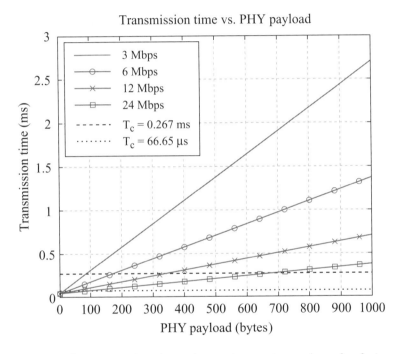

Figure 6.33 Plot of transmission times for various rates and payload sizes. Upper and lower bounds to the coherence time are superimposed.

The acting ASTM DSRC standard (ASTM 2003) states that compliant devices must sustain a PER < 10% when traveling at 85 mph and sending 1000 byte packets *and* when traveling at 120 mph and sending 64 byte packets. Based on Figure 6.33 and Table 6.3, we see that even at ordinary highway speeds (on average 62 mph for TX and 64 mph for RX), 1000 byte packets will always experience multiple fades and 64 byte packets are at risk unless higher data rates are used. However, higher data rates will be more sensitive to ICI and other channel distortions, so error rates might not necessarily be reduced. Furthermore, Doppler scales directly in proportion to speeds, so performance at the 120 mph limit will degrade further unless improvements, such as multiple equalizations over a packet transmission, are made.

A third point worth noting lies in the observation, made in Section 6.6.5, of significant signal attenuation when comparing urban NLOS environments to urban LOS environments at equivalent distances. Such differences in received power present problems in a form similar to the classic 'hidden terminal' problem. Specifically, given a receiver that has LOS to one transmitter and NLOS to another transmitter, the NLOS transmitter is hidden from the LOS transmitter. Consequently, at the receiver, transmissions will either collide, or the LOS transmission will predominate over the NLOS transmission. In an urban environment, one can imagine such events happening frequently as cars would receive transmissions more easily along their street rather than around corners; however, receiving data to avoid hazards could require NLOS communication (e.g. intersection collision warnings). Consequently, power control schemes should be pursued as an integral part of future DSRC development.

Overall, based on the measured data and the analysis above, the DSRC standard compensates adequately for most channel impairments and should function reasonably in its current form. That is not to say, however, that its signalling structures are perfectly optimized. Concerns about coherence time remain and must be addressed, especially since the high Doppler environments where it is smallest may be the most dependent on DSRC performance. Narrow coherence bandwidths with respect to pilot tone spacing also deserve some attention, though problems do not seem as severe as those regarding coherence time. Finally, the degree of signal attenuation in urban NLOS versus urban LOS environments suggests the need to examine power control methods for DSRC networks.

6.7 Future Directions

The data and analysis conducted here suggest that much remains to be done to improve the DSRC standard and to measure the wireless channel in more detail. For example, the coherence time problem provides one promising opportunity for innovation.

Various methods have been proposed to compensate for short coherence times, with each requiring various degrees of complexity. Some researchers, such as Sibecas et al. (2003) and Song and Lim (2003), have proposed pilot-based OFDM schemes for channel estimation. Modifications by Sibecas et al. (2003) target DSRC explicitly, but are limited in their tolerance to extreme Doppler due to the time delay incurred in estimating the channel response. The scheme from Song and Lim (2003) shows potential for Doppler spreads up to 200 Hz, but it was not simulated for kilohertz-order Doppler spreads and requires computationally intensive matrix inversions during channel estimation. Alternatively, other works (Choi et al. 2001; Hrycak and Matz 2006; Rugini et al. 2006; Schniter 2004; Shu et al. 2007) choose to correct for Doppler in the frequency domain by attempting frequency-domain equalization under various metrics. However powerful, equalization techniques suffer from the

PHYSICAL LAYER CONSIDERATIONS

need to conduct matrix operations and inversions, which make many algorithms computationally taxing. Furthermore, complexity scales not only with the number of subcarriers allocated, but also with OFDM symbol length and severity of Doppler spread.

In contrast, we have proposed in Zhang et al. (2008) a straightforward, robust method for compensating for short coherence times. Using the principles of time-domain differential OFDM (TDOFDM), it shortens the coherence time threshold of concern to 8 μs, the transmission time for an OFDM symbol. This widens the corresponding Doppler spread to 125 kHz – substantially larger than any of the Doppler spreads in our measurements. To accomplish this, TDOFDM differentially codes between successive subcarriers in two adjacent OFDM time symbols. Such differential coding amounts to a continuous equalization across an entire packet, leading to a tolerance of short coherence times. Additionally, TDOFDM is not computationally taxing, and requires minimal changes to the receiver and transmitter. Implementation details, along with verifying simulations, may be found in Zhang et al. (2008).

In terms of measurement, the use of antenna arrays would provide the ability to deduce angles of arrival in conjunction with Doppler and delay spreads. Such measurements would help verify the nature of scattering in heavily cluttered environments (e.g. urban NLOS). It would also provide insight into how multiple antennas could improve receiver sensitivity or aid in power control, another issue highly deserving of attention. Next, considering that the DSRC spectrum at 5.9 GHz allocates multiple channels, cross-channel interference is a concern (Kenney 2007). This stems from preliminary measurements that we conducted with prototype DSRC radios, though the data was not extensive or complete enough to present here. Finally, the opening of the 700 MHz spectrum by the FCC and its projected usage in Japanese DSRC systems provides a renewed opportunity to analyze vehicular wireless performance in another band with potentially different characteristics.

6.8 Conclusion

While much emphasis has and will deservedly be given to DSRC as a mechanism for increasing safety and efficiency on public roadways, its viability rests on the usage of a robust, low-latency physical layer. Decisions regarding DSRC's projected benefits should always take into account the capabilities of its physical layer technology and the ground truth of the environment under which it operates.

Measurements of this ground truth, by the authors and others, have shown the current DSRC standard to be sufficient, but not necessarily optimal, for its intended environment. Although the proposed standard may perform acceptably for short transmissions, longer transmissions may be subject to higher error rates in the absence of further processing. Analysis of our measurements also suggests the need to examine topics such as reduced coherence bandwidths, power control, and angle-of-arrival.

All of these items impart challenges that will affect higher layers of the DSRC protocol stack. Conversely, their presence also provides fertile ground for further research and innovation. The authors hope that this chapter has brought sufficient attention to these issues and has encouraged readers to further appreciate how physical layer issues play a driving role in DSRC's capabilities and future applications.

Acknowledgments

We could not have written this chapter without the aid and support of many individuals. First of all, we wish to thank Dr Carl Chun and Dr Wanbin Tang of the UC Berkeley Connectivity Lab for the integral role they played in the collection and analysis of the data presented above. Likewise at Toyota, we are indebted to Dr Ken Laberteaux for his insightful contributions and timely advice regarding our work, as well as for his tireless efforts in organizing both material and manpower to support our data collection campaigns. Two of his colleagues, Dr John Kenney and Lorenzo Caminiti, also deserve recognition for their significant inputs of time and effort to this chapter's underlying work. Finally, many thanks are due to Janet Chao, the chapter's illustrator, for her work in professionally transforming our many rough sketches into clear, communicative figures.

6.9 Appendix: Deterministic Multipath Channel Derivations

6.9.1 Complex baseband channel representation – continuous time

To derive the complex-baseband form $h_b(\tau, t)$, start with (6.8) and substitute for $y(t)$ and $x(t)$ with Equations (6.18) and (6.19):

$$y(t) = \int_{-\infty}^{+\infty} h(\tau, t) x(t - \tau) \, d\tau \tag{6.76}$$

$$\Re[y_b(t) e^{j2\pi f_c t}] = \int_{-\infty}^{+\infty} h(\tau, t) \Re[x_b(t - \tau) e^{j2\pi f_c (t - \tau)}] \, d\tau \tag{6.77}$$

$$= \Re\left[\int_{-\infty}^{+\infty} h(\tau, t) x_b(t - \tau) e^{j2\pi f_c (t - \tau)} \, d\tau\right] \tag{6.78}$$

$$= \Re\left[\int_{-\infty}^{+\infty} h(\tau, t) e^{-j2\pi f_c \tau} x_b(t - \tau) e^{j2\pi f_c t} \, d\tau\right] \tag{6.79}$$

$$= \Re\left[\left[\int_{-\infty}^{+\infty} h(\tau, t) e^{-j2\pi f_c \tau} x_b(t - \tau) \, d\tau\right] e^{j2\pi f_c t}\right]. \tag{6.80}$$

Comparing the left- and right-hand sides of Equation (6.80), note that equality holds when:

$$y_b(t) = \int_{-\infty}^{+\infty} h(\tau, t) e^{-j2\pi f_c \tau} x_b(t - \tau) \, d\tau \tag{6.81}$$

PHYSICAL LAYER CONSIDERATIONS

$$= (h(\tau, t)e^{-j2\pi f_c \tau}) * x_b(t). \tag{6.82}$$

Based on the above, conclude that

$$y_b(t) = h_b(\tau, t) * x_b(t) \tag{6.83}$$

and

$$h_b(\tau, t) = h(\tau, t)e^{-j2\pi f_c \tau} \tag{6.84}$$

$$= \sum_i a_i(t)\delta(\tau - \tau_i(t))e^{-j2\pi f_c \tau} \tag{6.85}$$

$$= \sum_i a_i(t)\delta(\tau - \tau_i(t))e^{-j2\pi f_c \tau_i(t)} \tag{6.86}$$

$$= \sum_i a_{i,b}(t)\delta(\tau - \tau_i(t)), \tag{6.87}$$

with

$$a_{i,b}(t) = a_i(t)e^{-j2\pi f_c \tau_i(t)}. \tag{6.88}$$

The complex-baseband channel is effectively the passband channel after a frequency shift, by the carrier frequency, down to the baseband.

6.9.2 Complex baseband channel representation – discrete time

To convert a continuous-time channel model to discrete time, begin by assuming that the allocated bandwidth of the passband channel is $B = 1/T_B$ Hz with sinc pulse-shaping and ideal filtering in use at the transmitter and receiver, as illustrated in Figure 6.6. Beginning with the baseband convolution relationship in Equation (6.20), make the effects of the transmitter's filters explicit:

$$y_b(t) = x_b(t) * h_b(\tau, t) \tag{6.89}$$

$$= \left(\sum_n x_b[n] \operatorname{sinc}(\pi B(t - nT_B))\right) * h_b(\tau, t) \tag{6.90}$$

$$= \left(\sum_n x_b[n] \operatorname{sinc}(\pi B(t - nT_B))\right) * \left(\sum_i a_{i,b}(t)\delta(\tau - \tau_i(t))\right) \tag{6.91}$$

$$= \sum_n \sum_i x_b[n] \operatorname{sinc}(\pi B(t - \tau_i(t) - nT_B))a_{i,b}(t). \tag{6.92}$$

Next, sample at every symbol time T_B:

$$y_b[m] = \sum_n x_b[n] \sum_i \operatorname{sinc}(\pi B(mT_B - \tau_i(mT_B) - nT_B))a_{i,b}(mT_B) \tag{6.93}$$

$$= \sum_n x_b[n] \sum_i \operatorname{sinc}(\pi(m - B\tau_i(mT_B) - n))a_{i,b}(mT_B) \tag{6.94}$$

with $p = m - n$, then:

$$y_b[m] = \sum_p x_b[m-p] \sum_i \text{sinc}(\pi(p - B\tau_i(mT_B)))a_{i,b}(mT_B) \qquad (6.95)$$

$$= x_b[m] * \left(\sum_i a_{i,b}(mT_B) \text{sinc}(\pi(p - B\tau_i(mT_B))) \right) \qquad (6.96)$$

$$= \sum_p x_b[m-p] h_b[p, m]. \qquad (6.97)$$

Therefore, provided the channel is LTV, conclude that the sampled impulse response is:

$$h_b[p, m] = \sum_i a_{i,b}(mT_B) \text{sinc}\left(\pi\left(p - \frac{\tau_i(mT_B)}{T_B} \right) \right) \qquad (6.98)$$

$$= \sum_i a_i(mT_B) e^{-j2\pi f_c \tau_i(mT_B)} \text{sinc}\left(\pi\left(p - \frac{\tau_i(mT_B)}{T_B} \right) \right). \qquad (6.99)$$

The quantity $h_b[p, m]$ should be interpreted as the pth tap of the channel at the time instant m. If the channel was LTI, then the quantities $\tau_i(t)$ and $a_i(t)$ would no longer be time-varying. Thus, the relationships in Equation (6.95) could be simplified to

$$y_b(t) = \sum_p x_b[m-p] \sum_i a_{i,b} \text{sinc}(\pi(p - B\tau_i)), \qquad (6.100)$$

which changes Equation (6.99) into

$$h_b[p] = \sum_i a_i e^{-j2\pi f_c \tau_i} \text{sinc}\left(\pi\left(p - \frac{\tau_i}{T_B} \right) \right). \qquad (6.101)$$

6.10 Appendix: LTV Channel Response

The presence of two 'time' variables, τ and t, in LTV channel representations complicates their response to an input $x(t)$. The τ variable refers to delay in the channel while t refers to the progression of actual time. Thus, the impulse response of a LTV channel evolves as t changes. LTI channels do not evolve with time and drop their dependence on t.

The first point to note is that a Fourier transform can be taken with respect to τ and t individually. The dual variables will be designated as f_τ and f_t, respectively. A Fourier transform with respect to τ yields the frequency response familiar from standard signals and systems curricula.

To understand how the output $y(t)$ of a LTV channel $h(\tau, t)$ changes as a function of its input $x(t)$, begin by writing the Fourier transform of $h(\tau, t)$ with respect to τ:

$$H(f_\tau; t) = \int_{-\infty}^{+\infty} h(\tau, t) e^{-j2\pi f_\tau \tau} \, d\tau. \qquad (6.102)$$

PHYSICAL LAYER CONSIDERATIONS

The above equation states that the LTV channel's frequency response (with respect to τ) changes over time. The inverse transform is given by

$$h(\tau, t) = \int_{-\infty}^{+\infty} H(f_\tau; t) e^{j2\pi f_\tau \tau} \, df_\tau. \tag{6.103}$$

Next, recall the LTV input–output relationship between $x(t)$, $h(\tau, t)$, and $y(t)$ given in Equation (6.8):

$$y(t) = \int_{-\infty}^{+\infty} h(\tau, t) x(t - \tau) \, d\tau. \tag{6.104}$$

The first goal is to convert this relationship into a form involving frequency-domain instantiations of $h(\tau, t)$ and $x(t)$. Substituting the contents of Equation (6.103) into Equation (6.104) yields

$$y(t) = \int_{-\infty}^{+\infty} \left[\int_{-\infty}^{+\infty} H(f_\tau; t) e^{j2\pi f_\tau \tau} \, df_\tau \right] x(t - \tau) \, d\tau, \tag{6.105}$$

$$= \int_{-\infty}^{+\infty} H(f_\tau; t) \int_{-\infty}^{+\infty} x(t - \tau) e^{j2\pi f_\tau \tau} \, d\tau \, df_\tau. \tag{6.106}$$

Substituting $\tilde{\tau} = t - \tau$ gives

$$y(t) = \int_{-\infty}^{+\infty} H(f_\tau; t) \int_{-\infty}^{+\infty} x(\tilde{\tau}) e^{j2\pi f_\tau (t - \tilde{\tau})} \, d\tilde{\tau} \, df_\tau, \tag{6.107}$$

$$= \int_{-\infty}^{+\infty} H(f_\tau; t) \int_{-\infty}^{+\infty} x(\tilde{\tau}) e^{-j2\pi f_\tau \tilde{\tau}} e^{j2\pi f_\tau t} \, d\tilde{\tau} \, df_\tau, \tag{6.108}$$

$$= \int_{-\infty}^{+\infty} H(f_\tau; t) X(f_\tau) e^{j2\pi f_\tau t} \, df_\tau. \tag{6.109}$$

The time-domain output $y(t)$ thus depends on the multiplication of the frequency responses $H(f_\tau; t)$ and $X(f_\tau)$ followed with an inverse Fourier transform with respect to f_τ. This is intuitively reasonable – the channel's output is the product of the Fourier transforms of the input and channel response at a specific instant t. A time instant t must be specified because the channel is evolving. If the channel were LTI and dropped all dependence on t, $H(f_\tau; t) = H(f_\tau)$ and the conclusion that $Y(f_t) = H(f_\tau)X(f_\tau)$ would follow. Obtaining a similar frequency-domain result for the LTV channel requires slightly more effort. As a side note, the transform of $x(t)$ is denoted as $X(f_\tau)$ and not $X(f_t)$ due to Equation (6.108).

To obtain $Y(f_t)$, apply the Fourier transform with respect to t directly to Equation (6.109):

$$Y(f_t) = \int_{-\infty}^{+\infty} \int_{-\infty}^{+\infty} [H(f_\tau; t) X(f_\tau) e^{j2\pi f_\tau t} \, df_\tau] e^{-j2\pi f_t t} \, dt \tag{6.110}$$

$$= \int_{-\infty}^{+\infty} \int_{-\infty}^{+\infty} H(f_\tau; t) X(f_\tau) e^{-j2\pi (f_t - f_\tau) t} \, dt \, df_\tau \tag{6.111}$$

$$= \int_{-\infty}^{+\infty} X(f_\tau) \int_{-\infty}^{+\infty} H(f_\tau; t) e^{-j2\pi(f_t - f_\tau)t} \, dt \, df_\tau \quad (6.112)$$

$$= \int_{-\infty}^{+\infty} X(f_\tau) H(f_\tau, f_t - f_\tau) \, df_\tau. \quad (6.113)$$

Equation (6.113) shows that to obtain the frequency response of the output, a second convolution, involving f_t, of the input and channel spectra is required. In terms of the channel, the Fourier transform of $h(\tau, t)$ with respect to t for fixed τ yields the per-tap Doppler spectra, as it encapsulates the change in a channel tap over time. It is through this additional convolution that the LTV effects of the channel manifest themselves in the output's frequency response.

6.11 Appendix: Measurement Theory Details

6.11.1 PN sequence bits

The 511 bits are listed below with 24 bits per line.

```
1 0 0 0 0 0 0 0    0 1 0 0 0 0 1 1    0 1 1 1 0 1 0 0
1 0 1 1 1 0 0 1    1 1 0 0 1 1 0 0    0 1 1 1 1 0 0 0
0 1 1 0 0 1 1 0    1 1 1 1 1 0 0 1    1 0 1 0 1 1 1 0
1 0 1 1 0 1 1 0    1 0 0 0 0 0 1 1    1 1 0 1 0 0 1 1
1 1 1 0 1 0 1 0    1 1 1 0 0 0 1 1    0 0 0 0 0 1 1 0
1 0 1 0 1 0 1 0    0 0 0 0 0 1 1 1    0 0 1 0 0 0 0 1
0 0 1 1 1 1 0 0    1 0 1 1 0 1 0 1    1 0 0 1 0 1 0 1
1 0 1 0 0 1 0 0    1 1 0 0 0 0 1 1    1 0 1 1 0 0 0 1
0 0 1 0 0 1 0 0    0 0 0 0 0 1 1 0    0 0 1 0 1 1 0 0
1 1 1 0 1 1 1 0    0 1 0 1 0 0 1 0    1 0 1 0 0 1 0 0
0 1 0 0 0 1 0 1    0 1 0 1 1 0 0 0    0 1 0 1 0 1 1 1
1 0 0 1 1 1 1 0    1 1 0 1 1 1 0 0    0 0 1 0 0 0 1 1
1 0 1 0 0 0 0 1    1 1 1 1 1 0 0 1    0 0 1 0 1 0 0 0
0 1 0 1 1 1 1 1    1 1 1 0 0 0 0 0    0 1 0 0 1 0 1 1
0 0 0 1 1 0 1 0    0 0 1 0 1 1 1 0    1 1 1 1 0 1 0 1
1 1 1 1 0 1 1 1    0 1 1 0 1 0 1 0    0 0 1 0 0 1 1 0
1 0 0 1 1 0 1 1    0 1 1 0 0 0 0 0    0 1 0 1 0 0 1 1
1 0 1 0 1 0 0 1    1 0 0 1 0 1 1 1    1 0 1 1 1 1 1 1
0 1 1 0 0 1 1 0    0 1 1 1 1 1 1 1    0 1 0 0 0 1 1 1
1 1 0 0 0 1 0 1    0 0 0 1 1 0 1 1    0 0 1 0 0 0 1 1
0 0 1 0 0 1 1 1    0 0 0 1 0 0 0 0    0 1 0 1 1 0 1 1
1 1 0 0 0 1 1
```

6.11.2 Generation of LTI channel estimates

A single LTI estimate of the channel, $\widehat{h_b[p]}$, may be calculated with knowledge of the sampled baseband received signal, $y_b[n]$. This signal is obtained via the system shown in Figure 6.15. The transmitted PN sequence, $x_{\text{MLS}}[n] = x_I[n]$, is pulse shaped

PHYSICAL LAYER CONSIDERATIONS

by $g_{rrc}(\tau)$ to produce $x_{MLS}(t) = x(t)$, which is then up-converted and sent over the air. At the receiver, the signal is down-converted, match filtered with $g_{rrc}(\tau)$, and sampled with sampling period T_s. Following sampling, a timing recovery algorithm calculates the optimum sampling point and produces a decimated sequence of approximate symbol values with period T_{sym}. Based on this process and Figure 6.15, $y_b[n]$ may be written as:

$$y_b[n] = y_b(nT_s) \tag{6.114}$$

$$= h_b(\tau) * x(t) * g_{rrc}(\tau)|_{t=nT_s} \tag{6.115}$$

$$= h_b(\tau) * \left(\sum_m x_{MLS}[m]g_{rrc}(t - mT_{sym})\right) * g_{rrc}(\tau)\Big|_{t=nT_s} \tag{6.116}$$

$$= h_b(\tau) * \left(\sum_m x_{MLS}[m]g_{rc}(t - mT_{sym})\right)\Big|_{t=nT_s}, \tag{6.117}$$

where $g_{rc}(t)$ is the raised cosine pulse. Under the assumption that the timing recovery algorithm performs perfectly, the output of estimated symbol values, $\tilde{y}[n]$, will obey $\tilde{y}[n] = y[(nT_{sym})/T_s]$. Proceeding to determine $\tilde{y}[n]$ and adjusting the sampling time for t appropriately,

$$\tilde{y}[n] = h_b(\tau) * \left(\sum_m x_{MLS}[m]g_{rc}(t - mT_{sym})\right)\Big|_{t=nT_{sym}} \tag{6.118}$$

$$= h_b(\tau) * \left(\sum_m x_{MLS}[m]g_{rc}(nT_{sym} - mT_{sym})\right) \tag{6.119}$$

$$= h_b(\tau) * \left(\sum_m x_{MLS}[m]g_{rc}((n - m)T_{sym})\right), \tag{6.120}$$

$$\text{where} \quad g_{rc}(kT_{sym}) = \begin{cases} 1 & \text{if } k = 0 \\ 0 & \text{otherwise,} \end{cases} \tag{6.121}$$

$$= h_b(\tau) * x_{MLS}[n]. \tag{6.122}$$

To further simplify Equation (6.122), rewrite $x_{MLS}[n]$ in a continuous-time form utilizing impulses and again evaluate at $t = nT_{sym}$:

$$\tilde{y}[n] = h_b(\tau) * \left(\sum_k x_{MLS}[k]\delta(t - kT_{sym})\right)\Big|_{t=nT_{sym}} \tag{6.123}$$

$$= \sum_k h_b(\tau) * x_{MLS}[k]\delta(t - kT_{sym})\Big|_{t=nT_{sym}} \tag{6.124}$$

$$= \sum_k h_b(t - kT_{sym})x_{MLS}[k]\Big|_{t=nT_{sym}} \tag{6.125}$$

$$= \sum_k h_b((n - k)T_{sym})x_{MLS}[k] \tag{6.126}$$

$$= h_b[n] * x_{MLS}[n]. \tag{6.127}$$

Equation (6.127) indicates that the sampled received signal is, as expected, the convolution of the sampled baseband channel and the original PN sequence. Recovery of $h_b[p]$ is the goal, so the next step is to cross-correlate $\tilde{y}[n]$ with $x_{\text{MLS}}[n]$:

$$R_{\tilde{y}x_{\text{MLS}}}[k] = \sum_n \tilde{y}[n]x^*_{\text{MLS}}[n-k] \qquad (6.128)$$

$$= \sum_n (h_b[n] * x_{\text{MLS}}[n])x^*_{\text{MLS}}[n-k] \qquad (6.129)$$

$$= \sum_n \left(\sum_l h_b[l]x_{\text{MLS}}[n-l]\right)x^*_{\text{MLS}}[n-k] \qquad (6.130)$$

$$= \sum_l h_b[l]\sum_n x_{\text{MLS}}[n-l]x^*_{\text{MLS}}[n-k] \qquad (6.131)$$

$$= \sum_l h_b[l]R_{x_{\text{MLS}}}[k-l] \qquad (6.132)$$

$$\approx \sum_l h_b[l]\delta[k-l] \qquad (6.133)$$

$$= \widehat{h_b[k]}. \qquad (6.134)$$

Consequently, after the autocorrelation operation, an estimate of $h_b[p]$, $\widehat{h_b[p]}$, is obtained.

6.11.3 Generation of Ricean K-factor estimates

To estimate the K-factor of a channel, select a single tap in the channel, $h_b[p_0, n]$, whose estimate is given by $\widehat{h_b[p_0, n]}$. If this tap is Ricean faded, then it can be written in the form

$$h_b[p_0, n] = C + v_{p_0}[n], \qquad (6.135)$$

where C denotes the mean value of the channel attributed to the dominant path, and $v_{p_0}[n]$ denotes the zero-mean fluctuations due to scattered paths. The K-factor is defined as the ratio of power in the dominant path to the total power in the scattered paths. Power in the dominant path is

$$\text{Pow}_{\text{dom}} = |C|^2, \qquad (6.136)$$

while power in the scattered paths is given by

$$\text{Pow}_{\text{diff}} = E[|v_{p_0}[n]|^2]. \qquad (6.137)$$

Greenstein et al. (1999) showed that estimates $\widehat{\text{Pow}}_{\text{dom}}$ and $\widehat{\text{Pow}}_{\text{diff}}$ of these two quantities may be made from measured data by calculating

$$\widehat{\text{Pow}}_{\text{dom}} = \sqrt{G_a^2 - G_v^2}, \quad \text{and} \qquad (6.138)$$

$$\widehat{\text{Pow}}_{\text{diff}} = G_a - \sqrt{(G_a^2 - G_v^2)}, \qquad (6.139)$$

where G_a and G_v are the first and second moments of $|h_b[p_0, n]|^2$ determined with empirical expected values:

$$G_a = E[|h_b[p_0, n]|^2] \tag{6.140}$$

$$G_v = \sqrt{E[(|h_b[p_0, n]|^2 - G_a)^2]}. \tag{6.141}$$

The interested reader may refer to Greenstein et al. (1999) for additional details on this estimation method.

References

Abdi A, Tepedelenlioglu C, Kaveh M and Giannakis G 2001 On the estimation of the K parameter for the Rice fading distribution. *IEEE Communications Letters* **5**(3), 92–94.

Acosta G and Ingram M 2006 Doubly selective vehicle-to-vehicle channel measurements and modeling at 5.9 GHz. *Proc. Wireless Personal Multimedia Communications Conference (WPMCC '06)*.

Acosta-Marum G 2007 *Measurement, Modeling, and OFDM Synchronization for the Wideband Mobile-to-Mobile Channel*. PhD thesis, Geogia Institute of Technology.

ASTM 2003 Standard specification for telecommunications and information exchange between roadside and vehicle systems – 5 GHz band dedicated short range communications (DSRC) medium access control (MAC) and physical layer (PHY) specifications.

Bahai AR, Saltzberg BR and Ergen M 2004 *Multi-Carrier Digital Communications*, second edn, Springer.

Bradaric I, Dattani R, Petropulu A, Schurgot, FLJ and Inserra J 2003 Analysis of physical layer performance of IEEE 802.11a in an ad-hoc network environment *Proc. IEEE Military Communications Conference (MILCOM '03)*, vol. 2, pp. 1231–1236.

Choi YS, Voltz P and Cassara F 2001 On channel estimation and detection for multicarrier signals in fast and selective Rayleigh fading channels. *IEEE Transactions on Communications* **49**(8), 1375–1387.

Cox D 1972 Delay Doppler characteristics of multipath propagation at 910 MHz in a suburban mobile radio environment. *IEEE Transactions on Antennas and Propagation* **20**(5), 625–635.

Durgin G, Rappaport T and Xu H 1998 Measurements and models for radio path loss and penetration loss in and around homes and trees at 5.85 GHz. *IEEE Transactions on Communications* **46**(11), 1484–1496.

Greenstein L, Michelson D and Erceg V 1999 Moment-method estimation of the Ricean K-factor. *IEEE Communications Letters* **3**(6), 175–176.

Hrycak T and Matz G 2006 Low-complexity time-domain ICI equalization for OFDM communications over rapidly varying channels. *Proc. of the 40th Asilomar Conf. on Signals, Systems, and Computers 2006 (ACSSC '06)*, pp. 1767–1771.

IEEE 1999 Supplement to IEEE standard for information technology telecommunications and information exchange between systems – local and metropolitan area networks – specific requirements. Part 11: wireless LAN medium access control (MAC) and physical layer (PHY) specifications: high-speed physical layer in the 5 GHz band.

IEEE 2003 Information technology – telecommunications and information exchange between systems – local and metropolitan area networks – specific requirements – Part 11: wireless LAN medium access control (MAC) and physical layer (PHY) specifications.

IEEE 2006a IEEE std. 1609.1 – 2006 IEEE trial-use standard for wireless access in vehicular environments (WAVE) – resource manager.

IEEE 2006b IEEE trial-use standard for wireless access in vehicular environments – security services for applications and management messages.

IEEE 2006c IEEE trial-use standard for wireless access in vehicular environments (WAVE) – multi-channel operation.

IEEE 2007a IEEE p802.11p/d4.0 draft standard for information technology – telecommunications and information exchange between systems – local and metropolitan area networks - specific requirements – Part 11: wireless LAN medium access control (MAC) and physical layer (PHY) specifications amendment 8: wireless access in vehicular environments.

IEEE 2007b IEEE standard for information technology – telecommunications and information exchange between systems – local and metropolitan area networks – specific requirements. Part 11: wireless LAN medium access control (MAC) and physical layer (PHY) specifications.

IEEE 2007c IEEE trial-use standard for wireless access in vehicular environments (WAVE) – networking services.

Kenney J 2007 Cross-channel interference test results: A report from the VSC-a project. Slides presented at standards meeting.

Matolak D, Sen I, Xiong W and Yaskoff N 2005 5 GHz wireless channel characterization for vehicle to vehicle communications. *Proc. IEEE Military Communications Conference (MILCOM '05)*, vol. 5, pp. 3016–3022.

Rappaport T 2002 *Wireless Communications: Principles and Practice*, second edn, Prentice Hall.

Rugini L, Banelli P and Leus G 2006 Low-complexity banded equalizers for OFDM systems in doppler spread channels. *EURASIP Journal on Applied Signal Processing* **2006**, 1–13. Article ID 67404.

Schniter P 2004 Low-complexity equalization of OFDM in doubly selective channels. *IEEE Transactions on Signal Processing* **52**(4), 1002–1011.

Schwengler T and Gilbert M 2000 Propagation models at 5.8 GHz – path loss and building penetration. *Proc. IEEE Radio and Wireless Conference (RAWCON 2000)*, pp. 119–124.

Shu F, Bi YF, Wang JX and Cheng SX 2007 Channel estimation and equalization for OFDM wireless system with medium doppler spread. *Proc. IEEE Wireless Communications, Networking, and Mobile Computing Conference 2007 (WiCom '07)*, pp. 403–407.

Sibecas S, Corral C, Emami S and Stratis G 2002 On the suitability of 802.11a/RA for high-mobility DSRC. *Proc. 55th IEEE Vehicular Technology Conference (VTC Spring '02)*, vol. 1, pp. 229–234.

Sibecas S, Corral C, Emami S, Stratis G and Rasor G 2003 Pseudo-pilot OFDM scheme for 802.11a and R/A in DSRC applications. *Proc. 58th IEEE Vehicular Technology Conference (VTC Fall '03)*, vol. 2, pp. 1234–1237.

Song WG and Lim JT 2003 Pilot-symbol aided channel estimation for OFDM with fast fading channels. *IEEE Transactions on Broadcasting* **49**(4), 398–402.

Tan IL 2008 *Broadband Channel Sounding: Characterizations and Implications for Vehicular Wireless Communications*. Master's thesis, University of California, Berkeley.

Tranter WH, Shanmugan KS, Rappaport TS and Kosbar KL 2004 *Communications Systems Simulation With Wireless Applications*, first edn, Prentice Hall.

Tse D and Viswanath P 2005 *Fundamentals of Wireless Communication*, first edn, Cambridge Univ. Press, New York, NY.

Zhang Y, Tan IL, Chun C, Laberteaux K and Bahai A 2008 A differential OFDM approach to coherence time mitigation in DSRC. *VANET '08: Proceedings of the Fifth ACM International Workshop on VehiculAr Inter-NETworking*, pp. 1–6. ACM, New York, NY, USA.

Zhao X, Kivinen J, Vainikainen P and Skog K 2002 Propagation characteristics for wideband outdoor mobile communications at 5.3 GHz. *IEEE Journal on Selected Areas in Communications* **20**(3), 507–514.

Zhao X, Kivinen J, Vainikainen P and Skog K 2003 Characterization of Doppler spectra for mobile communications at 5.3 GHz. *IEEE Transactions on Vehicular Technology* **52**(1), 14–23.

7

MAC Layer and Scalability Aspects of Vehicular Communication Networks

Jens Mittag,[1] Felix Schmidt-Eisenlohr,[1] Moritz Killat,[1] Marc Torrent-Moreno[2] and Hannes Hartenstein[1]

[1]Karlsruhe Institute of Technology (KIT), Karlsruhe, Germany
[2]Barcelona Digital Centre Tecnològic, Barcelona, Spain

7.1 Introduction: Challenges and Requirements

Communication in VANETs is intrinsically of broadcast nature: when a vehicle transmits a message, this message can be received typically by many neighboring vehicles – due to radio characteristics – and, indeed, should be receivable by all surrounding vehicles since the message might be of importance for safety and/or efficiency reasons. A central challenge of VANETs, however, is that no communication coordinator can be assumed. Although some applications will likely involve infrastructure (e.g. traffic signal violation warning, toll collection), several applications will be expected to reliably function using decentralized communications. Since no central coordination or handshaking protocol can be assumed, and given that many applications will be broadcasting information of interest to many surrounding cars, the necessity of a single shared control channel can be derived (even when multiple channels are available using one or more transceivers, at least one shared control channel is required). This 'one channel paradigm', together with the requirement for distributed control, leads to some of the key challenges of

VANET design. Clearly, medium access control (MAC) is a key issue in the design of VANETs. In this chapter, we first survey various MAC approaches proposed previously for vehicular networks. We then focus on the IEEE 802.11p draft of standard that, at the time of writing, is considered as the most promising MAC candidate due to availability and cost considerations. In addition to the key facts of the draft, modeling and simulation issues are addressed. We present performance results for IEEE 802.11p, particularly for probability of packet reception, and outline corresponding performance models that represent performance indicators depending on vehicular traffic density, packet sizes, and transmission rate. Finally, we address the issue of how to control the load on the channel to allow robust communication even under heavy load conditions. We conclude the chapter by pointing to open issues and future research directions.

Medium access control protocols for general wireless networks are surveyed, for example, in Gummalla and Limb (2000) and Kumar et al. (2006). In this chapter we focus on distributed MAC protocols since we assume that there will be no central communication coordinator in VANETs. As in general wireless networks, MAC issues such as half-duplex operation, time-varying channel, higher bit error rates compared to wire-line networks, and location-dependent carrier sensing that leads to hidden and exposed terminals as well as to packet capture opportunities, play a prominent role in VANETs. Also, standard MAC performance metrics such as throughput, delay, fairness, stability, robustness against channel fading, and support for QoS can and should be applied to assess MAC proposals for VANETs. Due to the specific nature of VANETs, however, those standard MAC issues and performance metrics need to be reconsidered with respect to their impact or specification, respectively. The following key issues have to be kept in mind when discussing MAC approaches for VANETs:

- **Hidden terminals.** When two wireless transceivers cannot sense each other due to radio propagation characteristics, they might send packets at the same time, causing a packet collision at a receiver that is within reach of both senders. This very well-known problem of hidden nodes, of course, significantly affects VANETs since the senders are not coordinated by a central entity and are not within a single 'cell' where each vehicle can sense the other vehicles.

- **System dynamics.** Mobility and rapid changes in environmental conditions lead to time (and frequency) varying channels with strong fading effects. Therefore, the problem of hidden (and exposed) terminals comes in a specific flavor.

- **Scalability.** The bandwidth of the frequency channels currently assigned or foreseen for VANET applications ranges from 5 to 20 MHz. With a high vehicular traffic density, those channels can easily suffer from channel congestion. Therefore, with high vehicular traffic density, packet capture will be standard and not an exception.

- **Types of communication.** For active safety applications, it can be assumed that each vehicle will periodically transmit status information to surrounding vehicles. Therefore, the basic type of communication will be one-hop broadcast messages. In addition, emergency messages will require quick and efficient information dissemination. Thus, there is also a need for quality of service support with respect to robustness of communication and differentiation of various types of communication.

By taking those observations into account, the following MAC metrics and requirements should be considered when evaluating MAC proposals for VANETs:

- **Probability of successful packet reception.** The key figure for one-hop broadcast messages is given by the average number of received packets depending on the distance to the sender. This metric represents the reception ratio of one-hop broadcast messages without retransmission schemes.

- **Channel access time.** For one-hop broadcast messages, channel access time is used as a key performance indicator for delay.

- **Congestion control.** Improvements of MAC approaches that can deal with highly varying channel load are typically labeled as 'congestion control' mechanisms in VANET literature. We address congestion control in Section 7.5.

- **Robustness against fading.** Both slow and fast fading can vary strongly in typical VANET environments. Thus, MAC approaches have to be evaluated for a wide range of fading conditions. We outline how to model and simulate various fast fading conditions in Section 7.3.3 and provide corresponding simulation results in Section 7.4.

- **Prioritization of messages.** Most likely, various types of messages will be transmitted within a single (control) channel. Since some might be more urgent than others, prioritization of messages might be an important feature of a MAC approach. We outline some prioritization mechanisms in Sections 7.3.2 and 7.5.

The above-mentioned requirements and metrics are taken into account in detail for the analyses of approaches based on IEEE 802.11p. In the following Section 7.2, we first present a survey of alternatives to IEEE 802.11p that are discussed on the basis of the results given in the original papers.

7.2 A Survey on Proposed MAC Approaches for VANETs

Various medium access control approaches were proposed, analyzed, modified, and extended for use in VANETs. While currently the focus of research, development, and standardization is on an appropriate derivative of the CSMA-based IEEE 802.11a wireless LAN standard, this section presents a survey on alternative approaches, their basic ideas and corresponding specifics as well as their common challenges.

7.2.1 Time-division multiple access based approaches

With time-division multiple access (TDMA), the available frequency band is slotted in time. Ideally, each time slot is used by a single sender only to avoid packet collisions. In ad hoc networks, however, the lack of centralized control leads to challenges of how to assign time slots to senders, of how to perform slot synchronization, and of how to deal with hidden and exposed nodes.

In Crowther et al. (1981) the Reservation-ALOHA (R-ALOHA) protocol was proposed to improve the throughput of the purely contention based protocol slotted ALOHA. Though it was initially developed for satellite communication, it has been extended and analyzed by Mann and Rückert (1988), Zhu et al. (1991), Liu et al. (1995), and Ma et al. (2005) for the use in inter-vehicle communication or distributed packet radio networks in general.

In R-ALOHA, the channel is divided into frames, and frames themselves are subdivided into N consecutive slots. In addition, R-ALOHA requires sufficiently large dimensioned time slots to cover the transmission of a single data packet and to dominate the maximum channel propagation delay. By using a simple sensing, each station should be able to determine whether a slot is *unused* or *used*. A slot is declared unused if no packet has been transmitted or if a collision has been detected, while a slot is declared used if exactly one packet was observed in this slot and was successfully received by the observer. Consequently, if a station wants to transmit a packet, it randomly selects an unused slot and contends for it, as it would do in slotted ALOHA. If the initial transmission attempt in the unused slot succeeds, i.e. no other station has selected the same slot, the station will become the owner of the slot and can use it as long as needed. However, if a collision with another station occurs, a different unused slot is probed during the next frame.

Mann and Rückert (1988) extended R-ALOHA to account for the mobility of stations in vehicular environments. They propose the Concurrent Slot Assignment Protocol (CSAP) for traffic information exchange to take care of the hidden terminal problem and the fact that stations cannot detect collisions they are involved in. They divide each frame into N data slots and N collision slots, where data slot i corresponds to collision slot i. A station which is currently the owner of a slot i uses collision slot i to broadcast its perspective, as a bitmap vector, on the reservation status of all other slots. Since every station that owns a slot is broadcasting such a bitmap vector, each station is able to merge the vectors and obtain a slot assignment in the two-hop surroundings. The authors state that CSAP is suited for limited mobility in the system and suggest using a purely contention-based solution if mobility may be high.

Based on CSAP, Zhu et al. (1991) propose a similar approach called DCAP, which further divides the channel according to the movement direction of vehicles and includes a fast channel handover in case of increasing co-channel interference. Due to mobility it can happen that an initially unused slot is interfered by other stations. If the packet loss ratio increases above a specific threshold, the station initiates a handover request and switches to the new logical channel once it has been established. An analysis and simulation of the protocol states that handovers are performed very quickly and collision probabilities are very small.

The performance of R-ALOHA for inter-vehicle communication has been evaluated using a discrete Markov chain by Liu et al. (1995) and Ma et al. (2005). Liu et al. (1995) use a metric called *deadline failure probability* (DFP) to describe the reliability of R-ALOHA. They state that the probability of not receiving a status update from a station within a specific time, e.g. within the next four frames, decreases with an increasing number of stations or error rates. Ma et al. (2005) investigate also the impact of multipath and shadowing, including the near–far effects and capture effects, on the performance of R-ALOHA. They state that capturing can significantly improve the stability of R-ALOHA.

The hybrid approach of Lott et al. (2001), which uses TDMA and CDMA, is a carry-on of the DCAP proposal by Zhu et al. (1991). Comparing with DCAP they not only distinguish between used and unused slots, but also between slots that cannot be used due to an existing reservation and slots that cannot be used due to interference. This distinction is used to implicitly signal negative acknowledgments within the slot assignment bitmap vector. For instance, in the case where a vehicle cannot decode a packet successfully, it will indicate this fact by tagging the slots as being interfered and thereby tell the sender that it did not receive the packet.

Recently, Bilstrup et al. (2008) evaluated the ability of self-organizing TDMA (STDMA) to support inter-vehicle communication. Similar to the proposals described above, STDMA applies the principle of R-ALOHA, but since it has been developed and standardized for the automatic identification system (AIS) for communication between ships and is part of the ITU-R Recommendation M.1371-1 (ITU 1998), with the focus on broadcast communication. In AIS, each ship transmits periodic heartbeat messages, which for instance contain information about its current position and heading, to establish a similar mutual awareness as envisioned by inter-vehicle communication. To support this objective, STDMA divides time into frames with fixed duration and frames themselves further into equally sized slots. However, no global frame synchronization is required and only time slots have to be aligned. Based on this definition, each ship performs the following steps to determine the time slots within a frame for the transmission of its heartbeat messages: i) directly after joining the network, each ship monitors the channel for the duration of at least one frame to determine the current reservation and usage of slots; ii) given the heartbeat rate r per frame, each ship then selects r 'random' slots per frame, such that the slots are equally distributed in time and, if possible, unused. If there are no unused slots available, slots which are used by the ship located furthest away will be selected; iii) each reserved slot is used during the following three to eight frames and has to be reselected afterwards using the same reservation scheme again. According to Bilstrup et al. (2008) STDMA is able to support safety-related inter-vehicle communication if no channel fading is considered. However, the channel characteristics of the allocated frequency spectrum for inter-vehicle communication will be subject to severe fading and whether STDMA remains robust or not in such an environment is an open question.

7.2.2 Space-division multiple access based approaches

With space-division multiple access (SDMA) based medium access schemes for inter-vehicle communication, access to the medium is controlled dependent on the current location of a vehicle. Instead of a distributed assignment of spreading codes or time slots among vehicles, the specific time-slot, the to-be-used frequency, or the spreading code of a transmission is derived from the geographical position of a vehicle. Of course such a scheme relies on the availability of user location information such as provided by GPS or magnetic positioning systems. Note, that the accuracy of the position information provided by GPS or magnetic positioning systems might be insufficient for an effective SDMA-based medium access scheme.

Bana and Varaiya (2001) introduced the first SDMA-based system model in which the geographical space and the available bandwidth are partitioned into N divisions, e.g. equally sized cells for space division and N TDMA-, CDMA-, or FDMA-channels for bandwidth division. By using a 1:1 map between the space and bandwidth divisions and by obtaining its current location on the road, a vehicle is then able to determine its channel assignment. The authors claim to provide delay-bounded access, since vehicles are mobile and changing their location over time. Further, the bandwidth assignment is fair as long as channels are of equal size. However, this scheme is static and inefficient. Since their model requires a space partitioning in which only one vehicle per division is allowed, a large number of channels are left unused if not all divisions are occupied. To deal with this issue, the authors propose to allow more than one vehicle per space division and use a contention-based scheme within a space division.

A similar idea has been published by Katragadda et al. (2003). Their location-based channel access (LCA) protocol is very similar to the proposal of Bana and Varaiya (2001) in the sense that space is divided into cells and mapped to channels. They also discuss the tradeoff between cell size, the corresponding time for which a cell-to-channel mapping can be used and the efficiency in case of a small number of vehicles. They conclude that it is necessary to allow more than one vehicle per cell and use a CSMA/CA approach inside each cell.

7.2.3 Code-division multiple access based approaches

In a code-division multiple access (CDMA) based system, concurrent access to the wireless medium is provided by the usage of different spreading codes among several senders. In principle, each sender multiplies its data signal with a spreading code before transmitting the signal to the wireless channel, thereby increasing the transmitted signal bandwidth. In order to successfully decode the signal, a receiver has to perform the reverse process and 'divide' the received signal by the spreading code. With this technique, it is possible to decode multiple incoming transmissions simultaneously, given that the individual streams arrive with equal signal strengths. In a distributed CDMA system, the challenging task is then (i) how to assign spreading codes to stations, such that codes are not used by two senders at the same time, and (ii) how to equalize the received signal strengths to mitigate the well-known near-far problem.

A promising study that considered CDMA for inter-vehicle communication has been published by Nagaosa and Hasegawa (1998). They propose a multicode sense (MCS)/CDMA system in which each vehicle senses the currently used spreading codes in the network in order to determine unused codes. For this to work, they assume the usage of equal transmission powers by all vehicles and the capability to demodulate all possible spreading codes in parallel. The challenge of equalizing the received powers is then simply ignored by arguing that low received signal strengths will correspond to packets from far distances and high signal strengths will correspond to packets from close distances. Since packets originated from close distances are of greater importance, this may be tolerated.

In Lott et al. (2001) the adoption of UMTS terrestrial radio access (UTRA-TDD) was investigated for its use in VANETs. UTRA-TDD incorporates elements of code-division multiple access (CDMA) and of time-division multiple access (TDMA). It was observed that transmit power control represents a challenging problem when applying CDMA in ad hoc networks due to the variety of senders and receivers and their respective locations. The authors of Lott et al. (2001) concluded that the different codes cannot be used for multiple access but could allow a single sender to transmit independently to various receivers simultaneously. However, for one-hop broadcast communication, this feature might be of less importance, although it could be used to make transmissions robust against noise and interferences.

7.3 Communication Based on IEEE 802.11p

For vehicle-to-X communication it is inevitable that all participating parties agree on a common standard. The currently foreseen standard for this specific type of communication is IEEE 802.11p (IEEE 2008). Basically it is one amendment within the family of IEEE 802.11 (IEEE 2007) standards that define the widely used technology for wireless local area networks (WLAN). IEEE 802.11p is adapted to the specifics that have to be respected in vehicle-to-X communications.

In Section 7.3.1, a brief overview of the IEEE 802.11 standard family is given, and a closer look at the technical functionality, details, and specifics of IEEE 802.11 is given in Section 7.3.2. In Section 7.3.3, modeling and simulation of IEEE 802.11p-based networks is presented, and the important parameters are explained; we also consider the modeling of the radio channel.

7.3.1 The IEEE 802.11 standard

The first version of the IEEE 802.11 standard was published in 1997 and it specifies the medium access control (MAC) and physical layer (PHY) for wireless local area networks (WLANs). Over the years the standard has continuously been developed and has grown, so that numerous amendments have been created in order to i) extend the functionality (e.g. in terms of security, quality of service, or interoperability), ii) support advanced transmission techniques and higher data rates (e.g. orthogonal frequency division multiplexing (OFDM)), and iii) operate in several frequency bands (e.g. within the ISM bands at 2.4 GHz and 5.8 GHz). Several of these amendments were aggregated in one version to form the up-to-date standard IEEE 802.11-2007 (IEEE 2007). The following paragraphs briefly describe the functional blocks defined in the standard, particularly focusing on the parts that have to be adapted for the operation in vehicle-to-X communications. An overview of the standard is given, for example, in Gast and Loukides (2005).

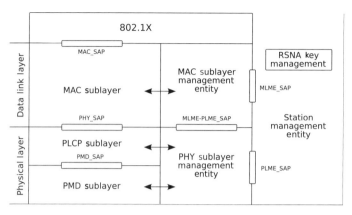

Figure 7.1 Reference model of IEEE 802.11 – layers, sublayers, and service access points. Adapted from Figure 5-10 in IEEE 802.11 Standard for information technology – Telecommunications and information exchange between systems – Local and metropolitan area networks – Specific requirements – Part 11: Wireless LAN medium access control (MAC) and physical layer (PHY) specifications. (Reproduced by Permission of © 2007 IEEE.)

In Figure 7.1, the reference model of IEEE 802.11 is shown. Several functional blocks are specified that interact among each other over a set of service access points (SAPs). The left column contains the sublayers that establish wireless communication: the medium access control (MAC) sublayer includes the methods for accessing the medium in a coordinated fashion, the physical layer convergence protocol (PLCP) sublayer transforms the MAC frame into a medium-independent physical frame structure by adding preamble, headers, and trailers, and finally the physical medium dependent (PMD) layer encapsulates all functionality to transmit the data bits over the air and is individual for each transmission technology. Although the generic concept is the independence of higher layers from the characteristics of the physical transmission, both MAC and PLCP have some dependencies; for example, MAC parameters (slot times, inter-frame spaces) are adapted depending on medium characteristics and application scenario, or the PLCP frame format differs with the transmission technique used. The middle and right columns of Figure 7.1 illustrate the different management entities that are needed to manage and configure the different layers and the station as a whole. IEEE 802.11 has two external interfaces: one is the wireless interface, the other one is a SAP to the next higher protocol layer, logical link control (LLC), and provides the request of frame transmission to, and signaling of frame reception from the wireless interface.

Part of MAC and PHY's fundamental functionality is the definition of station addresses, the grouping of stations to connected sets and the addressing scheme in the exchanged frames. IEEE 802.11 offers different opportunities to build such so-called basic service sets (BSS). For instance, nodes can form an independent BSS (IBSS) with no central coordination authority, or, as in environments with infrastructure, i.e. an access point, be part of an infrastructure BSS which is identified by an individual identification number, the BSSID. An announced BSS may be joined by first scanning for an available BSS, followed by an authentication, and finally an association process.

IEEE 802.11 defines the frame structures at different layers. At MAC layer, a generic 802.11 MAC frame is defined that builds the basis for all existing frames. It includes a bit field for frame control, a duration field, several addresses, the frame body and a frame control sequence (FCS) for error detection. Within the frame control bit field the type and subtype of a frame is specified, so that specific frame formats for management, control, and data transmission can be distinguished. Each subtype is derived from the generic format and adapted for the specified usage, i.e. specific fields and data elements are added or left out.

IEEE 802.11 provides several approaches for medium access control: point coordination function (PCF), that is only applicable if a central coordinating station like an access point is available, and distributed coordination function (DCF). Later, the standard was extended by more enhanced coordination functions that support the distinction of different service qualities, i.e. by the enhanced distributed channel access (EDCA) that will be explained in the next section. Here we first concentrate on DCF, as centralized approaches like PCF might not apply for typical vehicle-to-X communication.

The DCF follows the principle of carrier sense multiple access with collision avoidance (CSMA/CA), i.e. the channel is only accessed if the physical layer does not observe any ongoing activity on it and collision avoidance is provided by several additional technologies on the MAC layer described in the following. To allow medium access strategies, the physical layer has to notify the channel status to the MAC layer, called clear channel assessment (CCA). It is not specified how exactly a wireless card should identify the status of the medium, instead, the medium should be indicated busy if the received power level is higher than a certain threshold in case a valid frame transmission is observed. It should also be indicated busy in the absence of a valid transmission if the received power level exceeds a second, higher threshold. An important mechanism are inter frame spaces (IFSs), which are time durations that the medium has to be indicated as idle before the station may transmit. IFSs of different length for different frame types allow prioritized access. For example, important control packets such as acknowledgments are sent after a short inter frame space (SIFS), whereas regular data packets are not transmitted before the medium was sensed idle for the duration of a distributed IFS (DIFS), that exceeds the length of SIFS by two so-called slot times. In the case where the medium is determined busy, Figure 7.2 illustrates the medium access strategy: the station selects a random number of backoff slots within a certain range, the contention window. The slots are counted down after the medium was sensed idle for the duration of a DIFS; the countdown is interrupted whenever the medium is determined busy. Whenever the countdown reaches zero the frame is transmitted. When there are unicast packets for which no acknowledgment is received, a retransmission is scheduled after a newly selected number of backoff slots under the use of an increased contention window (exponential backoff). Retransmission limits are defined that restrict the number of transmission retries.

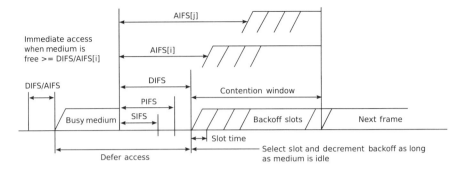

Figure 7.2 Distributed coordination function for medium access in IEEE 802.11. Adapted from Figure 9-3 in IEEE 802.11 Standard for information technology – Telecommunications and information exchange between systems – Local and metropolitan area networks – Specific requirements – Part 11: Wireless LAN Medium Access Control (MAC) and Physical Layer (PHY) Specifications. (Reproduced by Permission of © 2007 IEEE.)

Several physical layer specifications exist in IEEE 802.11; here we focus on the amendment IEEE 802.11a that serves as the basic technology used for vehicle-to-X communication. The adaptions specifically necessary for vehicle-to-X communication are to be defined in the amendment IEEE 802.11p. IEEE 802.11a operates in the 5.2–5.8 GHz frequency band and uses orthogonal frequency division multiplexing (OFDM) as transmission technology (see Chapter 6). The channel bandwidth is separated into 52 orthogonal subcarriers, i.e. subcarrier spacing is designed so that at the center frequency of one subcarrier all others have an amplitude of 0. By using coding schemes, data bits are redundantly spread over 48 of these subcarriers so that erroneously transmitted bits in one subcarriers do not necessarily damage a frame reception. Depending on the modulation scheme each subcarrier encodes a specific number of bits in each symbol; for example, using the relatively simple binary phase shift keying (BPSK) modulation scheme, each subcarrier encodes 1 bit. By applying an inverse fast Fourier transform (IFFT) the signals of all subcarriers are transformed into the time domain as symbols of fixed length. Subsequent symbols are separated by a guard interval of fixed duration in order to avoid interferences between the distinct symbols (inter-symbol interference, ISI). At a receiver the processing is applied in reversed order; of course, the IFFT is replaced by an FFT. As OFDM is a transmission technology that gives higher robustness against changing and varying channel conditions compared to other spread spectrum approaches it is a reasonable choice for vehicle-to-X communications.

7.3.2 IEEE 802.11p: towards wireless access in vehicular environments

IEEE 802.11p is a variant of IEEE 802.11a that additionally covers the specifics of vehicle-to-X communication: highly dynamic and mobile environment, message

MAC LAYER AND SCALABILITY ASPECTS

transmission in an ad-hoc manner, low latency, and operation in a reserved frequency range. These specifics require several adaptations to the standard. Historically, the IEEE 802.11p standard evolved out of the 'ASTM E2213-03 Standard Specification for Telecommunications and Information Exchange Between Roadside and Vehicle Systems' (ASTM 2003) that was transformed into an IEEE compliant style in 2004. In the following years standardization discussion continued, and the first version that successfully passed the IEEE 802 working group letter ballot was draft version 4.0. At the time of writing (March 2009), the improved draft version 5.0 of IEEE 802.11p passed the IEEE 802 working group recirculation letter ballot. Thus, IEEE 802.11p is still within the standardization process and further prepared in order to be brought towards sponsor ballot before being approved and published (also see Chapter 10). Though related to an earlier status of the IEEE 802.11p draft, a further description of IEEE 802.11p development and design decisions is given by Jiang and Delgrossi (2008).

Medium access layer

A fundamental difference of IEEE 802.11p compared to 'normal' IEEE 802.11 networks is the ability to communicate outside the context of a basic service set to enable communication in an ad-hoc manner in a highly mobile network. The IEEE 802.11 authentication and association processes preceding a first frame exchange would last too long, e.g. in the situation of communication between two vehicles with opposing driving direction. Consequently, authentication and association are not provided by the IEEE 802.11p PHY/MAC, but have to be supported by the station management entity (SME) or a higher layer protocol. In the vehicle-to-X use case the protocols of the IEEE 1609 (IEEE 2006) standard family contain the necessary procedures. IEEE 802.11p adds the mode of communication outside a BSS into the standard as this way of operation was not foreseen.

The communication outside of a BSS reduces the functionality of MAC to the basic needs. All unnecessary frame formats are removed and only a small number of the necessary frames remain: data is transmitted using the *QoS Data* frame format to allow prioritization of frames on a packet level basis according to the EDCA mechanisms described in the next paragraph. Unicast frames are acknowledged and may be preceded by an optional RTS/CTS frame exchange. A special management frame is introduced, the *timing and information frame*. It is suggested that roadside units are allowed to advertise information on the services provided in a rapid fashion. Such information may contain timestamp and time synchronization information, supported data transmission rates or information on enhanced station coordination (EDCA), and the possibility of announcing services of higher layers as, for example, specified in IEEE 1609.

An important aspect for vehicular communications concerning safety will be the prioritization of important safety- and time-critical messages over the ones that do not directly concern safety. IEEE 802.11p therefore specifically adapts enhanced distributed channel access (EDCA) that was originally proposed in the IEEE 802.11e amendment of the standard, introducing quality of service (QoS) support. The medium access rules defined by the DCF are replaced by the ones of EDCA, where

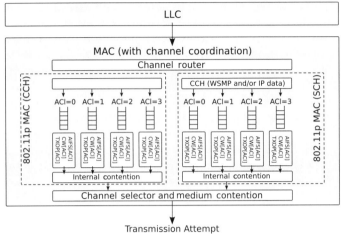

Figure 7.3 Scheme of enhanced distributed channel access (EDCA). Adapted from Figure 4 of IEEE Trial-Use Standard for Wireless Access in Vehicular Environments (WAVE) – Multi-channel Operation. (Reproduced by Permission of © 2007 IEEE.)

four different access categories (AC) are defined. Each frame is assigned one of the four access categories by the application creating the message, depending on importance and urgency of its content. Each AC is identified by its access category index (ACI), holds its own frame queue and has an individual set of parameters coordinating the medium access. Figure 7.3 gives an overview of the EDCA architecture in vehicle-to-X communications: two types of supported channels are shown, the control channel (CCH) and the service channels (SCH); a description of multi-channel operation can be found in Chapter 10. For each channel four separate queues are provided, each of them having a specific setting of contention parameters, shown in Table 7.1. The arbitration inter frame space number (AIFSN) replaces the fixed DIFS time defined in the DCF. The time for which the medium has to be sensed idle before it can be accessed has to exceed the time of an SIFS by AIFSN slot times. Also, the contention window minimum and maximum values are individual for each AC. For example, regarding the values given in Table 7.1, frames sent with ACI=3 have a high chance of accessing the medium earlier due to the lower AIFSN value and the lower contention window borders in case of a backoff. To summarize, frames with an ACI of 0 have regular access, an ACI of 1 is foreseen for non-prior background traffic, while ACIs of 2 and 3 are reserved for prioritized messages, e.g. critical safety warnings. Yet, there is no strict prioritization: contention between the access categories is resolved internally; only the frame having the lowest waiting time actually contends with other stations on the medium. Note that 'internal' collisions are possible; in this case the frame of the higher access category (having a lower ACI) is preferred.

As already mentioned, all functionality that is omitted to allow an instantaneous message exchange has to be addressed on a higher layer of abstraction.

Table 7.1 Default EDCA parameter set for the operation outside of a BSS.

Access category index	AIFSN	CW_{min}	CW_{max}
0	6	7	15
1	9	15	1023
2	3	3	7
3	2	3	7

For vehicle-to-X communication the IEEE 1609 trial-use standards (IEEE 2006) are developed providing the necessary services on a higher layer within the payload of IEEE 802.11p frames; a short overview is given here, more details can be found in Chapter 10.

- IEEE 1609.1 defines a resource manager that should allow multiple applications run on roadside units to communicate with the on-board units of multiple vehicles. It serves on the application layer.

- IEEE 1609.2 defines security services for the vehicle-to-X communication, such as authentication of stations and encryption of messages.

- IEEE 1609.3 specifies networking services for vehicle-to-X communication, including a specific stack and protocol to handle WAVE short messages (WSM).

- IEEE 1609.4 defines how the operation of multiple channels is organized and has a strong relation to the EDCA mechanisms described above.

Physical layer

At the physical layer IEEE 802.11p is similar to IEEE 802.11a, with some adaptations for the specific application domain. The operation takes place in a separate and reserved frequency band. In the USA the Federal Communications Commission (FCC) has allocated a 75 MHz wide frequency spectrum from 5.85 to 5.925 GHz in 1999. In Europe, a 30 MHz wide frequency spectrum was allocated by the Electronic Communications Committee (ECC) in August 2008, with a possible extension to 50 MHz. IEEE 802.11a specifies operation for 5, 10 and 20 MHz channels; while 'classic' wireless networks typically use 20 MHz channels, 10 MHz channels are envisioned for vehicle-to-X networks due to robustness issues and the possibility to reuse existing wireless chipsets. Several measurements by Alexander et al. (2007); Cheng et al. (2008, 2007); and Tan et al. (2008) showed a Doppler spread (caused by the fast moving nodes) up to 2 kHz and RMS delay spread (caused by multipath propagation) of up to 0.8 μs. In a 20 MHz channel of IEEE 802.11a the guard interval between subsequent symbols has a length of 0.8 μs and thus is critical, being too short to mitigate inter-symbol interferences (ISI). A longer guard interval of 1.6 μs is achieved when using half the bandwidth, as it is done in IEEE 802.11p. The duration of a data symbol also doubles to 6.4 μs. Thus, the measured delay spread is shorter than the guard interval, mitigating inter-symbol interferences (ISI).

Inter-carrier interferences (ICI) are mitigated as well because the Doppler spread is much smaller than half the subcarrier separation distance of 156.25 kHz. By only using half the bandwidth the capacity of the channel also reduces to the half, i.e. only 3 Mbps instead of 6 Mbps in the most basic mode. Owing to multi-path propagation and the high vehicular mobility, the channel coherence time may be shorter than the duration of a data frame so that the channel estimation performed during preamble reception may become invalid at the end of a frame. However, solutions exist that overcome these limitations, e.g. by using an advanced receiver as proposed in Alexander et al. (2007) in which a time-domain channel estimation and a frequency-domain channel tracking are performed to equalize the channel, or by using differential OFDM modulation proposed in Zhang et al. (2008).

IEEE 802.11p specifies more adaptions that hardware devices have to fulfill, e.g. with regard to the operating temperature ranges and the allowed tolerances of frequencies and clocks. Low bit error rates support highly reliable communication, and IEEE 802.11p therefore (optionally) specifies more stringent regulations with respect to adjacent and non-adjacent channel rejection and transmit spectrum masks. This should reduce the influence of neighboring channels on each other.

7.3.3 Modeling and simulation of IEEE 802.11p-based networks

In research and development, simulations are often used in order to investigate and evaluate systems that do not yet exist, or to determine the behavior of a system under various parameterizations. In the case of vehicle-to-X communication, simulation models and parameterization corresponding to the IEEE 802.11p definitions have to be used. One widely used simulator in research is NS-2 (NS-2 2008). The default simulator supports the simulation of wireless networks following the IEEE 802.11 standards. Yet, the provided models and implementation have their deficiencies and restrictions in accuracy, extendibility, and the structure and clarity of the implementation. The work of Chen et al. (2007) proposes ways to overcome the mentioned problems. Their functional and structural overhaul is illustrated in Figure 7.4.

In this approach, each functional block of the MAC and PHY layers is implemented as a separated module with clear functionality and interfaces for a logical and ordered structure. On the MAC layer the following main blocks are modeled: the transmission and reception of frames, the coordination of frame transmission and reception, the management of the backoff procedure, and the management of the physical and virtual channel status. Additional functionality can be extended to the simulator, e.g. handling of management frames, multi-channel support or quality of service support following EDCA.

A frame that is forwarded to the MAC to be transmitted is first handled within *transmission coordination*. Here, all processes before the channel is accessed are modeled, like waiting for inter-frame spaces, waiting for the right amount of backoff slots (coordinated in detail by the *backoff manager*) or performing a preceding RTS/CTS exchange. Also, retransmissions are processed by the *transmission coordination*. Then, when finally accessing the channel the *transmission coordination* module forwards the frame to the *transmission* module that interacts with the *physical layer*.

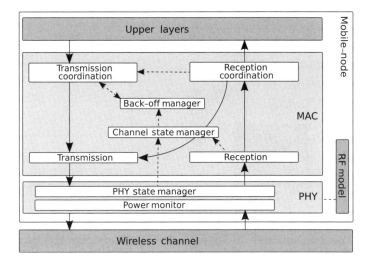

Figure 7.4 Architecture of the overhauled IEEE 802.11 implementation of the network simulator NS-2.

The *transmission* module also accepts control frames from the *reception coordination* and management frames, if such modules are incorporated. On packet reception the *reception* module is informed by the physical layer whenever a frame has arrived. First it 'virtually' performs the frame control sequence check to determine whether the frame was received correctly; in simulations, this information is provided by the *physical layer* depending on frame reception conditions as described below. Further, it applies the address filtering and, if applicable, forwards the packet to *reception coordination*. In the case of frames with errors the necessary actions are taken (usage of extended IFS). The *reception coordination* transmits CTS and ACK to the *transmission* module as a response to incoming RTS and DATA frames; it also signals CTS and ACK reception to the *transmission coordination* module. Finally, the channel state is managed by the *channel state manager*, which obtains the physical channel state from the *physical layer*, as well as information from the duration field of received frames that keep the channel virtually busy by the so-called network allocation vector (NAV).

While building a simulation model of the MAC layer mainly consists of correctly mapping the definitions, functionalities, and procedures defined in the standard into a simulation implementation, the effort to be done on the physical layer is far more complicated. The level of detail with which frame transmission and reception is represented by the simulation models can be selected from a wide range of possibilities. For OFDM systems, for example, the entities modeled could be continuous waveforms, different subcarriers, distinct OFDM symbols, data bits, or complete data frames.

Here, we present our approach of simulations based on packet/frame level models. Such a model implies several model assumptions that are described in the following. First, we assume that at each node r each frame f has one individual

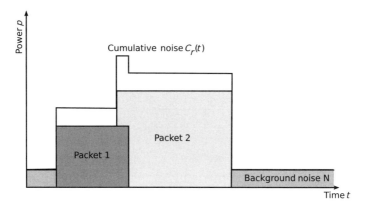

Figure 7.5 Illustration of the cumulative noise concept: the signal strengths of packet 1 and packet 2 are both added to the background noise, yielding a cumulative noise power level $C_r(t)$.

but uniform reception power $P_r(f)$ over the complete time of packet reception that is derived by the *RF models* described later in this section. Second, the reception powers of different frames received in parallel at node n_r at time t additively sum up to a cumulative power level $I_r(t)$ as illustrated in Figure 7.5. Additionally, there is a constant power level, the background noise N. An interference and noise level, the cumulative noise $C_r(t)$, can be calculated for each point in time:

$$C_r(t) = N + I_r(t) = N + \sum_{i \in F_r(t)} P_r(i), \qquad (7.1)$$

where $F_r(t)$ is the set of all frames that interfere with each other at node r at time t. Subsequently, the signal-to-interference-and-noise ratio (SINR) of a frame f can be determined as:

$$SINR(f, t) = \frac{P_r(f)}{C_r(t) - P_r(f)} = \frac{P_r(f)}{N + I_r(t) - P_r(f)} = \frac{P_r(f)}{N + \sum_{i \in F_r(t), i \neq f} P_r(i)}. \qquad (7.2)$$

A third assumption claims that successful reception of a frame is possible if the SINR experienced during the frame reception is always above a specific threshold value. The SINR thresholds are specific for each combination of modulation and coding scheme and can differ for the preamble/header portion of a packet (always transmitted with the lowest modulation and coding scheme) and its payload (where other schemes can be chosen). Thus, these two parts of a frame are modeled separately. Table 7.2 represents the SINR thresholds for different modulation and coding schemes as recommended by Jiang et al. (2008). Finally, a successful reception is possible only if the receiver is able to lock onto a packet, i.e. successfully decode its preamble and header. The 'capturing capability' is supported by radio chipsets to different extents. Extended capturing allows the receiver to stop the reception process of a currently received frame and start it for a newly arriving one.

The functionality is normally supported during the reception of a preamble, and modern radio chipsets also allow capturing within the payload reception phase.

Table 7.2 SINR thresholds used for vehicle-to-X communication simulations (from Jiang et al. (2008)).

Data rate	Modulation scheme	Coding rate	SINR threshold [dB]
3	BPSK	1/2	5
4.5	BPSK	3/4	6
6	QPSK	1/2	8
9	QPSK	3/4	11
12	16-QAM	1/2	15
18	16-QAM	3/4	20
24	64-QAM	2/3	25
27	64-QAM	3/4	N/A

In the simulation model the capturing functionality is substituted into two modules, a *power monitor* module and the *physical state manager*. The *power monitor* is responsible for keeping track of all frames that arrive at a certain node, and updates the interference and noise level at a station on every change, allowing the calculation of the SINR for every packet at every point in time.

The *physical state manager* includes the modeling of transmission and reception of frames. The frame transmission model is simple. Each frame is assigned a specific transmission power and has a duration that depends on its length and the used modulation and coding scheme. The frame itself, combined with power and duration information, is handed to the *wireless channel* object that disseminates the frames to all nodes and also models the propagation delay. Frame reception is modeled on packet level in the *physical state manager* and under the assumptions described above: successful reception is only achievable in the case where the SINR of a frame never falls below a certain threshold and if the receiver was able to lock onto or capture the frame.

Modeling of radio channel characteristics in a network simulator

A key influencing factor for the probability of packet reception is given by the assumed radio wave propagation model. The most detailed view on the radio wave propagation is likely given by a ray tracing simulation. However, the required computational effort does not allow simulation results in acceptable run-time. Hence, researchers have often applied simplifying models in order to represent radio propagation characteristics. The literature distinguishes between deterministic and probabilistic models, both accordingly characterizing the attenuation of the radio signal strength over distance. Deterministic models often used are the free space and the two-ray ground model. For one specific distance between transmitter and receiver of a message these models always return one received radio signal strength. Such modeling has been shown to strongly differ from what occurs in reality. An often cited representative for the more realistic probabilistic models is the Nakagami-m model. It covers fading effects and estimates the received signal strength for a multipath environment. The model is represented as a function with two parameters Ω and m, where Ω is the average received radio signal strength and m identifies the fading intensity. For accurately chosen parameters Taliwal et al. (2004) and Yin et al. (2006) have shown that it suitably agrees with empirical data measured under highway conditions. The resulting probability of reception over distance of such modeling is shown in Figure 7.6 for a scenario where reception is not influenced by other transmissions in parallel, i.e. for a single transmitter in the scenario. For a deterministic communication model we can identify the communication range, that is the maximum distance at which receptions are still possible. Thus, when declaring transmission powers, we can also use the distance in meters that would correspond to the communication range in the two-ray ground model. In the following, we sometimes refer to the deterministic radio propagation model when declaring the chosen transmission power.

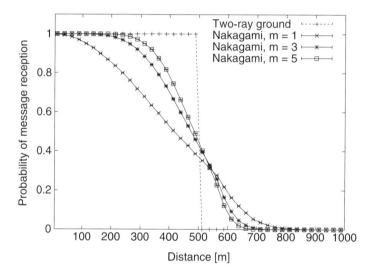

Figure 7.6 Probability of successful message reception when only one node is sending. As illustrated, the deterministic two-ray ground model provides a 100% reception probability up to a distance depending on the chosen transmit power (in this example, the transmit power was chosen to allow a communication range of 500 m). When probabilistic radio propagation models are used, e.g. Nakagami-m, the reception probability drops below 100% already at close distance. Moreover, a higher fading intensity, i.e. lower m values, provides less reliable reception.

Simulation parameter settings

For our studies, several simulation parameters have to be set to specific values, Table 7.3 provides a detailed overview. For radio propagation we choose moderate radio channel conditions which are represented by the fading parameter m = 3 of the Nakagami model and Ω is modeled to follow the path loss given by the two-ray ground model.

All nodes in the communication network are assumed to communicate according to the IEEE 802.11p standard. The standard considers various data rates ranging from 3 Mbps up to 27 Mbps. Since Maurer et al. (2005a) emphasized the benefits of lower data rates for a robust message exchange, and Jiang et al. (2008) showed the optimality of 6 Mbps in wide application range, we also choose 6 Mbps while having safety applications in mind. We assume operation on a 10 MHz wide channel as it is foreseen for the common control channel. The values for preamble, header and symbol lengths on the PHY layer as well as for minimum contention window, SIFS and slot time on MAC layer are set according to the definitions in the standard draft, while noise floor, carrier sense threshold, and the SINR thresholds have been measured in test environments and proposed by the chipset vendors.

Concerning the data traffic we assume that nodes periodically broadcast messages to their one-hop neighborhood as it is envisioned by beacon messages, for instance. For each transmitted packet we assume a size of 400 bytes comprising approximately 200 bytes of payload and 200 bytes for the indispensably required security protection. Note that today's considerations assume beacon messages to contain information on a node's identifier (might be pseudonyms), a timestamp, the position (longitude, latitude), the speed, and the direction of a vehicle.

Table 7.3 Simulation configuration parameters.

Parameter	Value
Radio propagation model	Nakagami m = 3
IEEE 802.11p data rate	6 Mbps
Channel bandwidth	10 MHz
Preamble length	32 µs
PLCP header length	8 µs
Symbol duration	8 µs
Noise floor	−99 dBm
Carrier sense threshold	−94 dBm
SINR for preamble capture	5 dB
SINR for frame body capture	10 dB
Minimum contention window	15
Slot time	13 µs
SIFS time	32 µs
Packet size	400 bytes

7.4 Performance Evaluation and Modeling

Based on the simulation methodology as described in the previous section, this section presents performance evaluation results for IEEE 802.11p when used for the most basic type of communication: one-hop broadcast messages. As described in the introduction to this chapter and in the previous section, various factors can lead to failure of packet reception, most notably the hidden node problem and fading of the channel.

In the first part of this section, we present results for the probability of successful packet reception. As the underlying scenario, a highway is assumed where all vehicles send periodic messages ('beacons') as one-hop broadcasts. We evaluate the probability of reception depending on the distance between sender and receiver and depending on other influencing factors such as the configured transmission power and transmission rate.

In the second part of this section, we present an approach as to how the simulation results obtained via the network simulator NS-2 can be transformed into an analytical empirical model that covers the dependencies with respect to vehicular traffic density, transmission power, and transmission rate. Since for an application designer it might be a huge burden to be dependent on exhaustive simulation runs when appropriate configurations have to be determined, the empirical model can facilitate the design process. In addition, the empirical model can be used 'online' for parameter optimization. We provide an overview of proposed analytical modeling attempts in the literature showing that a comprehensive mathematical or information-theoretic model which considers all important influencing factors has not yet been proposed. Then this section continues with the empirical approach: we generate numerous simulation traces on basis of our simulation results. The evaluation of the simulation traces and the succeeding application of linear least squares curve fitting techniques enables the derivation of a closed-form analytical expression which gives the probability of one-hop packet reception in dependency of various parameters.

7.4.1 Performance results of IEEE 802.11p-based active safety communications

The application of inter-vehicle and vehicle-to-infrastructure communications in the future will enable a lot of different kinds of applications. Probably one of the most promising applications is covered by the term 'active safety' systems in which intelligent vehicles will act cooperatively to avoid dangerous situations and accidents. By a *periodic* exchange of one-hop broadcast status messages, which will contain the vehicle's current geographic position, velocity, driving direction, etc., it is envisioned that vehicles will be able to establish a mutual awareness. This awareness can then be used to detect dangerous traffic situations, e.g. the end of a traffic jam or an overtaking vehicle. Additionally, vehicles are envisioned to warn each other in case a dangerous situation is detected, e.g. an icy road or the explosion of the airbag. Since these warnings are sent out in case of an emergency situation, these messages are commonly called *event-driven* messages.

In this section we will focus on the exchange of periodic status messages, also called beacons, and identify the primary parameters that determine the communication performance of those messages by simulation. For the simulations we use a simple and straight highway scenario with three lanes per direction and different vehicle densities. The vehicles are not moving and are configured to send periodic beacon messages with the setup described in Table 7.3.

The first parameter which has a significant impact on the performance of successful one-hop broadcast reception is the number of vehicles frequently sending beacons within a certain area, i.e. the vehicle density. For instance, in scenarios with a small vehicle density the probability of successful beacon message reception is likely to be higher compared to scenarios with a high vehicle density. Indeed, as illustrated in Figure 7.7(a), the observable probability is decreasing with increasing vehicle densities, while nothing else has been changed in the scenario configuration.

The second parameter that significantly influences the results is the transmission rate (or 'packet generation rate'). It should be intuitive that an increasing transmission rate, i.e. a higher rate at which packets are generated, will also reduce the probability of successful beacon reception, since more packets will contend for the channel and possibly overlap and collide. The impact of increasing the transmission rate from 2 Hz to 5 Hz and 10 Hz is illustrated in Figure 7.7(c). Again, only the rate was altered for the different curves – configurations such as vehicle density or transmission power have not been modified.

The third and last primary parameter for the performance of periodic one-hop broadcast communication is the used transmission power. In principle, an increase of the transmission power increases the number of vehicles that are within each others' communication range and competing for the channel. Thus, a change of transmission power is comparable to a change of vehicle density, except that transmission power can be adjusted intentionally to increase or decrease the communication range. As a result, as shown in Figure 7.7(e), a decrease of the transmission power reduces the chance of receiving messages from vehicles located further away. However, in situations where the wireless channel is congested, e.g. in Figure 7.7(f) where, compared to Figure 7.7(e), the transmission rate was doubled to 10 Hz, a reduced transmission power will help to successfully receive messages from close distances.

From the comparison of the curves in Figure 7.7(e) and 7.7(f) we can conclude that the impact of a transmission power adjustment depends on the used transmission rate. Similarly, though it is not shown here, the impact of the adjustment depends also on the actual vehicle density, i.e. a reduced transmission power will have a different effect in low-density traffic situations from that in high-density traffic situations. In the same way, an adjustment of the transmission rate will cause different results when performed either at low or high configured transmission powers, cf. Figure 7.7(c) and 7.7(d). Also, a change of vehicle density will influence the communication performance more or less extensively if different transmission rates are used, see Figure 7.7(a) and 7.7(b). Therefore, if we want to tweak or predict the performance of the vehicular communication system, we have to consider all three parameters simultaneously.

The same set of parameters have been reported and merged into a single performance metric by Jiang et al. (2007) to describe and compare the quality of the wireless communication channel. Of course, there are definitely many more parameters which can influence the performance of one-hop broadcast reception,

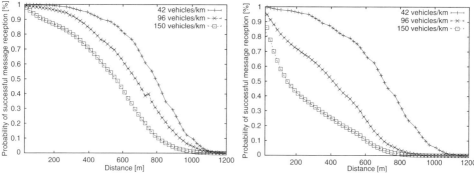

(a) Fixed transmission power of 20 dBm and fixed packet generation rate of 5 packets/sec.

(b) Fixed transmission power of 20 dBm and fixed packet generation rate of 10 packets/sec.

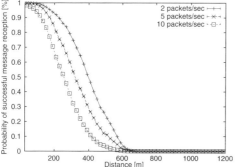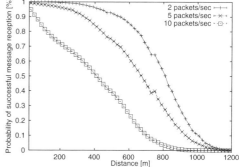

(c) Fixed vehicle density of 96 vehicles/km and fixed transmission power of 10 dBm.

(d) Fixed vehicle density of 96 vehicles/km and fixed transmission power of 20 dBm.

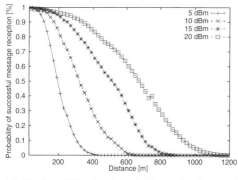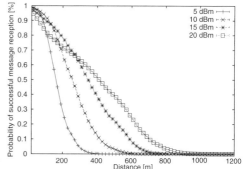

(e) Fixed vehicle density of 96 vehicles/km and fixed packet generation rate of 5 packets/sec.

(f) Fixed vehicle density of 96 vehicles/km and fixed packet generation rate of 10 packets/sec.

Figure 7.7 Probability of successful beacon reception in dependence of the distance to the sender for different vehicle densities, Figure (a) and (b); different beacon generation rates, Figure (c) and (d); and different transmission powers, Figure (e) and (f). Confidence intervals are shown but are hardly visible since variance of the results is very low. Simulation parameters are as given in Table 7.3.

MAC LAYER AND SCALABILITY ASPECTS

e.g. the carrier sense threshold used by the PHY layer to indicate when to block transmission requests from the MAC layer, or the average size of beacon messages.

7.4.2 Computational costs of simulation

The computational costs of the simulations as outlined in the previous sections are as follows. When assuming a discrete event-based simulation model which incorporates effects such as additive white noise and capture capabilities of modern hardware chipsets and which treats the preamble and the payload of a frame separately, the number of events E per transmission that have to be handled and processed by the simulator engine is then given by

$$E = 3(n-1), \quad (7.3)$$

where n is the number of vehicles in the scenario. For each possible receiver, one event has to be scheduled which signals the arrival of a new frame at the receiver. Even if the frame arrives with a very low signal strength, it has to be considered in order to accurately model the additive interference level. In addition, two events have to be scheduled, which signal the end of a frame's preamble and the end of the frame itself.

If we extend Equation (7.3) to account for all vehicles in a scenario sending beacon messages and also for different message frequencies, the number of events per simulation, as a function of number of vehicles n, the transmission rate r in Hertz, and the simulated time t in seconds, is given by

$$E(n, r, t) = t(nr)3(n-1). \quad (7.4)$$

Thus, the computational costs are scaling linearly with respect to the simulated time t and the transmission rate r, but nonlinearly with the number of nodes n. If we ignore the dimension of varying frequencies, we are able to plot the number of simulated events as a function of n and t, which is shown in Figure 7.8. The extent of the quadratic increase becomes more impressive if we compare the absolute number of events from scenarios with 200 and 1000 vehicles. For instance, if we assume a transmission rate of 5 Hz and a simulation time of 500 seconds, which, together with the 200 or 1000 vehicles, still reflect a small number of events in Figure 7.8, we have to deal with only 5 980 000 events at PHY and MAC layer in the 200-vehicle scenario and already 149 900 000 events in the 1000-vehicle scenario – an increase by a factor of 25. Obviously, a detailed simulation of the MAC and PHY protocols is not practicable and too time-consuming when simulating networks with a large number of vehicles to investigate large-scale effects of inter-vehicle communications. For such simulations, it would be better to have a MAC and PHY model which abstracts the small-scale effects while still providing, from a statistical point of view, a sufficient degree of accuracy.

7.4.3 Analytical models for performance of IEEE 802.11 networks

The majority of proposed models on IEEE 802.11 networks refer to a paper by Bianchi which studies the maximum achievable throughput of the IEEE 802.11

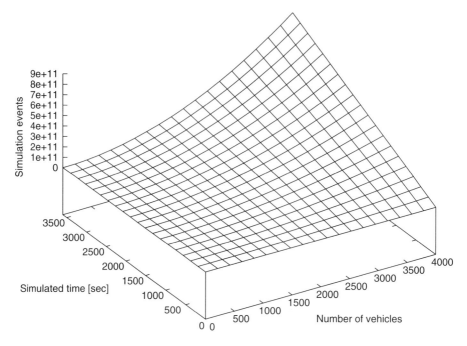

Figure 7.8 Computational costs for a detailed 802.11 simulation, in terms of simulation events, as a function of the number of vehicles in the network and the simulated time. For this illustration, the packet generation rate was set to five beacon messages per second and per vehicle.

protocol. Bianchi assumed IEEE 802.11 networks in general and considered assumptions which hardly adhere to conditions found in a vehicular environment. His key approximation is a constant and independent probability of collision for transmission attempts, regardless of the number of suffered retransmissions. In particular, the highly mobile environment which allows rapidly changing communication conditions puts this assumption into question. Additionally, he considers simplified conditions on the physical layer and limited capabilities of network cards that will be discussed later on. Nevertheless, the solid analysis provided in Bianchi's paper establishes the basis for succeeding papers releasing taken assumptions.

Bianchi's paper assumes saturated conditions, i.e. all nodes in the network always have packets ready for transmission. Due to this assumption, a node in the network is either transmitting a message or has chosen a backoff slot in its contention window. Bianchi applied two-dimensional Markov chains to model the contention window of a single node: one dimension considers the backoff slots and the second dimension takes account of differing backoff stages in which the number of backoff slots varies. The transitions between the system states are taken by a node according to the following transition law: *a)* whenever the medium is sensed idle for a DIFS period and the backoff counter is larger than zero, the node decrements its backoff slot. When the backoff counter equals zero the node starts a transmission. In case *b)* when the transmission is successful the node resets the contention window and starts processing the next package to be transmitted. If *c)* the transmission collides because another node in the network has chosen the same slot in its contention window, the node increases its backoff stage and reenters the contention phase by choosing a new backoff slot. For this Markovian process Bianchi determines the steady-state probability distribution over system states and, thus, derives the probability that a node starts a transmission in a generic time slot. The generalization of this transmission probability for all nodes in the network eventually leads to the probability of a successful transmission and, thus, to the system's throughput. For a detailed discussion on this approach we refer the reader to Bianchi's paper (Bianchi 2000).

Bianchi has validated his results by comparison to a simulation study. The conformance of both approaches, however, also results from implemented models in the simulator that comply with the assumptions taken by Bianchi. Today's advanced simulators comprise models that have been shown to significantly change a simulation's outcome as shown, for example, by Schmidt-Eisenlohr and Killat (2008). Likewise, many follow-up papers have proposed extensions that relax some assumptions taken in Bianchi's approach. In the following we give a short but nonexhaustive overview of papers proposing refinements of the above-mentioned model.

First follow-up papers dealt with Bianchi's modeling of backoff stages. In conformance with the IEEE 802.11 standard Bianchi considers a finite number of backoff stages, i.e. the size of the contention window is at most doubled a finite number of times. In contrast to the standard, however, Bianchi does not consider packet drops due to a failed transmission in the last backoff stage. He assumes a node to stay in the last backoff stage and to retry unsuccessful transmissions (potentially) an infinite number of times. Chatzimisios et al. (2004) addressed this issue and proposed a modeling in line with the standard. Furthermore, as a second contribution, the authors digressed from the assumption of inevitably successful transmissions in the presence of only a single sender and, thus, introduced transmission errors modeled by a *bit error rate (BER)*. In their evaluation the authors compare analytical with simulation results, infer the suitability of their considerations, and highlight the impact of an increasing bit error rate on traffic throughput and delay. Both of their evaluation approaches, analysis and simulation, however, consider an ideal channel in the sense that a node is able to detect any ongoing transmission regardless of its (geographical) distance from the sender. This inaccuracy has often been discussed in papers that address the problem of *hidden terminals*.

Papers referring to the hidden terminal problem commonly identified the chosen time steps in Bianchi's Markov model as the key difficulty. From the perspective of a single node's contention window succeeding system states are taken when the communication channel has been sensed idle (for sufficiently long) or when the node transmits a message. During a transmission, however, system conditions are assumed to be stable, i.e. the initial decision on a successful or failed reception is not changed while a transmission is going on. In case of the hidden terminal problem, however, some nodes are not able to detect an ongoing transmission, might therefore start a new (simultaneous) transmission and thus turn a firstly assumed successful packet reception into a failed one. A revised modeling takes asynchronous transmissions into account and digresses from considering packet transmission time as atomic time unit. Instead Hou et al. (2003), Tsertou and Laurenson (2006), and Ekici and Yongacoglu (2008), for example, assume finer granular time steps whereas the former two primarily focused on the RTS/CTS and the latter, more generally, also on the basic IEEE 802.11 access mechanism. All three papers additionally cluster neighboring nodes according to their distance into 'contending' and 'hidden' nodes with the latter covering all potential hidden terminals. This enhanced modeling requires modifications to Bianchi's analysis: while the probability of transmission in a generic slot remains unaffected by hidden nodes the probability of collision is indeed impaired. The analysis in all papers is backed by comparison with simulation studies which depend on certain assumptions. In contrast to the first two papers, however, Ekici and Yongacoglu (2008) consider an earlier work that relaxes Bianchi's assumption of *saturated load conditions*.

Duffy et al. (2005) digressed from the assumption of saturated conditions by introducing a (constant) probability reflecting the likeliness of packet arrival per time slot and node. The exploitation of the configuration interval of the introduced probability from 0 to 1 therefore allows us to model the entire spectrum from lightly loaded to saturated conditions on the communication channel. Since under this modeling a node is not necessarily contending for access to the communication channel all the time, the introduction of additional system states to Bianchi's model were required. Almost at the same time Engelstad and Østerbø (2005) published a paper that analyzes performance measures of the priority scheme in the EDCA mechanism of IEEE 802.11e networks. In the special case of only a single priority class, however, the analysis complies with previously used models. Engelstad and Østerbø (2005) proposed a comprehensive Markov chain model that extends Bianchi's model by the consideration of non-saturated conditions, packet drops due to finite retransmissions, and post-backoff timer. Again, both these papers validate their analysis via NS-2 simulations assuming that packets arrive in non-saturated conditions according to a Poisson distribution.

All previously mentioned papers focused on the IEEE 802.11 MAC mechanism not taking into account effects resulting from the underlying PHY layer. Indeed, performance metrics such as the system throughput certainly depend on the environment that influences, for instance, the radio wave propagation. Pham (2005), for instance, addressed this issue and studied the performance of IEEE 802.11 under idealistic radio conditions and under the probabilistic Rayleigh channel. In this way, fading effects of the radio signal are introduced that Pham modeled by means of a second Markov chain representing the 'up' and 'down' time of the communication channel. His analysis also comprises an advanced Markov chain for the IEEE 802.11 standard and covers its basic and RTS/CTS access mechanism. His concluding comparison to NS-2 simulations mostly agree with the analytically derived results but the paper does not cover, e.g. a discussion on the hidden terminal problem or on the *capturing effect*.

The capturing effect results from advanced network cards that can lock onto the most powerful packet in the case of multiple simultaneous receptions. Thereby the system's throughput can be significantly increased since not all simultaneously arriving packets are automatically discarded. Li and Zeng (2006) proposed a mathematical analysis that considers the capturing effect, Rayleigh fading and log-normal shadowing of the wireless channel. The paper assumes all nodes to be uniformly distributed in a circular area and determines, depending on the number of simultaneously transmitting nodes, the probability distribution of the radio strength of the joint signals. Thereby, a comparison to the strength of a single signal can determine whether one out of all the simultaneously transmitted messages could have been captured. Li and Zeng (2006) incorporated the determined capturing probability to an IEEE 802.11 mechanism description by Tay and Chua (2001), who carried out a similar analysis of the wireless standard as Bianchi without modeling the problem by means of Markov chains. The assumptions of saturated conditions, for instance, are likewise taken and thus influence Li and Zeng (2006)'s results when including the capturing effect. On the other hand their approach assumes ideal carrier sensing and the concluding fact of strictly synchronized communicating nodes. Thereby fading effects, or the hidden terminal problem which give reason for asynchronously started transmissions (and so collisions), have not been considered in the paper. Also, aforementioned contributions with respect to the hidden terminal problem would complete the modeling process only partly since a thoroughly modeled carrier sensing needs to consider the joint signal strengths of surrounding transmissions that has been introduced as *cumulative noise* in Section 7.3.3.

The presented overview certainly does not cover all the papers on modeling issues in IEEE 802.11 networks. Our selection of papers is meant to sketch the efforts of many researchers to analytically capture communication conditions in wireless networks following the IEEE 802.11 protocol. Regarding our aim of modeling vehicular networks in which multiple complex factors have a joint impact on communication, we conclude that a comprehensive analytical representation does not exist yet. The identified key challenges for the IEEE 802.11 MAC mechanism – configurable load on the communication channel, bit errors during transmission, the hidden terminal problem, a probabilistic radio wave propagation, the capturing effect and cumulative noise – have, to our knowledge, never been jointly discussed by analytical means. Finally, Table 7.4 summarizes our discussion and gives an overview of the identified key challenges as addressed by the presented papers.

7.4.4 An empirical model for performance of IEEE 802.11p networks

An analytical model which covers all known particularities in IEEE 802.11 networks is not yet available. On the other hand, however, advanced simulators have been proposed that can jointly consider various effects on the system's performance. In this section we follow Killat and Hartenstein (2009) and propose an empirical approach: First, we apply an advanced simulator and generate an exhaustive set of data points which cover our problem space in a uniform manner. Second, we make use of general linear least squares curve fitting techniques in order to seek an analytical expression that suitably represents the generated data set. The advantages gained from this procedure are at least twofold. On the one hand, we

Table 7.4 Discussed issues of the referenced papers on modeling communication conditions in IEEE 802.11 networks.

Paper	IEEE 802.11 MAC	Configurable load	Bit errors	Hidden terminal	Prob. radio wave propagation	Capturing effect	Cumulative noise
Bianchi (2000)	✓	–	–	–	–	–	–
Tay and Chua (2001)	✓	–	–	–	–	–	–
Hou et al. (2003)	✓	–	–	✓	–	–	–
Chatzimisios et al. (2004)	✓	–	✓	–	–	–	–
Duffy et al. (2005)	✓	✓	–	–	–	–	–
Engelstad and Østerbø (2005)	✓	✓	–	–	–	–	–
Pham (2005)	✓	✓	✓	–	✓	–	–
Li and Zeng (2006)	✓	–	–	✓	✓	✓	–
Tsertou and Laurenson (2006)	✓	–	–	✓	–	–	–
Ekici and Yongacoglu (2008)	✓	✓	–	–	–	–	–

obtain a compact representation of the generated data set that replaces a lookup table on the simulated data points and can, to some extent, estimate non-simulated data points. On the other hand, an analytical expression may be used when optimal communication protocol parameters need to be determined. For example, an optimal transmission power configuration may depend on various parameters and might be given by an optimization problem. Where an analytical expression on the communication performance exists, numerical procedures can determine the optimization's problem solution.

The capabilities of the proposed empirical model are certainly limited to the information captured in the generated data set. In other words, the analytical expression gained from the curve fitting process can only interpolate between known data points. Appropriate predictions beyond the known data points, however, will not be covered. Hence, thorough considerations of the assumptions subject to the generated data set are indispensably required in advance. From now on, our focus is on inter-vehicle communications to be used for road safety related applications. We investigate a key metric in vehicular ad-hoc networks, namely the probability of one-hop broadcast packet reception, depending on changing conditions of the environment. Our study is restricted to highway scenarios in order to exclude radio reflection effects from surrounding buildings that has hardly been explored so far. In the remainder of this subsection we firstly outline further assumptions underlying the conceived empirical model before we elaborate on its generation and conclude with its validation.

Assumptions

The empirical model is built on the simulation assumptions presented in Section 7.3.3. Briefly, we assume a 10 MHz channel and a data transmission rate of 6 Mbps which has been shown to be most suitable for safety-related applications. All nodes are assumed to communicate according to the IEEE 802.11p standard and to periodically transmit 400 byte packets where the transmission rate is considered as a variable input parameter to the desired empirical model.

Regarding the radio wave propagation, we assume the probabilistic Nakagami-m model and choose moderate radio channel conditions represented by a fading parameter $m = 3$. However, when we state chosen transmission powers in the following, we will make use of deterministic radio wave propagation models as introduced in Section 7.3.3. When deterministic models are assumed, one can identify a fixed communication range which is the largest distance at which packet receptions are still possible. Thus, when referring to the transmission power, we will state the distance ψ in meters that would correspond to the communication range in a deterministic model.

Table 7.3 provides a detailed overview of the assumptions subject to the following simulation study.

Model building

Multiple factors influence the probability of one-hop packet reception. However, in the most simple case of only a single sender in the network, the influencing factors are reduced to the propagation of the radio signal, since interferences can naturally be ruled out. Assuming a radio propagation that follows the Nakagami $m = 3$ distribution Killat et al. (2007) have shown that the probability of receiving one-hop broadcast packets can be analytically derived by

$$\mathscr{P}_R^{\text{single}}(x, \psi) = e^{-3(\frac{x}{\psi})^2} \left(1 + 3\left(\frac{x}{\psi}\right)^2 + \frac{9}{2}\left(\frac{x}{\psi}\right)^4\right) \tag{7.5}$$

where x denotes the distance between sender and receiver, and ψ states the chosen transmission power in meters as introduced before.

With many transmitters, however, additional effects need to be considered (see Section 7.4.3) that are not reflected in Equation (7.5). For this purpose, Killat et al. (2007) suggested utilizing simulators that generate plenty of data points, at best, uniformly covering the problem space. Clearly, the problem space itself needs to be identified and, with increasing complexity, suitably restricted to manageable dimensions. In this scope we conceive the following problem space: we assume a varying vehicular traffic density δ [veh/km] in which all vehicles regularly transmit messages at a rate of f [Hz] using a configured transmission power of ψ [m]. Now, the objective is the probability of one-hop packet reception in dependency of the problem parameters over distance, i.e. $\mathscr{P}_R(x, \psi, \delta, f)$.

By making use of the advanced simulator proposed by Chen et al. (2007) we simulated more than 600 scenarios, each running for 100 s and 30 seeds. In each scenario a unique data point in the problem space (tx power, vehicle density, tx rate) is considered. All scenarios are evaluated by focusing on the transmissions triggered by one chosen reference vehicle. In order to avoid correlations between subsequent transmissions of the reference vehicle, its transmission interval is, in contrast to the other nodes, fixed to a relaxed configuration of 1 s. For all transmissions of the reference vehicle the number of potential and actual receptions over distance are recorded, thus yielding a maximum of 6000 captured packets per unit distance (note that each distance exists on both sides of the sender). Concluding, the simulation results give a lookup table on the probability of reception over distance for some data points in the problem space of transmission power, vehicular density, and transmission rate.

Instead of a lookup table, however, we are looking for an analytical expression for further usage in simulations or parameter configurations. For this purpose, Killat et al. (2007) suggested applying general linear least squares curve fitting techniques using Equation (7.5) as a starting point which is extended with linear and cubic terms; additionally the fitting parameters a_1 through a_4 are introduced, i.e.

$$\widetilde{\mathscr{P}}_R(x, \psi) = e^{-3(\frac{x}{\psi})^2} \left(1 + \sum_{i=1}^{4} a_i \left(\frac{x}{\psi}\right)^i\right). \tag{7.6}$$

We applied the *Levenberg–Marquardt*[1] curve fitting method and confirm a visual perfect match for all results of our simulation study. For an empirical model,

---

[1] In this work we utilized the open source software *GRETL (Gnu Regression, Econometric, and Time-series Library)* version 1.7.1.

however, a law is needed that gives the respective fitting parameters depending on the scenario configuration. Mathematically speaking, we seek functions

$$h_i : (\text{tx power, tx rate, vehicle density}) \rightarrow \mathbb{R}$$

which give, for any combination in the inputs, the corresponding fitting parameters a_i. In fact, a polynomial of sufficiently high degree would approximate the conceived functions arbitrarily well. However, due to the three-dimensional nature of the input, the number of coefficients in the polynomial rapidly increases and thus stress the complexity of the resulting empirical model. Hence, we look for a reduction of the complexity and thus make use of the *communication density* which has been introduced by Jiang et al. (2007).

The communication density has been conceived as a metric for assessing the load on the communication channel and simply states the number of sensible events per unit of time. It is calculated as the product of communication range, vehicle density and transmission rate, yielding a value expressed in packet transmissions per unit time. Jiang et al. (2007)'s results indicated that broadcast transmissions can expect very similar performance behavior in differing environments when the communication densities coincide. Indeed, our numerous simulations exhibit concurring probability of receptions for various scenarios with the same communication density as illustrated in Figure 7.9.

Inspired by this coherence, we study the determined fitting parameters of Equation (7.6) based on the respective communication density. However, Figure 7.10(a) shows significant deviations of the fitting parameters a_i from polynomials Φ_i of fourth degree based on the communication density. Obviously, there exists another relationship of the fitting parameters apart from the communication density. In order to reveal this parameter, we study the deviations of the fitting parameters a_1 through a_4 from the polynomials Φ_1 through Φ_4 applied in Figure 7.10(a). Table 7.5 lists the correlation coefficients of the residuals $\rho_i = a_i - \Phi_i$ to various input combinations. In particular residuals ρ_2 and ρ_3 show a noticeable correlation to the transmission power. Therefore we choose a two-dimensional polynomial on the communication density and the transmission power as fitting function to the parameters a_1 through a_4. Regarding the degree of the polynomial, we compare the coefficient of determination R^2 of the fitting process for various degrees of the polynomial and, according to Figure 7.10(b), assess a degree of four as a suitable choice for all parameters.

Table 7.5 Correlation coefficients of residuals ρ_1 through ρ_4 to various input combinations.

	ψ	δ	f	$\psi \cdot \delta$	$\psi \cdot f$	$\delta \cdot f$	$\psi \cdot \delta \cdot f$
ρ_1	**0.5910**	-0.2645	-0.2506	0.2748	0.3111	-0.5779	0.0000
ρ_2	**-0.7310**	0.2989	0.2562	-0.2938	-0.3857	0.5747	0.0000
ρ_3	**0.7055**	-0.2879	-0.2386	0.2844	0.3786	-0.5487	0.0000
ρ_4	0.1263	-0.0435	0.0208	-0.0800	0.0176	**0.1549**	0.0001

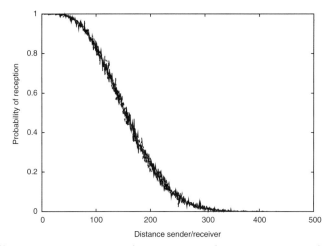

(a) Five differing scenario setups with $\psi = 200$ m and a communication density of 120.

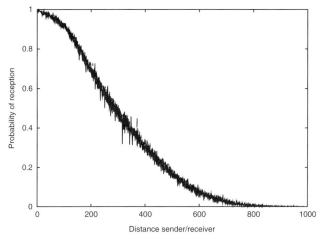

(b) Four differing scenario setups with $\psi = 500$ m and a communication density of 500.

Figure 7.9 Differing scenarios with common transmission power and communication density exhibit very similar behavior with respect to probability of reception over distance.

We can now declare the desired functions h_i as:

$$h_i(\xi, \psi) = \sum_{j,k \geq 0} h_i^{(j,k)} \xi^j \psi^k, \qquad i = 1 \ldots 4$$

with $\xi = \psi \cdot f \cdot \delta$ and $j + k \leq 4$

and list the involved coefficients $h_i^{(j,k)}$ in Table 7.6. Based on the obtained relationship between scenario variables and fitting parameter we can finally formulate the

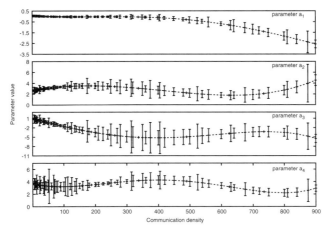

(a) Curve fit of fourth degree polynomials Φ_i to the fitting parameters a_1 through a_4 based on the communication density (with 95% confidence interval).

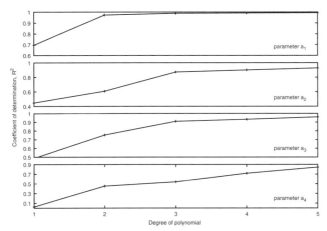

(b) Accuracy of fitting two-dimensional polynomial with varying degree to the parameters a_1 through a_4.

Figure 7.10 Approximating fitting parameters a_1 through a_4.

empirical model as:

$$\mathscr{M}: \quad \widetilde{\mathscr{P}}_R(x,\delta,\psi,f) = e^{-3\left(\frac{x}{\psi}\right)^2}\left(1 + \sum_{i=1}^{4} h_i(\xi,\psi)\left(\frac{x}{\psi}\right)^i\right). \tag{7.7}$$

Model validation and use case

In the beginning of Section 7.4.4 we claimed two beneficiaries from the closed-form representation of the empirical model: a convenient replacement of the lookup table

Table 7.6 Coefficients $h_i^{(j,k)}$ subjected to the polynomials h_1 through h_4. Although some values seem to be negligible all values significantly influence the resulting probability of reception.

	(j,k)				
	(0,0)	(1,0)	(2,0)	(3,0)	(4,0)
$h_1^{(j,k)}$	0.0123679	−2.25450e−06	6.36982e−12	−2.09306e−17	1.05684e−23
$h_2^{(j,k)}$	2.99714	2.53145e−05	−5.63148e−11	9.81719e−18	3.06358e−23
$h_3^{(j,k)}$	−0.610698	−6.96673e−05	1.95332e−11	1.80545e−16	−1.39711e−22
$h_4^{(j,k)}$	4.15044	−9.01791e−06	1.49252e−10	−2.44958e−16	1.09573e−22
	(3,1)	(2,1)	(2,2)	(1,1)	(1,2)
$h_1^{(j,k)}$	−7.55774e−21	1.07606e−14	4.35680e−18	4.18407e−09	−2.95060e−12
$h_2^{(j,k)}$	−8.01474e−20	1.36395e−13	−1.66585e−17	−4.08656e−08	−7.91283e−11
$h_3^{(j,k)}$	1.06503e−19	−3.27537e−13	8.08115e−17	1.96554e−07	−4.45038e−11
$h_4^{(j,k)}$	1.12769e−19	−4.38097e−14	−9.45343e−17	−1.18915e−07	2.42917e−10
	(1,3)	(0,1)	(0,2)	(0,3)	(0,4)
$h_1^{(j,k)}$	−7.17582e−15	0.00109906	−8.53786e−06	2.19213e−08	−1.70152e−11
$h_2^{(j,k)}$	8.52923e−14	−0.0152642	0.000105275	−2.42757e−07	1.86228e−10
$h_3^{(j,k)}$	1.49057e−14	0.0604508	−0.000411583	9.25967e−07	−6.90747e−10
$h_4^{(j,k)}$	−1.22398e−13	−0.0389028	0.000307395	−7.01965e−07	5.05531e−10

and the possibility of estimating non-simulated data points in the problem space, on the one hand, and the usability for simulations and parameter adjustments on the other hand. With respect to the replacement of the lookup table we can confirm that a comparison of the model's 'prediction' with the simulation traces used in the model-building process yielded a wide conformance: across all investigated scenarios we investigated the sum of squared errors (SSEs) over all distances. The maximum SSE found in a scenario ($\psi = 600$ m, $\xi = 150$) aggregated to 0.393 (cf. Figure 7.11(b)). The largest observed deviation between model and simulation results amounts 5.1% and is illustrated in Figure 7.11(a) at a distance of 139 m.

Regarding the ability of the model to predict scenarios which have not been used in the model-building process we refer to a problem that likewise emphasizes the benefit of the model for solving parameter configuration problems. Assuming that an application runs on all vehicles, to work properly, the application requires certain constraints to be fulfilled. These restrictions, for instance, could be expressed in given probabilities of packet reception q_i to be guaranteed at given distances x_i. At best, each node has chosen a communication configuration which meets the

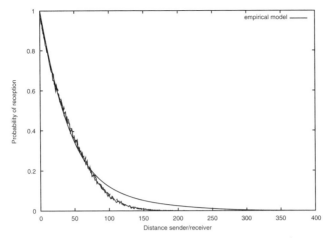

(a) Scenarios ($\psi = 200$ m, $\xi = 900$) for those where the maximum deviation has been determined.

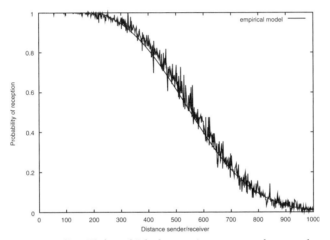

(b) Scenario ($\psi = 600$ m, $\xi = 15$) for which the maximum sum of squared errors (SSE) has been determined.

Figure 7.11 Comparison of simulation results with empirical model.

application's constraints with a minimum amount of occupied resources (of the communication system). If we consider, for the sake of simplicity, transmission power adjustment as the only means of influencing communication conditions we may utilize the model $\mathcal{M}$ to determine a vehicle's optimal transmission power ψ, that is the solution to

$$\min \ \psi \qquad (7.8)$$
$$\text{subject to} \ \ q_i - \mathcal{P}_R(x_i, \delta, \psi, f) \leq 0, \quad \forall i. \qquad (7.9)$$

In our sample scenario we assume a traffic density of $\delta = 125$ veh/km and the conceived application to transmit messages at a rate of $f = 2$ Hz; we confirm that this combination has not been used in the model-building process. Additionally, we assume three constraints subject to the application: a required probability of reception of $q_1 = 95\%$ at $x_1 = 100$ m, of $q_2 = 85\%$ at $x_2 = 200$ m and of $q_3 = 75\%$ at $x_3 = 300$ m. By making use of numerical solving algorithms we obtain from the optimization problem given in Equation (7.9) $\psi_1 = 245$ m, $\psi_2 = 364$ m, and $\psi_3 = 444$ m as the minimum transmission power configurations to meet constraints one to three, respectively. As illustrated in Figure 7.12 the numerically gained power values obviously closely coincide with the corresponding simulation results. Figure 7.12 depicts the probability of packet reception at distances 100 m, 200 m, and 300 m based on various transmission power configurations. Since the third constraint is dominating the others in the sense that a transmission power lower than ψ_3 would already be sufficient to adhere to constraints one and two, ψ_3 is likewise the solution to the above-mentioned optimization problem when all constraints need to be fulfilled at the same time.

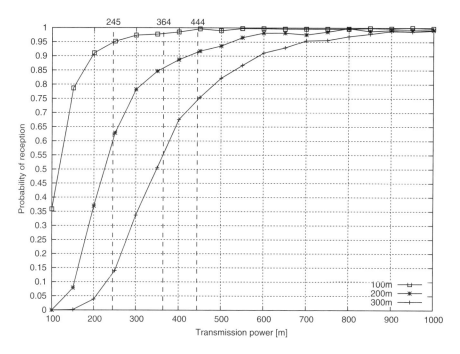

Figure 7.12 Probability of packet reception with resect to the chosen transmission power. Simulation results are shown as curves and numerically computed values to meet given constraints are shown as numbers.

Concluding, we summarize the model's benefits demonstrated in this subsection: i) the replacement of a corresponding lookup table, ii) the ability to 'predict' the outcome of non-simulated scenarios, and iii) the closed-form expression that enables computations on the model to be used in parameter configuration problems.

7.4.5 Conclusion

In this section we have pointed to the difficulties of implementing safety applications in a vehicular environment operating over a shared and limited communication channel. We demonstrated the influence of the scenario configuration on the performance of the communication system which might provoke the violation of performance thresholds required by safety applications. Countermeasures are expected to be taken by application designers who therefore require a good understanding of the communication system. A tool which may give insights into communication characteristics has been derived by means of advanced simulations and provides an analytical expression for the probability of packet reception over distance for the input parameters transmission power, transmission rate, and vehicular density. The usability of this tool is at least twofold: i) it replaces extensive simulation studies to determine one key figure of merit in (vehicular) wireless communication systems, namely the probability of packet reception; ii) it allows to solve analytically given problems as, for example, used for optimal parameter configuration.

7.5 Aspects of Congestion Control

One of the main concerns with future vehicular ad-hoc networks is to avoid the degradation of the wireless communications caused by the envisioned amount of data generated by the vehicles. Especially when high penetration rates of equipped vehicles are accomplished, the underlying wireless technology will be seriously challenged.

In the literature, many studies can be found that foresee the need of strategies to control the load on the wireless channel (Blum et al. (2004); Kosch et al. (2006); Torrent-Moreno et al. (2005a); Wischhof and Rohling (2005)). By controlling the load, saturated channel conditions and their negative impact on the performance of wireless communications will be avoided. Some of these studies, e.g., Wischhof and Rohling (2005) and Kosch et al. (2006), propose utility- or relevance-based approaches in order to prioritize the dissemination of relevant information, for instance by delaying or even dropping the least important messages in order not to saturate the channel. These approaches categorize the data traffic that is generated by the wide range of applications and assign different priorities dependent on the relevance of the safety related information, e.g. the distance to a dangerous location or the elapsed time since the detection of a dangerous situation.

However, although utility- or relevance-based strategies will be necessary, it is still not clear whether they are sufficient to avoid stressed wireless channel conditions. Indeed, as outlined by Torrent-Moreno et al. (2004, 2005b); Xu et al. (2004a) and Torrent-Moreno et al. (2006), even the amount of periodic status messages could saturate the wireless channel. Since those messages are the building block of VANETs and are necessary for the detection of potentially dangerous situations, simply dropping or delaying their transmission is not an acceptable solution.

Instead, different strategies which control the load of the periodic traffic without dropping the transmissions are required.

7.5.1 The need for congestion control

In the following, we present the different aspects that need to be considered when assessing the question whether IEEE 802.11p is suitable for supporting the required information exchange to enable active safety applications.

First, the limits of the wireless communication channel have to be sized. As defined by the FCC (Federal Communications Commission of the USA) [FCC (2004)], a spectrum of 75 MHz has been allocated at 5.9 GHz. The entire spectrum is divided into seven 10 MHz channels, out of which one channel is reserved exclusively for the exchange of safety-related information. The remaining six channels are to be used for non-safety applications. Independently, the data rates provided by IEEE 802.11p [IEEE (2008)] with such 10 MHz channels range from 3 to 27 Mbps.

Among the different data rates, the lower ones are to be preferred for safety applications due to their robustness against noise and interference by means of lower required signal-to-interference noise ratios – see Maurer et al. (2005b). While lower data rates imply greater transmission times and thus a higher probability of multiple overlapping transmissions, they require only a low SINR to successfully decode a packet. In comparison, higher modulation rates require an increased SINR but at the advantage of a reduced probability of multiple overlapping transmissions. Whether a higher data rate with a higher required SINR but shorter transmission times and a reduced probability of packet collisions improves communication is in principle a trade-off discussion. To investigate this trade-off, a simulation study was performed by Jiang et al. (2008) in order to determine the most robust data rate for broadcast communication. Their results state that a 6 Mbps data rate turns out to be the best selection for safety-related communication, although the 4.5 Mbps and 9 Mbps rates performed nearly as well as the 6 Mbps data rate or even slightly better in low saturated channel conditions.

Apart from the limitations of the wireless channel itself, the available bandwidth might be further reduced due to single transceiver systems and multi-channel operation. As described earlier, there will be one control channel for safety-related communication and six service channels for non-safety communication. If vehicles are equipped with one transceiver only, a switching between channels is necessary to support non-safety and safety communication. According to Wang and Hassan (2008) non-safety communication in such a set-up will only be possible in low-density traffic situations or if the amount of data traffic generated by periodic status messages is limited.

The second aspect that has to be considered is the offered channel load generated by periodic status messages. In principle, the load can be estimated by a simple multiplication of the expected message generation rate of each vehicle, the average message size, the configured transmission power of each vehicle, i.e. the communication range, and the number of vehicles within each other's communication range, i.e. the vehicle density. In the following we will therefore discuss these individual factors to get a glimpse of how much periodic beacon traffic will be generated to support active safety applications.

According to previous studies, e.g., Xu et al. (2004b) or Reumerman et al. (2005), and the VSC (Vehicle Safety Communications Project) final report [VSC (2004)], it is envisioned that several messages per second from each vehicle will be needed in order to provide the required accuracy for safety applications. Especially with applications with a huge potential safety benefit, such as cooperative forward collision warning, there could be a requirement for a periodic rate of up to ten messages per second. Regarding message size, security studies by Raya and Hubaux (2005) concluded that, depending on the used signing algorithm, periodic status messages will have a size between 250 and 800 bytes due to digital signatures and certificates. Also, vehicles will most likely be able to communicate up to a distance of 1 km or even more when using a maximum transmission power of about 20 dBm. Assuming that vehicles use a fixed transmission power, which corresponds on average to a fixed communication range, the number of vehicles within each other's communication range then depends on the vehicle traffic density on the highway. An overview of how many vehicles one can expect on highways is given by the Highway Capacity Manual [Board (2000)]. It classifies the quality of traffic flow or rather the level of service provided by a highway dependent on the vehicle density on the highway. As shown in Table 7.7, the vehicle density increases for lower quality levels and the average speed decreases. Assuming a communication range of 1000 m and a three lanes per direction highway, the maximum number of expected vehicles within communication range for these service levels will vary between 84 and 300.

Table 7.7 The capacity of multi-lane highways and the corresponding average speeds according to Board (2000). In addition, the number of vehicles within the communication range is listed for a three lane per direction highway and a 1000 m communication range.

Level of service	Max. density (vehicles/km/lane)	Average speed (km/h)	Vehicles within 1000 m comm. range
A	7	100.0	84
B	11	100.0	132
C	16	98.4	192
D	22	91.5	264
E	25	88.0	300

A back-of-the-envelope calculation easily shows that, for example with 200 neighboring nodes sending 10 packets per second each of size 400 bytes, the generated load will be 6.4 Mbps and already slightly higher than the maximum possible data rate of 6 Mbps which could be served by an optimally coordinating medium access protocol. In scenarios with a higher vehicle density, the situation gets even worse. Such a high load on a CSMA-controlled channel is likely to result in an increased amount of packet collisions and, consequently, in a decreased 'safety level' as seen by the safety application.

With the reasoning above, it is clear that strategies designed to control the channel load caused by the exchange of periodic status messages are required. Note that, up to now, the only congestion control mechanism proposed in the IEEE 802.11p draft [IEEE (2008)] is to prevent any message, except from the highest priority, to be transmitted if the measured channel occupancy is larger than 50%. This measure, however, would not solve the problem of the channel load resulting from periodic status messages.

Now, the question arises of how to control the load on the channel and what are the design criteria that have to be applied. Basically, there are two parameters to adjust in order to control the load on the medium (assuming the packet size is reduced to the minimum data set required): transmission power and periodic beacon generation rate.

Due to the technological limitations mentioned above, the existing trade-offs need to be appropriately balanced in order to find the optimal point of operation. While a higher packet generation rate can increase the information accuracy with frequent updates, an uncontrolled strategy can lead to a saturated medium with a high rate of message collisions. Likewise, although a message sent with higher transmission power can reach further distances, it will increase the level of interference on other transmissions. On the other hand, a higher transmission power could increase the robustness of a specific message transmission.

Moreover, in a communication network where safety is the main goal, fairness becomes a critical issue. Trying to optimize packet delivery ratio, achieved bandwidth, etc., without taking fairness into consideration can be a harmful approach. In other words, improving, for example, the overall packet delivery ratio of the system while not satisfying the safety requirements of a single node (transmission power, channel access opportunities, etc.) may become a danger to all surrounding nodes.

In the following, we describe relevant strategies, designed to avoid channel congestion in vehicular environments, which make use of transmission power adjustment or beacon generation rate control.

7.5.2 Congestion control by means of transmit power control

Transmit power control in mobile networks has been intensively studied in the past, but frequently with the objective of maximizing overall system capacity, energy consumption, or connectivity for point-to-point communications. Since the envisioned pattern of vehicle-to-vehicle communication will primarily be point-to-multipoint and energy consumption will in general not be an issue, those existing studies are not applicable to vehicular networks. For a description of the design principles behind those power control studies we refer to the work of Kawadia and Kumar (2005).

Nevertheless, encouraged by the increasing interest caused by the potential of vehicle-to-vehicle communications, some researchers applied the conventional power control goals to vehicular environments. For instance, Artimy et al. (2005) adjust the transmission power pursuing a high degree of connectivity in the vehicular network. Although in this section we are focusing on congestion control, their strategy is of interest since they estimate the local density around a vehicle based on its movement pattern and speed. However, it is not clear whether such an approach on its own is sufficient for congestion control, since, as we have seen in Section 7.5.1, there might also be the need to reduce the transmission power in free-flow traffic situations with high mobility due to the huge amount of offered beaconing load.

Strategies intended to avoid channel saturation conditions can also be classified based on the selected channel access mechanism. Assuming a TDMA reservation scheme Caizzone et al. (2005) suggest controling the transmission power of a vehicle to keep the number of neighbors between a predefined minimum and maximum value. The proposed algorithm makes use of the information carried on TDMA multiframes to estimate the number of vehicles that comprise its neighborhood. A vehicle A defines its neighborhood as the set of nodes from which it receives the indication that its time-slot (within the multiframe) is busy and not collided. Therefore, before each transmission, each vehicle increments or decrements its transmission power so that the estimated number of neighbors stays within the predefined range.

Caizzone et al. performed a simulation study in order to show the performance of their algorithm. Although the scenario set up shows no characteristics of a vehicular environment, an average of 3600 active nodes move randomly in a square network of 3 km × 3 km, they demonstrate how successful channel access opportunities can be several times higher when applying transmission power control.

Assuming IEEE 802.11p as the selected wireless technology, Torrent-Moreno et al. (2006) propose a distributed transmit power control method D-FPAV (distributed fair power adjustment for vehicular environments) to control the load of periodic messages on the channel and prioritize safety-related emergency messages. The authors justify the need of a power control strategy in a CSMA/CA scenario presenting the results of a simulation study performed for an environment configured similarly to the parameters specified in Section 7.3.3. Their setup simulates a straight and 6 km long highway with three lanes per direction and an average of 11 vehicles per km and lane driving at an average speed above 120 km/h. Each vehicle was configured to transmit 10 beacons per second with a size of 500 bytes each and a data rate of 3 Mbps. Despite the fact that the assumed data rate of 3 Mbps is not in line with today's findings, their proposal meets the requirements of a congestion control mechanism tailored to inter-vehicle communications.

As illustrated in Figure 7.13, for which a highway scenario with three lanes per direction and 16 vehicles per km and lane has been simulated, the benefit of reducing the transmission power can be expressed in terms of an increased probability that a beacon is successfully received by neighboring vehicles. More specifically, a reduction of the transmission power from 20 dBm to 10 dBm or 5 dBm leads to an increased probability of successful beacon reception for neighbors located in the close surrounding and to a decreased probability for neighbors located further away. In general, increasing the transmission power of one message increases its robustness against power fluctuations as well as interference and, thus, it is capable of reaching further distances. However, increasing the transmission power of all nodes in a network increases each vehicle's carrier sense range and the number of nodes sharing the channel at all locations, thus, reducing the spatial reuse of

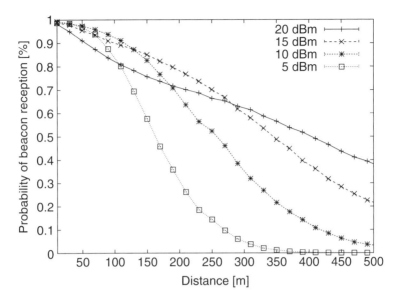

Figure 7.13 Probability of successful beacon reception with respect to the distance in three lane per direction highway scenario with 16 vehicles/km per lane. Comparison of the results obtained by using different transmission powers with a data rate of 6 Mbps, a contention window value of 15 and a beacon generation rate of 10 beacons/sec.

the wireless channel. Indeed, increasing the transmission power from 10 dBm to 20 dBm results in a significantly higher number of collisions at close distances from the sender due to the higher level of interfering signals. The reason for these low reception rates at close distances to the sender is the inability of the channel access mechanism to allocate and to coordinate the large number of transmissions from neighboring nodes in this scenario.

Torrent-Moreno et al. conclude from these results that a lower transmission power could be a better choice if it provides higher reception rates at close distances, since this area is more relevant from a safety perspective (due to the kinetic energy of the moving vehicles). D-FPAV is their proposed solution to adjust the transmission power of beacon messages and, thus, control the load on the channel in order to ensure a high probability of beacon reception at close distances to the sender.

The design goals and characteristics of D-FPAV are the following: i) it is fully distributed and able to quickly react to the dynamic topologies of vehicular networks; ii) it controls the beaconing load under a strict fairness criterion that has to be met for safety reasons. According to the authors, strict fairness must be guaranteed since it is very important that every vehicle has a good estimation of the state of all vehicles (with no exception) in its close surrounding. More specifically, a higher transmit power should not be selected at the expense of preventing other vehicles from sending/receiving their required amount of safety information; iii) D-FPAV allows

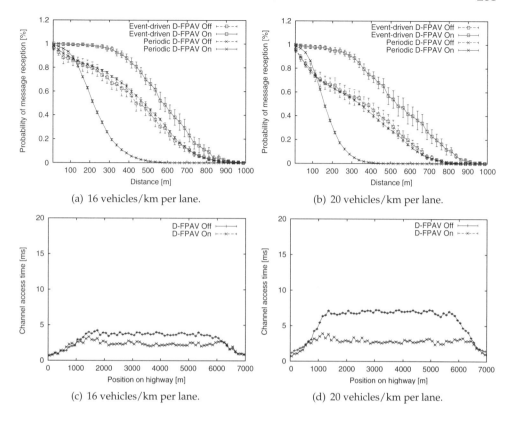

Figure 7.14 Probability of successful reception of periodic beacons and one-hop event-driven messages with respect to the distance to the transmitter as well as the average channel access times with respect to the position on the highway. Both with D-FPAV on/off, MBL = 5.0 Mbps, and vehicle densities of 16 vehicles/km per lane and 20 vehicles/km per lane.

a clear prioritization of event-driven over periodic messages, since the transmission power of event-driven messages is not reduced.

The objective of the D-FPAV strategy is to determine a power assignment for all vehicles in a distributed manner such that the minimum of the transmit powers used for beaconing is maximized and the network load (or bandwidth consumption) experienced by each vehicle remains below a predefined threshold called MBL (maximum beaconing load). By limiting the load to an MBL level below the limit of the communication channel, e.g. setting the MBL to 5.0 Mbps in a 6.0 Mbps channel, it is possible to reserve a portion of the bandwidth for emergency or other messages. To reach this objective, every vehicle runs FPAV (see Torrent-Moreno et al. 2005b), a localized algorithm based on a 'water filling' approach as proposed in Bertsekas and Gallager (1987), to calculate the maximum common transmit power level which should be used by neighboring vehicles. Since this calculation is performed on

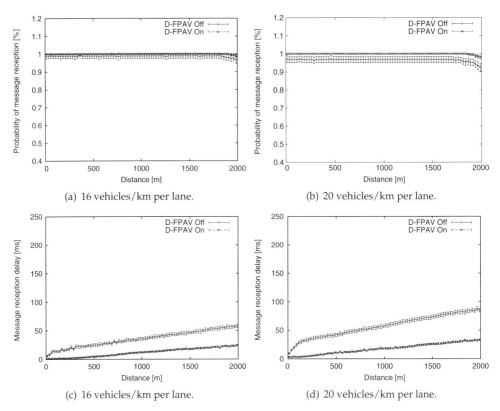

Figure 7.15 Probability of information delivery inside the dissemination area and reception delay with respect to the distance from the message originator (with multi-hop retransmissions) with D-FPAV on/off, MBL = 5.0 Mbps, and vehicle densities of 16 vehicles/km per lane and 20 vehicles/km per lane.

local and possibly incomplete information of the network topology, it might happen that some vehicles will experience an MBL violation with this transmit power. To prevent such violations each vehicle will periodically exchange its own transmit power computation with neighboring vehicles and, for the transmission of beacon messages, select the minimum power level amongst the one computed locally and those computed by surrounding vehicles. For an in-depth description of the protocol and algorithm see Torrent-Moreno et al. (2006; 2007; 2009).

Figure 7.14 illustrates the impact of D-FPAV on reception and channel access performance of periodic beacon and event-driven emergency messages when simulating 7 km long and straight highway scenarios with three lanes per direction and 16 or 20 vehicles/km per lane. Apart from the setup described in Section 7.3.3, the vehicles in these simulations have been configured to transmit beacon messages 10 times a second. Additionally, one vehicle located in the middle of the scenario was selected to start the dissemination of an event-driven message, targeting all vehicles 2 km behind itself, once every second. As can be seen in Figures 7.14(a) and 7.14(b), D-FPAV with an MBL threshold of 5.0 Mbps reaches its design goals in both scenarios and provides, compared to the case without D-FPAV, higher reception probabilities for beacon messages at close distances from the sender (at the expense of lower reception probabilities at further distances) and an increased reception probability for event-driven messages at all distances from the originator. Furthermore, as can be seen in Figures 7.14(c) and 7.14(d), the channel access times of periodic beacon messages, with respect to the position on the highway, are reduced to nearly half of the time as experienced without D-FPAV.

The effect of prioritization of event-driven emergency messages over beacon messages is further illustrated in Figure 7.15. The dissemination strategy EMDV (emergency message dissemination for vehicular environments) which is used here has also been proposed by Torrent-Moreno et al. and is basically a position-based forwarding approach, in which the message should generally be forwarded by a preselected optimal relay. However, in case of packet collisions and reception failure at the relay, a contention-based scheme, as proposed by Briesemeister and Schäfers (2000), is used to increase the reliability of the dissemination. The authors have shown that, combined with D-FPAV, the reliability of emergency message dissemination can be increased from 98.3%, or 96.5%, up to 100% (see Figures 7.15(a) and 7.15(b)). What looks like a marginal gain of the average turned out to be quite significant in a few scenario runs: without D-FPAV, it could happen that the initial emergency message from the originator was not successfully received by anyone at all – disastrous in a safety of life situation. Furthermore, the effectiveness and speed of disseminations, in terms of average number of retransmissions and dissemination delay, were benefiting: on the one hand, the average number of retransmissions required to cover the destination area (all vehicles located 2 km behind the originator) was reduced from 18.05 to 9.96 in the 20 vehicles/km scenario and from 11.18 to 8.48 in the 16 vehicles/km scenario. On the other hand, D-FPAV reduced the average reception delay with respect to the distance to the originator by nearly 50% (compare Figures 7.15(c) and 7.15(d)). Also regarding the worst case reception delay observed in all simulations, D-FPAV achieved a reduction from 294 ms to 204 ms or, for the 16 vehicles/km scenario, a reduction from 248 ms to 167 ms.

When we compare the results shown above with earlier simulations by Torrent-Moreno et al. we can see that the gain of transmit power control in a 6 Mbps channel is less dominant than in the case of a 3 Mbps channel. The performance and reliability increase for emergency messages and for beacons from the close surrounding is marginal in the 16 vehicles/km per lane scenario and starts to become significant when increasing to 20 vehicles/km per lane. Clearly, there seems to be a modulation-specific threshold of the offered beacon load beyond which transmit power control provides a significant performance gain. As simulations show, this threshold depends not only on the modulation scheme used, but also on the radio propagation conditions: in Figure 7.16, the probability of successful beacon reception is illustrated when simulating a scenario with either D-FPAV enabled or disabled and applying different channel fading conditions in the simulations. The results show that transmit power control provides an increased reception probability at close distances if channel fading is assumed to be non-severe (Nakagmi m = 3) and no significant improvement in case of a severe channel fading (Nakagami m = 1). The observed behavior of the probability of beacon reception without the use of D-FPAV can be explained as follows: the greater variation of the received signal strengths in severe conditions causes a reduced number of interferences from far distances and therefore the chance of successfully decoding a message increases for close distances and decreases for far distances. Since D-FPAV has already reduced the interference from far distances by adjusting the transmit powers, a greater variation of the received signal strengths does not improve the performance of beacons from close distances and rather reduces the chance of successfully decoding a message. Indeed, interference and resulting collisions are not the dominating factor anymore.

In addition, the probability of reception at close distances under severe fading conditions shows almost the same performance for D-FPAV on and off. Obviously, an intense fading (or path loss) in signal propagation has the same effect as a transmit power adjustment and provides comparable results with respect to the probability of successful message reception. Severe fading can therefore be seen as nature's approach of how to perform congestion control in VANETs.

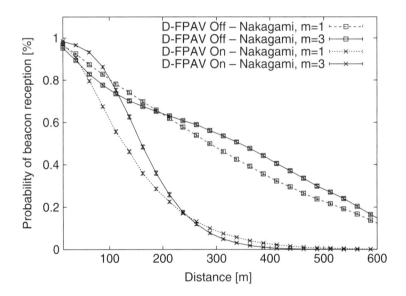

Figure 7.16 Probability of successful beacon reception with respect to the distance to the transmitter for D-FPAV on and off and different fading intensities (Nakagami m = 1 reflecting severe and Nakagami m = 3 reflecting medium fading conditions).

A different approach regarding transmit power control has been followed by Khorakhun et al. (2008) who suggest adjusting transmit power with respect to a network-wide desired channel busy time limit. The channel busy time is basically the fraction of time in which the channel, from the perspective of a single vehicle, is not idle, i.e. the vehicle is either sending/receiving packets or detecting an energy level above the clear channel assessment threshold of 802.11 MAC. Compared to the channel beacon load metric used by Torrent-Moreno et al. the channel busy time is not able to reveal the number of nodes sharing the channel or the applied priorities for periodic and event-driven messages. However, in their approach, each vehicle monitors its own channel busy time and either increases or decreases the transmit power by one step, depending on whether the measured channel busy time is above or below the desired threshold. In order to obtain a smoother and more stable transmit power adaptation among vehicles, i.e. neighboring vehicles selecting similar transmit power levels, the authors introduce a feature called *average transmit power*. With this feature enabled, every vehicle sets its determined transmit power level inside every packet header and calculates the average transmit power of its neighborhood using the indications of surrounding vehicles. Then, every time a vehicle wants to increase its transmit power by one step, it checks whether its own transmit power level is below the average power level or not. If its own transmit power level is above the average power level, it defers the increase.

The approach by Khorakhun et al. has been evaluated in a simulation study, for which a 4 km long highway with two lanes per direction, 16 vehicles/km per lane and an average vehicle velocity of 60 km/h has been set up. Further, the vehicles have been configured to transmit 1000 byte packets once every second using a data rate of 1 Mbps and a maximum transmission power of 15 dBm. Even though this configuration is not in line with current assumptions, e.g. a too-low packet generation rate, too-low data rate or too-large packet sizes, the authors could show that their approach is able to keep the channel busy time as experienced by each individual vehicle at a predefined desired level.

Compared to D-FPAV, which, based on an estimation, adjusts the transmit power in advance and thereby avoids congested channel conditions, the approach by Khorakhun et al. is based on a feedback-loop principle and is only able to react to changing channel conditions. It, therefore, cannot avoid congested situations. As stated by the authors, if the increase or decrease by one power step is very small, the transmit power adjustment might take several iterations before the desired channel busy time is achieved. By contrast, the use of large power steps reduces the number of iterations needed and thereby the reaction time, but, at the same time, increases the possibility of ending up with an unstable system oscillating around the desired channel busy time level. In addition, their approach does not distinguish between periodic beacons and event-driven messages and thus is not able to prioritize event-driven emergency messages in terms of higher transmit powers without affecting the feedback loop and thereby the transmit powers used for beacons.

7.5.3 Congestion control by means of rate control

Similar to the adjustment of transmit power levels it is possible to reduce the load on, or the congestion of, the wireless communication channel by adjusting the packet generation rate of periodic beacon messages. While a high beacon generation rate will increase the update rate of the received information, which might be necessary for safety applications if vehicles move at high velocities such as 30 m/s or more, a low beacon generation rate will increase the latency of the information, which might be higher than the interval between two successive transmissions owing to packet losses. The optimal beacon generation rate is therefore subject to the required information accuracy of the applications.

Without having safety in mind, Xu and Barth (2004) proposed an adaptive interval control broadcast protocol for traffic information distribution in VANETs, in which the beaconing rate is adjusted according to the current velocity of the vehicle, the transmission attempt failure and the packet reception success rate. Their semi-dynamic approach, which uses a simple lookup table, increases the transmission interval whenever the vehicle slows down sufficiently and whenever a maximum transmission attempt failure rate or a minimum packet reception rate is observed. However, whether and how statistics such as the reception success rate can be obtained in a fully distributed system has not been discussed.

Though ElBatt et al. (2006) are not explicitly proposing a strategy for controlling the beacon generate rate, they provide a detailed study of the broadcast reception performance and beacon inter-arrival time characteristics of different beacon generation rates. They simulated a straight 1-mile-long highway scenario with four lanes per direction and an inter-vehicle spacing of 10 m to investigate the suitability of inter-vehicle communication for the application of cooperative collision warning. In their simulations all vehicles were transmitting 100 byte beacons using a transmit power equal to 150 m communication range and a fixed beacon interval, which varied between 50 ms and 700 ms from run to run. The communication was fixed at 6 Mbps and the wireless channel was configured to account for path loss and fading. As one would expect, their results indicate that the probability of successful packet reception increases with a decreasing beacon generation rate, e.g. when moving from a 50 ms to a 700 ms beacon interval. However, the observed inter-arrival time of beacons is lowest for the 100 ms beacon interval. According to their explanation, a very short interval time of 50 ms lead to an overloaded channel and resulted in consecutive packet losses, which were dominating the inter-arrival time. With an increased beacon interval, packet losses due to collisions were reduced and inter-arrival times were dominated by the configured beacon interval. ElBatt et al. conclude that an optimal inter-arrival time depends on the joint consideration of consecutive packet losses and beacon generation rates.

An approach that adjusts the beacon generation rate with respect to accuracy requirements has been suggested by Rezaei et al. (2007). The authors propose a strategy in which every vehicle runs a so-called *self estimator* to estimate the position, speed and heading of the vehicle. The estimator actually uses an extended Kalman filter and integrates differential GPS positions and vehicle sensor information. In addition, each vehicle runs several *neighbor estimators* to estimate the position of each surrounding vehicle, using the information received through beacons, and a *remote estimator* to mimic the estimation of its own position from the perspective of neighboring vehicles. For this to work, the remote estimator is fed with the same information as the neighbor estimators of surrounding vehicles, i.e. the information contained in its own beacon messages. The objective is then to transmit beacons only when the difference between the calculations of the self estimator and the remote estimator exceeds a maximum deviation threshold, which can happen when the vehicle is changing direction or speed. However, in their studies the authors assume that the wireless channel has no loss and only a negligible delay.

Using the locally measured channel busy time as an indicator, Khorakhun et al. (2008) have proposed a feedback-loop-based strategy to decide whether to increase or decrease the transmission rate. Similar to their transmit power control approach described in Section 7.5.2, a vehicle increases its beacon generation rate whenever the channel busy time is below a desired channel busy time threshold and decreases the beacon generation rate if it is above. Furthermore, the average transmission rate is calculated by the exchange of current generation rate values used by the vehicles. By a comparison of the average transmission rate with their own rate and permitting an increase of their own rate only when it is below the average, the algorithm is then able to provide a fair rate adaptation among all nodes. However, as mentioned already in Section 7.5.2, their approach is not able to strictly differentiate between periodic beacons and event-driven emergency messages.

7.6 Open Issues and Outlook

To identify open issues and challenges related to MAC and scalability aspects, it is helpful to distinguish between fundamental and 'accidental/incidental', i.e., regulatory or technology-related, issues. As a fundamental issue one can observe the growing number of radios per unit area. We have assumed in this chapter that every motor vehicle is equipped with a radio, but in the future roadside sensors and pedestrians might also need to be considered as well. To obtain the required scalability, transmit power and rate control, as well as the joint optimization of both power and rate appear to be essential building blocks. Thus, the aspect of medium access is not strictly a layer-two issue but a cross-layer issue. An underlying challenge, the issue of dealing with the hidden terminal problem of decentralized MAC approaches and/or the efficient use of the available bandwidth will remain.

Results of performance evaluations in general cannot simply be taken as 'the truth' but depend on the chosen models and assumptions. Appropriate models to assess VANETs could change over time depending, for instance, on

- regulatory aspects and multi-channel issues
- insights into the wireless channels experienced in VANETs
- antenna, transmitter, and receiver technology used
- software-defined radio techniques
- security and privacy mechanisms.

Since VANETs are not reality yet, model validation is based on 'face validation', i.e., the agreement of experts on the validity of assumptions. A considerable amount of effort has been spent on exactly this issue of determining suitable models and parameters. Still, the proper modeling of interfacing layers zero, one, and two or the level of security provided in each beacon message might considerably evolve in the near future as a result of experiences gained by field operational tests and improved simulator frameworks like the one provided by NS-3.

Under the given model assumptions, the performance evaluation of IEEE 802.11p clearly shows that, as is typical for wireless local area networks based on carrier sense multiple access, the corresponding MAC approach has difficulties in dealing with stressed channel conditions. On the other side, it should be noted that over a wide range of conditions, IEEE 802.11p performs well. Since a safety-critical system has to operate well in basically all possible situations, further work can be expected that addresses the issue of stability and robustness under stressed conditions. Furthermore, consolidation efforts are required in order to provide performance comparisons of the various proposed approaches.

Bugs can occur in simulation code as well as in real world deployments. Therefore, analytical results for the probability of packet reception and reception delay would be beneficial for a baseline reference as well as for – potentially online – optimization of configurations. As surveyed in this chapter, there exist various results of analytical models that differ in the degree of realism, and those approaches require a significant amount of further adjustments to be able to accurately predict the performance of VANETs.

Finally, to assess the quality of the communication system, the quality of experience with respect to the target application should be ultimately used instead of a communication-centric quality of service. Thus, again, VANETs themselves prove to be intrinsically a cross-layer issue. To allow for time-efficient quality-of-experience-focused simulations, we presented an approach to build an empirical model for probability of packet reception. In the future, the empirical model could be replaced by an analytical one when those become available.

Acknowledgments

We acknowledge the support of the German Research Foundation (DFG) through the Research Training Groups 895 'Information Management and Market Engineering' and 1194 'Self-Organizing Sensor-Actuator-Networks'; the German Ministry of Education and Research (BMBF) and NEC Deutschland GmbH for the 'Network on Wheels' project; the Ministry of Science, Research and the Arts of Baden-Württemberg (Az: Zu 33-827.377/19,20), Klaus Tschira Stiftung, INIT GmbH, and PTV AG for support of the junior research group on Traffic Telematics; the PRE-DRIVE C2X (PREparation for DRIVing implementation and Evaluation of C-2-X communication technology) project which is funded under the Seventh Framework Programme of the European Commission.

We would also like to acknowledge the use of the HP XC4000 high performance computing system of the state of Baden-Württemberg, Germany, operated at the Steinbuch Centre for Computing (SCC), Karlsruhe Institute of Technology (KIT), for the simulation runs presented in this chapter.

References

Alexander P, Haley D and Grant A 2007 Outdoor Mobile Broadband Access with 802.11. *IEEE Communications Magazine* **45**(11), pp. 108–114.

Artimy M, Robertson W and Phillips W 2005 Assignment of Dynamic Transmission Range Based on Estimation of Vehicle Density. *Proceedings of the 2nd ACM International Workshop on Vehicular Ad Hoc Networks (VANET)*, Cologne, Germany, pp. 40–48.

ASTM 2003 ASTM E2213-03 Standard Specification for Telecommunications and Information Exchange Between Roadside and Vehicle Systems.

Bana S and Varaiya P 2001 Space Division Multiple Access (SDMA) for Robust Ad-hoc Vehicle Communication Networks. *Proceedings of the 4th IEEE Intelligent Transportation Systems*, San Francisco, USA, pp. 962–967.

Bertsekas D and Gallager R 1987 *Data Networks*. Prentice Hall.

Bianchi G 2000 Performance Analysis of the IEEE 802.11 Distributed Coordination Function. *Journal on Selected Areas in Communications* **18**(3), pp. 535–547.

Bilstrup K, Uhlemann E, Ström EG and Bilstrup U 2008 On the ability of the 802.11p MAC method and STDMA to support Real-Time Vehicle-to-Vehicle communication. *EURASIP Journal on Wireless Communications and Networking*.

Blum J, Eskandarian A and Hoffman L 2004 Challenges of Inter-Vehicle Ad Hoc Networks. *IEEE Transaction on Intelligent Transportation Systems* **5**, pp. 347–351.

Board TR 2000 *Highway Capacity Manual*. National Academy of Science.

Briesemeister L and Schäfers L 2000 Disseminating Messages Among Highly Mobile Hosts Based on Inter-Vehicle Communication. *Proceedings of the IEEE Intelligent Vehicles Symposium*, Dearborn, USA, pp. 522–527.

Caizzone G, Giacomazzi P, Musumeci L and Verticale G 2005 A Power Control Algorithm with High Channel Availability for Vehicular Ad-hoc Networks. *Proceedings of the 40th IEEE International Conference on Communications (ICC)*, Seoul, Korea, pp. 3171–3176.

Chatzimisios P, Boucouvalas A and Vitsas V 2004 Performance Analysis of IEEE 802.11 DCF in Presence of Transmission Errors. *Proceedings of the 39th IEEE International Conference on Communications*, Paris, France, pp. 3854–3858.

Chen Q, Schmidt-Eisenlohr F, Jiang D, Torrent-Moreno M, Delgrossi L and Hartenstein H 2007 Overhaul of IEEE 802.11 Modeling and Simulation in NS-2. *Proceedings of the 10th ACM Symposium on Modeling, Analysis, and Simulation of Wireless and Mobile Systems (MSWiM '07)*, New York, USA, pp. 159–168.

Cheng L, Henty B, Cooper R, Stancil D and Bai F 2008 Multi-Path Propagation Measurements for Vehicular Networks at 5.9 GHz. *Proceedings of the IEEE Wireless Communications and Networking Conference (WCNC)*, Las Vegas, USA, pp. 1239–1244.

Cheng L, Henty B, Stancil D, Bai F and Mudalige P 2007 Mobile Vehicle-to-Vehicle Narrow-Band Channel Measurement and Characterization of the 5.9 GHz Dedicated Short Range Communication (DSRC) Frequency Band. *IEEE Journal on Selected Areas in Communications* **25**(8), pp. 1501–1516.

Crowther W, Rettberg R, Walden D, Ornstein S and Heart F 1981 A System for Broadcast Communications: Reservation-ALOHA. *Proceedings of the 6th Hawaii International Conference on System Sciences*, Honolulu, USA, pp. 371–374.

Duffy K, Malone D and Leith D 2005 Modeling the 802.11 Distributed Coordination Function in Non-Saturated Conditions. *IEEE Communications Letters* **9**(8), pp. 715–717.

Ekici O and Yongacoglu A 2008 IEEE 802.11a Throughput Performance with Hidden Nodes. *IEEE Communication Letters* **12**(6), 465–467.

ElBatt T, Goel SK, Holland G, Krishnan H and Parikh J 2006 Cooperative Collision Warning using Dedicated Short Range Wireless Communications. *Proceedings of the 3rd International Workshop on Vehicular Ad Hoc Networks (VANET)*, New York, USA, pp. 1–9.

Engelstad PE and Østerbø ON 2005 Non-Saturation and Saturation Analysis of IEEE 802.11e EDCA with Starvation Prediction. *Proceedings of the 8th ACM International Symposium on Modeling, Analysis and Simulation of Wireless and Mobile Systems (MSWiM '05)*, New York, USA, pp. 224–233.

FCC 2004 Federal Communications Commission. FCC 03-324. FCC Report and Order.

Gast M and Loukides M 2005 *802.11 Wireless Networks – The Definitive Guide*, 2nd edn, O'Reilly.

Gummalla ACV and Limb JO 2000 Wireless Medium Access Control Protocols. *IEEE Communications Surveys and Tutorials*.

Hou TC, Tsao LF and Liu HC 2003 Analyzing the Throughput of IEEE 802.11 DCF Scheme with Hidden Nodes. *Proceedings of the 58th IEEE Vehicular Technology Conference (VTC-Fall)*, Orlando, USA, pp. 2870–2874.

IEEE 2006 IEEE 1609 Trial-Use Standard for Wireless Access in Vehicular Environments (WAVE).

IEEE 2007 IEEE 802.11-2007, Wireless LAN Medium Access Control (MAC) and Physical Layer (PHY) Specifications.

IEEE 2008 IEEE 802.11p/D5.0, Draft Amendment for Wireless Access in Vehicular Environments (WAVE).

ITU 1998 *Technical Characteristics for a Universal Shipborne Automatic Identification System Using Time Division Multiple Access in the Maritime Mobile Band*.

Jiang D and Delgrossi L 2008 IEEE 802.11p: Towards an International Standard for Wireless Access in Vehicular Environments. *Proceedings of the 67th IEEE Vehicular Technology Conference (VTC-Spring)*, Marina Bay, Singapore, pp. 2036–2040.

Jiang D, Chen Q and Delgrossi L 2007 Communication Density: A Channel Load Metric for Vehicular Communications Research. *Proceedings of the 4th IEEE International Conference on Mobile Ad-hoc and Sensor Systems (MASS)*, Pisa, Italy, pp. 1–8.

Jiang D, Chen Q and Delgrossi L 2008 Optimal Data Rate Selection for Vehicle Safety Communications. *Proceedings of the Fifth ACM International Workshop on VehiculAr Inter-NETworking (VANET)*, San Francisco, USA, pp. 30–38.

Katragadda S, Ganesh Murthy C, Ranga Rao M, Mohan Kumar S and Sachin R 2003 A Decentralized Location-Based Channel Access Protocol for Inter-Vehicle Communication. *Proceedings of the 57th IEEE Vehicular Technology Conference (VTC-Spring)*, Jeju, Korea, pp. 1831–1835.

Kawadia V and Kumar P 2005 Principles and Protocols for Power Control in Wireless Ad Hoc Networks. *IEEE Journal on Selected Areas in Communications (JSAC)*.

Khorakhun C, Busche H and Rohling H 2008 Congestion Control for VANETs based on Power or Rate Adaptation. *Proceedings of the 5th International Workshop on Intelligent Transportation (WIT)*, Hamburg, Germany.

Killat M and Hartenstein H 2009 An Empirical Model for Probability of Packet Reception in Vehicular Ad Hoc Networks. *EURASIP Journal on Wireless Communications and Networking*, Article ID 721301, 12.

Killat M, Schmidt-Eisenlohr F, Hartenstein H, Rössel C, Vortisch P, Assenmacher S and Busch F 2007 Enabling Efficient and Accurate Large-Scale Simulations of VANETs for Vehicular Traffic Management. *Proceedings of the Fourth ACM International Workshop on Vehicular Ad Hoc Networks (VANET)*, Montreal, Canada, pp. 29–38.

Kosch T, Adler C, Eichler S, Schroth C and Strassberger M 2006 The Scalability Problem of Vehicular Ad Hoc Networks and How to Solve It. *IEEE Wireless Communications* **13**(5), 22–28.

Kumar S, Raghavan VS and Deng J 2006 Medium Access Control Protocols for Ad-hoc Wireless Networks: A Survey. *Elsevier Ad-Hoc Networks Journal* **4**, 326–358.

Li X and Zeng QA 2006 Capture Effect in the IEEE 802.11 WLANs with Rayleigh Fading, Shadowing, and Path Loss. *Proceedings of the 2nd IEEE International Conference on Wireless and Mobile Computing, Networking and Communications (WiMob)*, Montreal, Canada, pp. 110–115. USA.

Liu TK, Silvester J and Polydoros A 1995 Performance Evaluation of R-ALOHA in Distributed Packet Radio Networks with Hard Real-Time Communications. *Proceedings of the 45th IEEE Vehicular Technology Conference (VTC)*, Chicago, USA, pp. 554–558.

Lott M, Halfmann R, Schultz E and Radimirsch M 2001 Medium Access and Radio Resource Management for Ad Hoc Networks Based on UTRA TDD. *Proceedings of the 2nd ACM International Symposium on Mobile Ad Hoc Networking & Computing (MobiHoc)*, New York, USA, pp. 76–86.

Ma X, Hrubik P, Refai H and Yang S 2005 Capture Effect on R-ALOHA Protocol for Inter-Vehicle Communications. *Proceedings of the 62nd IEEE Vehicular Technology Conference (VTC-Fall)*, Dallas, USA, pp. 2547–2550.

Mann A and Rückert J 1988 A New Concurrent Slot Assignment Protocol for Traffic Information Exchange. *Proceedings of the 38th IEEE Vehicular Technology Conference (VTC)*, pp. 503–508.

Maurer J, Fügen T and Wiesbeck W 2005a Physical Layer Simulations of IEEE802.11a for Vehicle-to-Vehicle Communications. *Proceedings of the 62nd IEEE Vehicular Technology Conference (VTC-Fall)*, Dallas, USA, pp. 1849–1853.

Maurer J, Fügen T and Wiesbeck W 2005b System Simulations Based on IEEE 802.11a for Inter-Vehicle Communications Using a Realistic Channel Model. *Proceedings of the International Workshop on Intelligent Transportation (WIT)*, Hamburg, Germany.

Nagaosa T and Hasegawa T 1998 Code Assignment and the Multicode Sense Scheme in an Inter-Vehicle CDMA Communication Network (Special Section on Spread Spectrum Techniques and Applications). *IEICE Transactions on Fundamentals of Electronics, Communications and Computer Sciences* **81**(11), 2327–2333.

NS-2 2008 Network Simulator NS-2, version 2.33, http://www.isi.edu/nsnam.

Pham PP 2005 Comprehensive Analysis of the IEEE 802.11. *Mobile Networks and Applications* **10**(5), 691–703.

Raya M and Hubaux J 2005 The Security of Vehicular Ad Hoc Networks. *Proceedings of the ACM Workshop on Security of Ad Hoc and Sensor Networks (SASN)*, Alexandria, USA.

Reumerman HJ, Roggero M and Ruffini M 2005 The Application-Based Clustering Concept and Requirements for Intervehicle Networks. *IEEE Communications Magazine* **42**(4), 108–113.

Rezaei S, Sengupta R and Krishnan H 2007 Reducing the Communication Required by DSRC-Based Vehicle Safety Systems. *Proceedings of the 10th IEEE Intelligent Transportation Systems Conference (ITSC)*, Seattle, USA, pp. 361–366.

Schmidt-Eisenlohr F and Killat M 2008 Vehicle-to-Vehicle Communications: Reception and Interference of Safety-Critical Messages. *it – Information Technology* **50**(4), 230–236.

Taliwal V, Jiang D, Mangold H, Chen C and Sengupta R 2004 Empirical Determination of Channel Characteristics for DSRC Vehicle-to-Vehicle Communication. *Proceedings of the 1st ACM International Workshop on Vehicular Ad Hoc Networks (VANET)*, Philadelphia, USA, p. 88.

Tan INL, Tang W, Laberteaux K and Bahai A 2008 Measurement and Analysis of Wireless Channel Impairments in DSRC Vehicular Communications. Technical Report UCB/EECS-2008-33, EECS Department, University of California, Berkeley.

Tay YC and Chua KC 2001 A Capacity Analysis for the IEEE 802.11 MAC Protocol. *Wireless Networks* **7**(2), 159–171.

Torrent-Moreno M 2007 *Inter-Vehicle Communications : Achieving Safety in a Distributed Wireless Environment – Challenges, Systems and Protocols (PhD Thesis)*. Universitätsverlag Karlsruhe.

Torrent-Moreno M, Jiang D and Hartenstein H 2004 Broadcast Reception Rates and Effects of Priority Access in 802.11-Based Vehicular Ad-Hoc Networks. *Proceedings of the 1st ACM International Workshop on Vehicular Ad Hoc Networks (VANET)*, Philadelphia, USA, pp. 10–18.

Torrent-Moreno M, Killat M and Hartenstein H 2005a The Challenges of Robust Inter-Vehicle Communications. *Proceedings of the 62nd IEEE Semiannual Vehicular Technology Conference (VTC-Fall)*, Dallas, USA, pp. 319–323.

Torrent-Moreno M, Mittag J, Santi P and Hartenstein H 2009 Vehicle-to-Vehicle Communication: Fair Transmit Power Control for Safety-Critical Information. *IEEE Transactions on Vehicular Technology*, **58**(7), pp. 3684–3707.

Torrent-Moreno M, Santi P and Hartenstein H 2005b Fair Sharing of Bandwidth in VANETs. *Proceedings of the 2nd ACM International Workshop on Vehicular Ad Hoc Networks (VANET)*, Cologne, Germany, pp. 49–58.

Torrent-Moreno M, Santi P and Hartenstein H 2006 Distributed Fair Transmit Power Assignment for Vehicular Ad Hoc Networks. *Proceedings of the IEEE Conference on Sensor, Mesh and Ad Hoc Communications and Networks (SECON)*, Reston, USA, pp. 479–488.

Tsertou A and Laurenson DI 2006 Insights into the Hidden Node Problem. *Proceedings of the 2006 International Conference on Wireless Communications and Mobile Computing (IWCMC)*, Vancouver, Canada, pp. 767–772.

VSC 2004 Task 3 Final Report – Identify Intelligent Vehicle Safety Applications Enabled by DSRC Public Document, Vehicle Safety Communications Project, Crash Avoidance Metrics Partnership.

Wang Z and Hassan M 2008 How much of DSRC is Available for Non-Safety use? *Proceedings of the Fifth ACM International Workshop on VehiculAr Inter-NETworking (VANET)*, San Francisco, USA, pp. 23–29.

Wischhof L and Rohling H 2005 Congestion Control in Vehicular Ad Hoc Networks. *Proceedings of the IEEE International Conference on Vehicular Electronics and Safety*, Xi'an, China pp. 58–63.

Xu H and Barth M 2004 A Transmission-Interval and Power-Level Modulation Methodology for Optimizing Inter-Vehicle Communications. *Proceedings of the 1st ACM International Workshop on Vehicular Ad Hoc Networks (VANET)*, Philadelphia, USA, pp. 97–98.

Xu Q, Mak T, Ko J and Sengupta R 2004a Vehicle-to-Vehicle Safety Messaging in DSRC. *Proceedings of the 1st ACM International Workshop on Vehicular Ad Hoc Networks (VANET)*, Philadelphia, PA, USA, pp. 19–28.

Xu Q, Sengupta R and Jiang D 2004b Design and Analysis of Highway Safety Communication Protocol in 5.9 GHz Dedicated Short Range Communication Spectrum. *Proceedings of the IEEE 57th Vehicular Technology Conference (VTC-Spring)*, Jeju, Korea, pp. 2451–2455.

Yin J, Holland G, ElBatt T, Bai F and Krishnan H 2006 DSRC Channel Fading Analysis from Empirical Measurement. *Proceedings of the First International Conference on Communications and Networking in China (ChinaCom)*, Beijing, China, pp. 1–5.

Zhang Y, Tan I, Chun C, Laberteaux K and Bahai A 2008 A Differential OFDM Approach to Coherence Time Mitigation in DSRC. *Proceedings of the 5th ACM International Workshop on VehiculAr Inter-NETworking (VANET)*, San Francisco, USA, pp. 1–6.

Zhu W, Hellmich T and Walke B 1991 DCAP, A Decentral Channel Access Protocol: Performance Analysis. *Proceedings of the 41st IEEE Vehicular Technology Conference (VTC)*, St Louis, USA, pp. 463–468.

8

Efficient Application Level Message Coding and Composition

Craig L Robinson

University of Illinois at Urbana-Champaign, IL, USA

As we move toward the large-scale deployment of wireless inter-vehicle communication systems, we must consider some real-world implementation issues and tradeoffs. For example, demands on the wireless channel (such as bandwidth or latency) will evolve as more vehicles become equipped, and new and evolving applications generate new data and transmission requirements. Yet, in order to ensure that these systems are interoperable, a standardized communication interface is required. Thus, one of the challenges we face is the deployment of a standard system that efficiently uses the finite wireless medium and yet can support future application requirements.

> "Standardization can only go so far, because we can't anticipate all possible future needs." (Berners-Lee et al. 2001)

This chapter will describe some of the important considerations that must be made in developing a wireless safety communication system. We will work mainly at the application level. Special effort is made to highlight implementation issues and restrictions as well as looking at how overall system objectives can be met. The principal method presented here is called the 'message dispatcher' (MD) (Robinson et al. 2006, 2007), which has subsequently become known as the 'message handler' (SAE 2008). The approach is based on two principal concepts:

1. Improve channel utilization by recognizing similarities in transmitted data.[1]

2. Separate message construction and communication from the application functionality.[2]

These concepts, as well as some mechanisms for achieving them, are described in this chapter. In the following section we give an overview of the wireless intervehicle communication environment and highlight some desirable features which a system architecture in this environment should posses. We then proceed immediately to introducing the message dispatcher (MD) in Section 8.2. To illustrate the functionality of the MD, we then present an example application and some sample data sets in Sections 8.3 and 8.4. Section 8.5 contains an extension of the general MD concept which uses linear predictive coding to reduce communication bandwidth requirements. Finally, Section 8.6 discusses how the original system architecture objectives have been met by the MD architecture. This material has also been presented in Robinson et al. (2006, 2007).

8.1 Introduction to the Application Environment

This section describes cooperative vehicular safety applications and their communication requirements. Further, it describes some attractive features which a data exchange system for vehicular safety should possess.

8.1.1 Safety applications and data requirements

Significant efforts, involving the vehicle industry and government agencies, have been made to identify which communication-enabled vehicular safety applications will provide the greatest benefits. The deliberations by the US National Highway Traffic Safety Administration (NHTSA), the US Department of Transportation (USDOT), and the Vehicle Safety Communications Consortium (VSCC) have identified eight such applications (Carter 2005; USDOT 2006), shown in Table 8.1 along with their *proposed* communication requirements.[3] We highlight the fact that these requirements have only been proposed. There has, to the authors' knowledge, been no *thorough* investigation or large-scale deployment of the systems on which these requirements are based. As a consequence, any communication protocol must be sufficiently flexible to enable changing requirements.

Considering Table 8.1, it appears that applications with the greatest safety potential rely heavily on single-hop broadcast communication with nearby vehicles and infrastructure. The eventual vehicular wireless communication environment is expected to have devices from various manufacturers implementing distinct (although ideally cooperative) applications. Moreover, many vehicles are likely to run multiple safety applications concurrently (e.g., an emergency brake light application, a lane change warning application, an intersection collision warning

---

[1] In a communication theoretic framework this can be considered as a manifestation of source coding.
[2] From an network architecture perspective this can be considered as the addition of a 'data coordination' element to the application layer in the OSI reference model.
[3] Similar deliberations are underway in Europe and Asia, with similar results.

Table 8.1 Eight high-priority vehicular safety applications.

Application	Comm. type	Freq.	Latency	Data transmitted	Range
Traffic signal violation	I2V one-way, P2M	10 Hz	100 msec	Signal status, timing, surface heading, light posn., weather	250 m
Curve speed warning	I2V one-way, P2M	1 Hz	1000 msec	Curve location, curvature, speed limit, bank, surface	200 m
Emergency brake lights	Vehicle-to-vehicle two-way, P2M	10 Hz	100 msec	Position, deceleration heading, velocity	200 m
Pre-crash sensing	Vehicle-to-vehicle two-way, P2P	50 Hz	20 msec	Vehicle type, yaw rate, position, heading, accel.	50 m
Collision warning	Vehicle-to-vehicle one-way, P2M	10 Hz	100 msec	Vehicle type, position, heading Velocity, acceleration, yaw rate	150 m
Left turn assist	I2V and V2I one-way, P2M	10 Hz	100 msec	Signal status, timing, posn. direction, road geom., vel. heading	300 m
Lane change warning	Vehicle-to-vehicle one-way, P2M	10 Hz	100 msec	Position, heading, velocity Accel., turn signal status	150 m
Stop sign assist	I2V and V2I one-way	10 Hz	100 msec	Position, velocity heading, warning	300 m

These applications were identified by NHTSA and VSCC (USDOT 2006). Note that communication frequencies are in the range of 1–50 Hz and maximum communication range span 50–300 meters. P2P represents 'point-to-point', P2M represents 'point-to-multipoint', I2V represents 'infrastructure-to-vehicle' and V2I represents 'vehicle-to-infrastructure'.

application, etc.). Each application is likely to have different, although overlapping, data element requirements, i.e., many safety applications may require the vehicle speed, vehicle location, current turning radius, etc.

Responding to these identified applications, the Society of Automotive Engineers (SAE) defines over 150 vehicle 'data elements' in their dedicated short range communications (DSRC) message set dictionary (SAE 2008). Examples of vehicle data elements are latitude and longitude, heading, acceleration (with varying precision: 4-bit, 8-bit, 16-bit), headlight status, and brake status. Of these data elements, 30 of the most frequently used elements are selected as the basis of a 'common message set'. The common message set is intended to provide a standardized set of messages which applications running on each vehicle could use to communicate. In earlier revisions of the SAE standard (prior to 2006), the messages were all of fixed size, structure, and content.

This approach had the potential to generate packets with redundant data. Some data in a fixed message might not be used at all, whereas in other cases, individual applications may require a different message, even though some data elements are common to both messages. Hence multiple applications within a single vehicle might separately (and redundantly) send the vehicle's current speed in different messages. Further, information that changes slowly or infrequently (e.g., windscreen wiper status) need not be sent frequently.[4] In addition, very few of the applications described in Table 8.1 have actually been fully developed. Subsequently, the exact usage characteristics of the safety messages are in flux at the very time they are being defined and standardized. This makes the process difficult, as these choices are likely to be refined over time. This leads to the conclusion that a message with fixed contents will either be very large, or not be able to meet all application requirements.

These observations lead to the development of the message dispatcher. The MD was subsequently integrated into the concept of operation of the SAE standard (SAE 2008). A similar approach has also been used in the Cooperative Intersection Collision Avoidance System (CICAS) project within the Collision Avoidance Metrics Partnership (CAMP) (CAMP 2008).

8.1.2 Desirable architectural features

In addition to simply providing information useful for the defined applications, there are several goals which a system architecture for wireless inter-vehicle safety communication should satisfy. Some of the environmental considerations were highlighted in the previous section. In this section we discuss some of the goals that a suitable system architecture should attempt to achieve. The goals are driven both from a technical standpoint and a business perspective, in that real world deployment and proliferation considerations need to be made. The goals include:

Future-proof: Vehicles will broadcast data that is probably valuable for multiple surrounding vehicles with multiple safety applications. However, creating and testing these safety applications is an ongoing effort. Therefore, a scheme must

---

[4]This statement ignores a potential requirement that new neighbours of a vehicle be promptly updated with the vehicle's current state.

be backward compatible as well as future-proof to newly defined, evolving, or upgraded applications.

Flexibility: It seems likely that in the heterogeneous marketplace for vehicles, different vehicles will be running different subsets of safety applications. A scheme should be sufficiently flexible to account for this.

Extensible: It is conceivable that not all safety applications will be universally standardized. Hence, a mechanism for adding support for non-standardized data elements (e.g. proprietary to one or more manufacturers) is desirable.

Unified interface: From an implementation perspective, it is attractive to construct an architecture where policy and self-policing between various applications, within a single vehicle, can be managed in a single entity. Further, authentication and other security primitives would ideally be managed collectively across safety applications.

Layered architecture: By providing a layered architecture that abstracts the message sending interface from the application designer, a separation of concerns for the application designer is achieved. This enables easier, faster and more modular development.

Low bandwidth usage: The available bandwidth is a finite resource and should be conserved wherever possible. Real-world testing by the VSCC (USDOT 2006) demonstrates that the channel capacity is an issue that will need to be addressed for large-scale deployment and in heavy traffic environments.

Information rate: Some data elements, such as headlight status, change infrequently. Thus, a solution should distinguish these properties and transmit information only when it is appropriate.

Recognize vehicle capabilities: Not all vehicles will be able (or willing) to measure and transmit certain pieces of information. This should be reflected in the message construction.

Enable product differentiation: Vehicle manufacturers desire the ability to provide unique applications and services to their customers. The functionality of these services should not be limited to the applications that are currently deployed or enabled by other vendors.

As described in the following sections, the message dispatcher architecture provides a sound and efficient architecture for the envisioned vehicular safety data exchange environment.

8.1.3 Broadcast characteristics

As illustrated in Table 8.1, safety messages tend to be locally broadcast with a maximum transmission range of 300 meters; messages sent by a vehicle will contain data elements useful to multiple vehicles in the nearby vicinity. As DSRC radios are required to communicate at least 300 meters, we assume that safety messages broadcast their messages in a single hop.[5]

As a result of the highly dynamic vehicular environment, it is likely that the immediate neighbors of a vehicle will change frequently. It is probably unnecessary,

---

[5]Extensions, such as dynamic power control (Kawadia and Kumar 2005) and geographical flooding (Ko and Vaidya 1999) are also possible. The message dispatcher concept readily extends to these situations.

and more difficult, to maintain an updated topology of 1–3-hop neighbors. Again, one-hop broadcasts seem appropriate. Further, coordinating transmission between vehicles will be difficult. Packet interference and loss seem likely. One method for dealing with this is to reduce channel load and use the channel infrequently, which is achievable through the message dispatcher architecture.

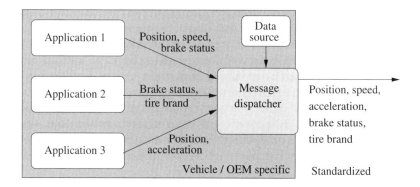

Figure 8.1 The message dispatcher assimilates data requirements from all the on-board applications and compiles a single message using a dictionary of defined data elements and standardized message construction guidelines. Some of the data may be obtained directly by the MD from an onboard data source (e.g. CAN bus).

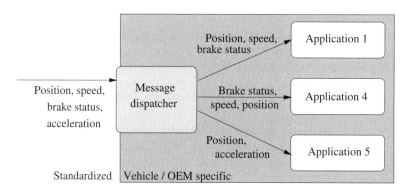

Figure 8.2 A receiving message dispatcher is responsible for separating and disseminating data elements from the received message to all on-board applications, as well as managing data requirements for surrounding vehicles.

8.2 Message Dispatcher

The basic architectural concept of the message dispatcher (MD) is illustrated in Figures 8.1 and 8.2. The Message Dispatcher's responsibility is to coordinate all

APPLICATION LEVEL MESSAGE CODING AND COMPOSITION

the data exchange requirements of the applications running on a vehicle. The MD accomplishes this by serving as an interface between the application layer and the communication stack.

Safety applications will register or send data elements to be broadcast to the MD. The MD then summarizes these data elements across applications and creates a single packet comprising the minimum set of the data elements to be transmitted (see Figure 8.1). In some implementations the MD might collect data from other data sources within the vehicle (e.g. through the CAN bus). The MD would also consider data requirements of other surrounding vehicles or roadside units, as described in Section 8.2.3. This combined message is then sent to the DSRC radio for broadcast. Any vehicle that receives a message would provide all on-board applications with the data elements they require, as shown in Figure 8.2.

The message dispatcher design can be divided into two broad topics. First, the definition of a data element dictionary (Section 8.2.1). Second, the specification of how these elements should be combined into a message (Section 8.2.2).

8.2.1 Data element dictionary

This section describes how data elements may be identified and formatted in a 'data element dictionary'. Each element in the dictionary is defined using the fields indicated in the example in Table 8.2.

Table 8.2 Vehicle latitude example data element.

Name	DE_VehicleLatitude
Unique ID	70
Unit	microdegrees
Accuracy	LSB is 1 microdegree
Range	−900000000 to 900000000
Size	32 bits
Description	The latitude position of the center of the vehicle, expressed in micro degrees and based on the WGS-84 coordinate system

Using this data dictionary, a message can be constructed by creating a string of unique identifiers followed by the value of the data element. Further, this unique ID overhead can be reduced when related data elements are grouped into a 'data frame'. For example, latitude is frequently updated and transmitted together with longitude. Overhead is reduced by formatting latitude and longitude into a single 'position' data frame with a single ID. Each data frame consists of data elements in a specified order and thus their unique IDs are not required within the data frame.

The SAE standard (SAE 2008) identifies over 150 data elements in its dictionary. Accordingly, several data frames have also been defined. The CICAS-V project uses a similar approach for data element definition (CAMP 2008).

Adding or modifying a data element under this architecture is relatively straightforward. The data dictionary would need to be updated and resubmitted to a central authority for updating the standard. While waiting for the standards body to act, new elements can be introduced by a lightweight tagging scheme, which is discussed in Section 8.2.2.

8.2.2 Message construction

Each message is constructed using the data elements and data frames specified by the data dictionary. The message dispatcher can choose to include (either in a frame or as an individual data element) elements in a particular message so as to meet latency, network loading or application demands. Thus, the process for constructing a message amounts to including data elements and frames prefixed by their unique identifier. A receiving MD can interpret this message based on the contents of its own data dictionary. In addition, the definition of an escape character can indicate the start of a data element, thus enabling transmission of variable length data, as well as enabling an out-of-date data dictionary to still interpret incoming data. The escape character will immediately be followed by a unique tag, of fixed length, that identifies the subsequent data. Using the tag, the message dispatcher would poll the subscribed applications for knowledge of the incoming data element.

As an example implementation of this message construction schema, the SAE standard (SAE 2008) divides messages into two sections. The first section is used to include data frames using their unique identifier followed by the series of data elements comprising the data frame. The second section is used to include individual data elements that have not already been included in the first section. Similarly, the CICAS-V project uses an efficient binary XML schema to construct messages (CAMP 2008) except that the data element size can be included in the message immediately after the ID.

8.2.3 What and when to send

Determining what elements are required to be sent by a vehicle in order to satisfy *surrounding* vehicles is a general problem in all mobile ad-hoc networks. To the authors' knowledge, there has been no robust methods proposed to obtain specific information from a newly encountered vehicle. Sending a large packet, comprising all defined data elements, at the maximum required rate among all elements is very inefficient. While not the main contribution of this chapter, we propose two possible solutions so as to illustrate the utility of the MD architecture.

- *Match received message*: If a message dispatcher receives a message with a data element it is not currently transmitting, it should include its own version of that data element in the following transmission. In this way it 'matches' the incoming message. To avoid a 'racing' situation shown in Figure 8.3, the original sender could include a data element indicating whether its message contents should be matched.

- *Request data elements*: Define and send a Data Element that specifically requests certain elements. This solution would be useful in probe-type applications where a roadside unit requests information from passing vehicles.

Figure 8.3   Vehicle A sends a message that the message dispatcher on vehicle B matches on subsequent transmissions. Although Vehicle C is beyond the range of interest of vehicle A, it too begins to match the message resulting in a 'racing' condition.

8.3   Example Applications

This section illustrates how the MD concept can be used to efficiently construct messages for multiple safety applications. Two safety applications are presented here: the Emergency Brake Warning (EBW) application and the Intersection Violation Warning (IVW) application. It is likely that all DSRC-equipped vehicles will send out a heartbeat message (HBM) comprising a minimal set of data elements and frames, as shown in Column 2 of Table 8.3. We will look at ways to reduce the heartbeat message channel load in a later section.

8.3.1   Emergency brake warning

The Emergency Brake Warning (EBW) application alerts the driver when a preceding vehicle performs a severe braking maneuver, as shown in Figures 8.4 and 8.9.

Method of operation: Consider Figure 8.4. When vehicle 1 performs severe braking, a request is passed to the MD to begin transmitting the data elements in Column 3 of Table 8.3 at the indicated frequency. Note that the data with ID *AH:EBWBreadcrumb*, which represents a sampled-path history of the vehicle, has not been defined in the data dictionary. It is appended to the message using the schema described in Section 8.2.2. On reception, vehicle 3 determines whether the braking car is in its forward path using the *AH:EBWBreadcrumb* data element. The vehicle can then take appropriate action. The path history is only used when a severe-braking event occurs and need not be sent until it is needed. A series of pictures illustrating the functionality of EBW are included in Figure 8.9.

8.3.2   Intersection violation warning

The Intersection Violation Warning (IVW) application warns the driver if violating a red light seems imminent – see Figure 8.5. Vehicles approaching the intersection are

Table 8.3 Sample application data element requirements.

Data element (DE) / frame (DF)	HBM	EBW	IVW	Message dispatcher	3 Hz	5 Hz	10 Hz
DF: PositionShort	3 Hz	5 Hz	10 Hz	In DE:PositionLong	•		
DF: AccelerationSet4Way	3 Hz	–	5 Hz	5 Hz	•	•	
DF: PositionLong	–	5 Hz	10 Hz	10 Hz	•	•	
DF: PositionConfidenceSet	–	3 Hz	3 Hz	3 Hz	•		
DF: SpeedandHeadingPrecision	3 Hz	5 Hz	5 Hz	5 Hz	•	•	
DE: Acceleration	–	3 Hz	–	In DF:AccelerationSet4Way			
DE: AntiLockBrakeStatus	–	3 Hz	–	3 Hz	•		
DE: BrakeAppliedStatus	3 Hz	3 Hz	–	3 Hz	•		
AH: EBWBreadcrumb	–	5 Hz	–	5 Hz and to EBW		•	
DE: TrafficLightID	–	–	–	To IVW			
DF: TrafficLightLocation	–	–	–	To IVW			
DF: TrafficLightPhases	–	–	–	To IVW			
AH: IVWMap	–	–	–	To IVW			
DE: IVWWarningFlag	–	–	10 Hz	10 Hz	•	•	•
DE: IVWWarningID	–	–	10 Hz	10 Hz		•	•
DE: IVWWarningVehPos	–	–	10 Hz	In DF:PositionLong			
...	...	...	...	...	...	...	...

A sample of the data elements (DE) and frames (DF) required by the two safety applications (emergency brake warning (EBW) and intersection violation warning (IVW)), as well as the heartbeat message (HBM). Column 5 represents the transmit requirements that the message dispatcher must satisfy when both applications are active as well as to which application an incoming data element should be routed. Columns 6, 7 and 8 represent the contents of a 3, 5 and 10 Hz message as described in Section 8.3.3.

APPLICATION LEVEL MESSAGE CODING AND COMPOSITION

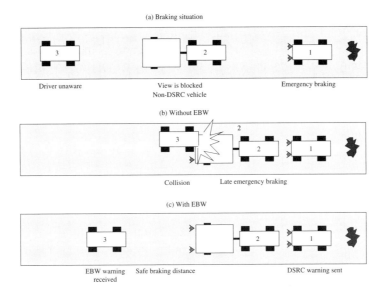

Figure 8.4 Brake lights are often difficult to see if there is a blocking vehicle. EBW provides a brake warning by using wireless communication to allow safe stopping.

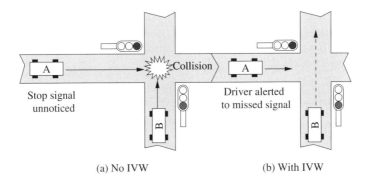

Figure 8.5 Without IVW vehicle A runs the light and causes a collision with vehicle B. When IVW is activated in Figure (b), both drivers are alerted allowing vehicle A to stop and vehicle B to proceed cautiously through the intersection.

also warned if an approaching vehicle has issued a warning. As shown in column 4 in Table 8.3, many of the required data elements are similar to the EBW application, although they may have different frequency requirements. Some data elements are unique to the IVW application and have not yet been defined in the data dictionary (e.g. *AH:IVWMap*).

Method of operation: Roadside units transmit traffic light information including its location, light status, time until color change, dimensions of intersection (called data element *AH:IVWMap*), etc. The IVW application registers with the MD to receive all incoming IVW related Data Elements. Vehicles then determine if a signal violation is imminent. If so, the driver is alerted and a message is sent to the traffic light and surrounding vehicles indicating that a violation is likely. Thus, the message dispatcher only sends IVW data when triggered by a violation event.

8.3.3 Message composition

Consider a vehicle that is speeding toward a red light. The driver has been alerted to a potential violation and is braking sharply. Thus, both the IVW and EBW systems are active. Table 8.3 lists a *subset* of the data elements and transmit frequencies which the message dispatcher must satisfy.

The message dispatcher combines the required data frames and data elements into a minimal set of messages. Duplicates are ignored, such as with *DE:Acceleration* and *DE:IVWWarningVehPos* in Table 8.3. Another example is data element *DF:PositionShort* which is not included in outgoing messages as it is a subset of the information already included in *DF:PositionLong*.

Since the applications have registered data at different frequencies, the MD constructs three different messages, called Msg10Hz, Msg5Hz, and Msg3Hz. The message contents are described in the final three columns of Table 8.3. Note that the data elements in the 10 Hz message also appear in the 5 Hz and 3 Hz messages. To meet the frequency requirements and minimize the utilized bandwidth, the MD will send the messages in the sequence illustrated in Table 8.4. When considering the table, note (A) indicates a time when the Msg10Hz is not sent since it is contained in Msg5Hz, and (B) a time when the Msg5Hz and Msg10Hz are not sent since they are both contained into Msg3Hz.

Table 8.4 Message dispatcher transmission timeline for the example in Table 8.3 and Section 8.3.3.

Time (sec)	Message sent
0.0	Msg3Hz
0.1	Msg10Hz
0.2	Msg5Hz (A)
0.3	Msg10Hz
0.4	Msg3Hz (B)
0.5	Msg10Hz
0.6	Msg5Hz
0.7	Msg10Hz
0.8	Msg5Hz
0.9	Msg3Hz

8.3.4 Implementation

The message dispatcher has been implemented by the Toyota Technical Center in two Toyota Prius cars. Each vehicle is retrofitted with a Linux-based miniature PC, an OBD-II vehicle interface, a DENSO prototype DSRC radio, and a commercial DGPS unit (see Figure 8.9(f)).

The MD implementation uses a callback mechanism to interface with applications. Upon initialization, applications register with the MD those data elements that it will **provide** and those it wants to **receive** from the MD. Data elements may be both provided and received. During data element registration, the application supplies the MD with the callback method to be invoked when it is time to send the data element, or when the MD has received an updated data element over the channel.

At the end of the registration process, the application specifies the frequency of transmission and instructs the MD to begin a periodic transmission of these data elements. The MD is then responsible for the message composition detailed in Section 8.3.3, using the provided callback method to get the data element values. Conversely, the transmission of other registered data elements may be event-driven. In this case, when the event is triggered by the application logic, the application is responsible for invoking a 'Send Now' API in the MD. With both periodic and event-driven communications possible, the MD includes an internal scheduler to decide when to send periodic data elements. When interrupted by an event-triggered transmission, the MD scheduler may reschedule future periodic transmission times.

The Toyota Technical Center has successfully implemented the MD described above in approximately 1000 lines of code. It has been extensively tested while running the EBW and IVW applications simultaneously. For these applications, a traffic light and two vehicles each run their own MD.

8.3.5 Analysis

A basic evaluation of the performance of the MD for the two applications in the TTC implementation is now given. Using the data element sizes specified in revision 9 of SAE J2735 (SAE 2008), the heartbeat message is 25.5 bytes, emergency brake warning message is 155.5 bytes, and the intersection violation warning message is 46.75 bytes. Message sizes are calculated without including data frame headers, data element identifiers, or any tagging schema. A common message set (CMS) is one that incorporates all the data elements necessary for the HBM, EBW, and IVW is 176.75 bytes. This CMS architecture was used by the SAE prior to the adoption of the MD architecture. In the following discussion we shall use the CMS as a baseline for evaluating the improvement offered by the MD architecture.

Assume that this common message set (CMS) is periodically transmitted with the highest frequency in Table 8.3, 10 Hz, in order to meet the requirements of all the applications. For one vehicle, this requires a channel usage of 14.1 kbps, as shown in Table 8.5. Conversely, a HBM being sent out by the MD at the 3 Hz frequency specified in Table 8.3 requires only 0.6 kbps. This is a reduction of 95% of channel load when neither of the applications is required to transmit. Channel usage rises to

3.7 kbps when IVW events occur (initiating a 10 Hz transmission) and to 6.2 kbps when EBW events occur (initiating a 5 Hz transmission). If both IVW and EBW events occur on one vehicle, the channel usage is equal to that of the CMS. However, this assumes that the elements for either application are sent continuously. Since the MD can dynamically manage message contents, the full EBW/IVW message need only be sent in EBW/IVW instances, which we assume to occur with the (overly generous) percentage frequencies shown in the Column 1 of Table 8.5. Thus, the expected channel load, shown in Column 3, is far less than the peak channel load.

Table 8.5 Bandwidth comparison using the message dispatcher.

Message type and (use freq.)	Bandwidth		E[bandwidth]
	CMS	MD	MD
Heartbeat (100%)	14.1	0.6	0.6
EBW (2%)	14.1	6.2	0.71
IVW (4%)	14.1	3.7	0.72
EBW & IVW (3%)	14.1	14.1	1.01

Comparison of bandwidth requirements (in kbps) under the message dispatcher (MD) and common message set (CMS) architectures. The 4th column represents the expected bandwidth usage assuming the frequency of usage denoted in parenthesis in the first column.

Further, it is significant to notice that additional saving will be achieved in overall bandwidth usage when there are multiple vehicles present. This is because, when using the MD, only a limited number of vehicles will need to transmit an EBW or a IVW message. The remaining vehicles continue to transmit the heartbeat message.

Although these results depend on several simplifying assumptions, it is clear that with a maximum DSRC channel capacity of 27 Mbps the reduction of channel load possible by employing the MD is relevant.

In Section 8.5 we shall present a further method for reducing the channel load which can readily be implemented in the message dispatcher framework.

8.4 Data Sets

We have described our MD deployment and vehicle testbed in Section 8.3.4. Using the testbed we collected several data sets under a variety of driving conditions. We have chosen three sets for further analysis. The sets were all recorded in and around Ann Arbor, MI. The trajectories are shown in Figure 8.6, and described in Table 8.6. The data can be downloaded at Robinson (2006).

Each data set contains samples of vehicle position (latitude and longitude), speed and acceleration (lateral and longitudinal). Other data such as brake status, brake pressure, and steering wheel angle were also recorded.

APPLICATION LEVEL MESSAGE CODING AND COMPOSITION

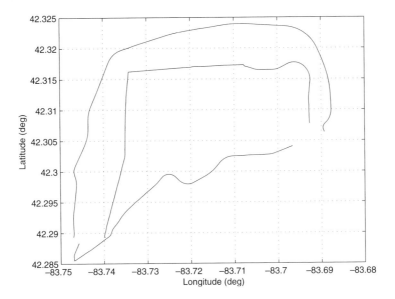

Figure 8.6 Traces of the three recorded data sets. The lower trajectory represents data set A, the middle data set B and the upper trajectory is the highway data set C.

Table 8.6 Data set parameters.

Parameter	Data set A	Data set B	Data set C
Environment	Urban	Urban	Highway
Sample freq.	5 Hz	5 Hz	5 Hz
Duration	8 min 24 sec	10 min 34 sec	6 min 17 sec
Length	6.1 km	7.6 km	9.1 km
# of stops	2	2	0
# of turns	7	5	0

8.5 Predictive Coding

The objective of this section is to show, using real-world data, that predictive coding can significantly reduce the channel load in vehicular safety applications.

The analysis thus far has basically been focused on two concepts:

1. Avoid duplication of data in multiple messages.

2. Compose sequences of messages so as to only send data at the minimum required update rate.

However, neither of these notions considers the uncertainty or variability associated with the underlying data. For example, analysis of the data sets reveals that brake status changes infrequently. When it does change, the status is often maintained.

So, compared to vehicle longitude which changes rapidly and continuously when a vehicle is moving, routine brake status can be transmitted less frequently. Only status changes need be transmitted. Of course, upon a change, it can and should be transmitted immediately. There is a difference between nominal state update transmission, and rapid reaction to changes.

Indeed, we can extend this notion to encompass data which change frequently, but which can be easily predicted. For example, given a particular vehicle position, velocity, and steering angle, the future trajectory of the vehicle can be predicted. Hence, to extend the example above, the longitude may not need to be transmitted frequently either, if it can be predicted well. For example, a stationary vehicle's position can easily be predicted.

The main idea presented in this section is to transmit data only when the error in estimating the state is 'sufficiently large'. We shall characterize the model used for the state estimation as well as the notion of 'sufficiently large error'. This idea is similar to conserving bandwidth while using computational power (Yook et al. 2002). We also consider the case where, when the error is small, only the least significant bits or some smaller correction update is sent rather than a complete data element with full resolution.

These types of transmission policies represent predictive coding (PC), which we describe in the following section.

8.5.1 Linear predictive coding

Linear predictive coding (LPC) (Elias 1955) is commonly used in the transmission, reproduction and generation of human voice over low data rate channels. It is a method of encoding signals in which the value of the signal at each sample time is predicted as a linear function of the past values of the signal (ATIS 2001). Predictive coding does not require the use of a linear prediction function.

The design of a predictive coding scheme can be divided into two parts. The first is to determine an appropriate model to be used to predict the signal which is being sampled. In voice applications, a linear model of a particular order is assumed. The model parameters are obtained using a least squares or auto-regressive fit on a window of sampled data. The model parameters (i.e. the coefficients of the characteristic polynomial and the model gain) are then transmitted across the channel. In general, the order of the model is chosen so as to be able to accurately reproduce the signal given a particular model input. This is the second part of LPC design: specifying the model input. In voice applications, the input is characterized by the voice pitch frequency. An appropriate pulse train and white noise signal is used as a model input.

The approach presented here differs from LPC in the following ways. First, the model is not linear. Second, we a priori define a Newtonian model to represent the behavior of the vehicle. The model is described in Section 8.5.2. By defining the model a priori, the model coefficients have been specified. Thus, in our update scheme, we shall only transmit state updates, and not the model coefficients. Third, we do not characterize the input function at every time instant, but rather assume a zero-order hold for intermediate values of the state which are not modeled. Finally,

8.5.2 System model

In this section we describe a simple first-order model used for state prediction. We use the discrete time index k. Transmission of a particular data element does not occur at fixed time intervals, but will occur at the discretized time instants. Define the time between transmission instants as Δ_k. We define the following velocity estimate update:

$$\hat{v}_k = \begin{cases} \hat{v}_{k-1} + \hat{a}_k \Delta_k & \text{when } v_k \text{ not transmitted,} \\ v_k & \text{when } v_k \text{ is transmitted} \end{cases} \qquad (8.1)$$

where $\hat{v}(k)$ and $\hat{a}(k)$ represent *estimates* of the velocity and acceleration at time k. The term v_k represents the *sampled* vehicle velocity. The estimate of the distance, $\hat{D}_k$, moved by the vehicle in time Δ_k, is computed using

$$\hat{D}_k = \hat{v}_k \Delta_k + 0.5 \hat{a}_k \Delta_k^2. \qquad (8.2)$$

The estimate of the vehicle heading, $\hat{\Phi}$, is updated using a non-slip tri-cycle model:

$$\hat{\Phi}_k = \hat{\Phi}_{k-1} - \frac{\hat{v}_k \Delta_k}{l} \tan \hat{\Psi}_{k-1}, \qquad (8.3)$$

where Ψ is the angle of the front wheels and l is the wheel base of the vehicle. The angle of the front wheels is linearly related to the steering wheel angle, which we are able to measure. The distance moved, (8.2), is resolved into a displacement using the most recent heading estimate. A longitude and latitude update can then be performed using this displacement. Estimates for any other data element not included in the model are given by a zero-order hold of the most recently transmitted value. Whenever a state observation is transmitted it replaces the state estimate at that time, as illustrated by the velocity update in (8.1).

Our estimation model represents a Kalman filter update scheme with noiseless observations. There is scope for improvement in this regard by deploying a full Kalman filter, and using a higher-dimensional model for state prediction and incorporating models for other parameters. This is beyond our purpose here, which is simply to demonstrate that using predictive coding yields a significant transmission data rate reduction for vehicular communication.

8.5.3 Tolerable error

In Table 8.3 we have presented transmission frequencies for various data elements in two example applications, as well as a 'heartbeat' message. We refer to these fixed frequency requirements as the 'regular transmission scheme'. To the best of the authors' knowledge, selection of these frequency values in practice has been based on three requirements:

1. to provide neighbouring vehicles with a sufficiently accurate estimate of a vehicle's current state, that is, to maintain the state estimation error within some 'tolerable error',

2. to ensure that vehicles entering an area receive a timely introduction from their new neighbours,

3. to ensure that neighbours are quickly updated about a state transition.

In the predictive coding scheme we propose here, there is no requirement that data elements be transmitted at a particular frequency. Hence, the state estimation error incurred between successive transmissions may be very large. To address the first item listed above, we shall use the state estimation error between successive sample times (under the regular transmission scheme), to define the tolerable error. This is the error measured at time $k+1$ when using a zero-order hold of the state at time k to predict the state at time $k+1$. Computing the expected value of this error yields what we call the 'expected error', shown in Table 8.7. The time between transmissions is 0.2 seconds (5 Hz transmission frequency). We use this error value in the following section as an indicator as to when a data element should be transmitted. This approach also gives an indication of when to transmit data so as to satisfy the last requirement in the list above.

In order to address the second requirement achieved by a regular transmission frequency, that is that new neighbouring vehicles are informed of the state, we stipulate that a data element must be transmitted at least as frequently as some lower bound. The frequency is chosen to be small compared to the regular transmit frequency, e.g., 0.25 Hz. We shall also investigate the effect of not having this requirement at all.

8.5.4 Predictive coding transmission policies

We now describe five transmission policies for reducing the number of bits transmitted over the channel:

1. Data elements are transmitted at the transmit frequency specified in Table 8.3. This is a reference policy.

2. Only transmit data elements when their estimation error exceeds the tolerable error bound in Table 8.7. A zero-order hold is used in between transmissions, i.e., the value of a data element is held constant until a new value is received. There is no predictive coding used. When an update is required, the full data element is sent. Each data element must be sent at a frequency of at least 0.25 Hz.

3. Same as Policy 2 except that the model described in Section 8.5.2 is used to predict the state in between sample transmissions, i.e., predictive coding. When an update is required, the full data element is sent. Each data element must be sent at a frequency of at least 0.25 Hz.

Table 8.7 Tolerable error values.

Data element	$E[\text{error}]$
Year	0
Month	0
Day	0
Hour	0
Minute	3.33E−003
Seconds	0.2
mseconds	398
Speed (m/s)	0.0716
Longitude (deg)	2.54E−005
Latitude (deg)	1.65E−005
Heading (deg)	2.42
Long. accel. (m/s^2)	0.0769
Lat. accel. (m/s^2)	0.0687
Yaw rate (m/s)	0.248
Steering wheel angle (deg)	1.04
Brake switch	5.70E−003
Brake torque (Nm)	5.89

Expected error immediately before a transmission for each of the data elements measured in the sample data sets and included in the heartbeat message. A zero-order hold is used. For reference, the expected error of the latitude and longitude amounts to approximately 1.9 m.

4. Same as in Policy 3, except that, when possible, a smaller state 'correction' is sent instead of the full data element. For example, when the error is sufficiently small, instead of transmitting the four-byte longitude data element, we transmit a one-byte correction. A *complete* data element must be sent at a frequency of at least 0.25 Hz.

5. Same as Policy 4, except that there is no minimum update frequency requirement.

In all the policies, whenever a data element is sent, the time in milliseconds must also be sent. We have used the data element definitions from the SAE standard (SAE 2008). For our purposes here we shall consider how these policies reduce the channel loading associated with a heartbeat message transmitted at 5 Hz, and containing all of the data listed in the left column of Table 8.7. Note that the data elements related to time (e.g., year, hour, etc.) are included as a time-stamp for the other data elements in the packet, and is not for clock synchronization.

8.5.5 Predictive coding results

We have summarized the results of the predictive coding analysis in several ways. Table 8.8 shows the average send frequency of each data element under the five policies. Several observations can be made. First, data elements that are highly predictable (e.g., year, month, and hour) do not need to be sent regularly. Hence, under policies 3 and 4 their send frequency is almost the minimum update frequency. Under policy 5 there is no minimum update frequency, and hence the required transmission frequency approaches zero (they must be sent at least once to start). Also note that since longitude and latitude change frequently while moving, they are required to be updated most frequently under policy 2. They can, however, be predicted fairly well, and hence under the remaining policies their frequency of transmission drops considerably. We also note that data elements that are not predicted (e.g. yaw rate) maintain a fairly constant transmit frequency across all policies, as expected. A more descriptive model would reduce their transmit frequency. Finally, since we assumed that the *msecond* data element is sent with any and every transmission, its high frequency of transmission (almost 5 Hz) indicates that some sort of data is sent at almost every transmission opportunity. Thus, there may be additional bandwidth savings realized by adding a data element to a packet already scheduled for transmission if it prevents the additional data element being sent in its own message later. This is especially applicable when there is significant message overhead.

Table 8.9 represents the percentage reduction in the transmitted rate for the four policies in comparison to the reference policy. Since highway driving has less abrupt changes in acceleration and direction as compared to urban driving, we can expect the state prediction for highway driving to be somewhat better than for urban driving. This is reflected by the larger reduction in message size associated

Table 8.8 Send frequencies for various data elements under the proposed policies.

Data element	Average transmission frequency (Hz)			
Policy #	Policy 1	Policy 2	Policy 3, 4	Policy 5
Year	5	0.25	0.25	0
Month	5	0.25	0.25	0
Day	5	0.25	0.25	0
Hour	5	0.25	0.25	0
Minute	5	0.25	0.25	0.02
msecond	5	4.86	4.49	4.35
Speed (m/s)	5	2.01	1.76	1.76
Longitude	5	2.90	0.26	0.07
Latitude	5	2.97	0.27	0.1
Heading	5	0.69	0.57	0.48
Accel. x (m/s^2)	5	1.83	1.83	1.8
Accel. y (m/s^2)	5	1.95	1.95	1.93
Yaw rate	5	1.39	1.39	1.36
Steering wheel	5	1.59	1.59	1.56
Brake switch	5	0.27	0.27	0.04
Brake torque	5	0.69	0.69	0.48

APPLICATION LEVEL MESSAGE CODING AND COMPOSITION

Table 8.9 Policy performance.

Data set	Policy 2	Policy 3	Policy 4	Policy 5
A	62%	77%	81%	84%
B	63%	77%	80%	83%
C	57%	78%	82%	84%

Percentage reduction in average data rate of transmitted messages compared to policy 1.

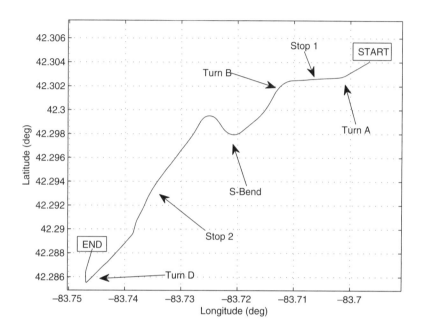

Figure 8.7 The route corresponding to Data Set A. The points marked correspond to those in Figure 8.8.

with the highway driving data set (Data Set C). Note that this is not the case under policy 2 because messages need to frequently include position updates, which change frequently during highway travel. The results also indicate that policy 5 does not perform that much better than policies 3 and 4. This is because, as shown in Table 8.8, some sort of message is sent at almost every opportunity.

Finally, Figure 8.8 presents a rolling average of the length of the messages transmitted along the trajectory followed in Data Set A (shown in Figure 8.7). Inspection of the data elements contained in the message during the 'peaks' indicates that they are correlated to changes in acceleration, such as when braking or accelerating before and after a stop. Acceleration is not modelled in our predictive model. Incorporating it indicates a future direction for extending the predictive coding method.

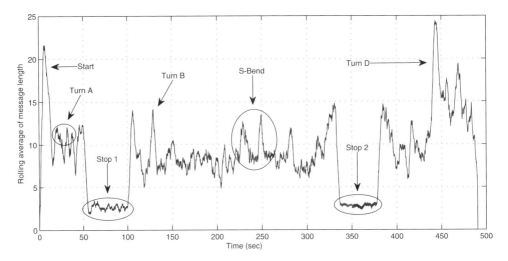

Figure 8.8 Representation of the length of messages transmitted along the route shown in Figure 8.7. A rolling average of 20 messages is used, and the length is defined as the number of characters transmitted in each message. The majority of the 'spikes' are caused by changes in acceleration, such as before and after a stop.

These results lead us to a few conclusions. First, predictive coding is highly effective at reducing the channel usage for vehicular safety applications. Even a simple state-triggered transmission policy such as the zero-order hold in policy 2 is significantly better than a regular high frequency transmission approach. Second, future message optimizations should focus on reducing the number of transmitted messages, as compared to trying to reduce the size of individual messages.

8.6 Architecture Analysis

In this section we describe how the goals identified in Section 8.1.2 are met by the MD architecture.

Each application stipulates its data requirements to a single location, the message dispatcher. Thus, any upgrades or modifications to the interface requirements are well-defined and localized (i.e., the *single interface* goal is met). The message dispatcher generates a combination message satisfying application requirements. Applications do not consider how data elements are shared between vehicles, achieving an effective *separation of concerns*. When the message format or protocol changes, only the MD implementation must change, and not the applications. This is an important abstraction for both the application and communication layer designers. Algorithms for avoiding channel overload, secure data transmission, and low latency message delivery can be implemented in the message dispatcher enabling *lower bandwidth usage*. Messages composed do not have to contain redundant information and thus can *recognize vehicle capabilities*.

(a) Three cars in transit. View of front vehicle is obscured.

(b) Middle vehicle swerves to avoid the braking front vehicle.

(c) Without EBW, emergency braking is required to stop the last vehicle.

(d) With EBW, the driver is alerted as the braking occurs and they start to brake.

(e) Safe stopping distance achieved.

(f) System hardware.

Figure 8.9 Illustration of the functionality of the emergency brake warning (EBW) system. Figures (a), (b), and (c) represent behavior without the EBW system. Three vehicles traveling at high speed are shown in Figure (a). The front vehicle begins to brake sharply (b), causing the middle vehicle to swerve at the last moment. The result is emergency braking and a potential collision by the rear vehicle in Figure (c). Alternatively, with EBW in Figure (d), as soon as the front car begins braking an EBW is transmitted via DSRC to the rear vehicle where the driver is alerted through the LCD screen, as well as via an alarm over the audio system. Ample time is then available for the tail vehicle to stop as shown in Figure (e). Figure (f) shows the hardware deployed in the rear of a Toyota Prius. The CAN network is accessed on the left-hand side, a GPS receiver is in the middle and the DSRC radio as well as the mini computer are on the right-hand side.

The message dispatcher can also evolve as application requirements change. If a unique or new set of data elements is required, this can be incorporated only when required by using the flexible terminal character with post-fixed identifier scheme. This illustrates the *flexibility*, *future-proofness*, and *extensibility* of the architecture. This enables the MD to effectively *exploit the communication ability*. Finally, since the applications have been successfully separated from the communication protocol, companies are able to develop their own products and thus *enable product differentiation*.

8.7 Conclusion

This chapter has provided some insight into some of the important considerations that need to be made in deploying a wireless inter-vehicle communication system. In particular, we presented a broad framework called the 'message dispatcher' which effectively deals with many of these considerations. Further, there are several interesting research avenues and application functionality enabled by this architecture. For instance, channel loading can explicitly be managed through dynamic message construction (e.g., based on vehicle traffic), packet collision avoidance algorithms can be implemented (e.g. transmit power modulation), specific 'important' data can be retransmitted in the case of loss, cooperation between MDs could be enabled to manage channel usage, enabling data delivery notifications, giving priority levels to information (e.g., low latency or application dependencies), filtering or other modifications to the raw incoming data can be performed and even multi-hop information passing schemes can be implemented.

Acknowledgments

The author would like to acknowledge those who have contributed greatly to prior publications of this work and without whom this chapter would not have been possible. In particular, D Caveney and K Laberteaux at the Toyota Technical Center (TTC), G Baliga currently at Google Inc, and PR Kumar at the University of Illinois at Urbana Champaign (UIUC). Funding and research resources were generously sponsored through contracts at TTC and UIUC.

References

ATIS 2001 *ATIS Telecom Glossary 2000* American National Standards Inc. http://www.atis.org/tg2k/t1g2k.html.
Berners-Lee T, Hendler J and Lassila O 2001 The semantic web. *Scientific American Magazine*.
CAMP 2008 CICAS-V dsrc message descriptions and examples. Technical report, Collision Avoidance Metrics Partnership.
Carter A 2005 The status of vehicle-to-vehicle communication as a means of improving crash prevention performance. Technical report, NHTSA. www-nrd.nhtsa.dot.gov/pdf/nrd-01/esv/esv19/05-0264-W.pdf.
Elias P 1955 Predictive coding i. *IEEE Transactions on Information Theory* **IT-1**(1), 16–24.

Kawadia V and Kumar PR 2005 Principles and protocols for power control in ad hoc networks. *IEEE Journal on Selected Areas in Communications* **23**, 76–88.

Ko YB and Vaidya N 1999 Geocasting in mobile ad hoc networks: Location-based multicast algorithms. *WMCSA*, New Orleans.

Robinson CL 2006 http://decision.csl.uiuc.edu/~testbed/TTCUIUCData.zip.

Robinson CL, Caminiti L, Cavney D and Laberteaux K 2006 Efficient coordination and transmission of data for vehicular safety applications. *The Third ACM International Workshop on Vehicular Ad Hoc Networks (VANET)*, pp. 10–19.

Robinson CL, Cavney D, Caminiti L, Baliga G, Laberteaux K and Kumar PR 2007 Efficient message composition and coding for cooperative vehicular safety applications. *IEEE Transactions on Vehicular Networks* **56**(6), 3244–3256.

SAE 2008 Dedicated short range message set (DSRC) dictionary (rev. 29). Technical Report SAE J2735, SAE.

USDOT 2006 Vehicle safety communications project – final report. Technical report, USDOT. www-nrd.nhtsa.dot.gov/departments/nrd-12/pubs_rev.html.

Yook JK, Tilbury DM and Soparkar NR 2002 Trading computation for bandwidth: Reducing communication in distributed control systems using state estimators. *IEEE Transactions on Control Systems Technology* **10**(4), 503–518.

9

Data Security in Vehicular Communication Networks

André Weimerskirch,[1] Jason J Haas,[2] Yih-Chun Hu[2] and Kenneth P Laberteaux[3]

[1]escrypt Inc., Ann Arbor, MI, USA
[2]University of Illinois at Urbana-Champaign, IL, USA
[3]Toyota Research Institute-North America, Ann Arbor, MI, USA

9.1 Introduction

Information technology has gained central importance for many new automotive applications and services. The costs for software and electronics are estimated to approach the 50% margin level in car manufacturing in 2015. Perhaps more importantly, there are estimates that already today more than 90% of all vehicle innovations are centered around IT software and hardware. Safety applications based on vehicular network communication are a major aspect of future innovation. Vehicle ad-hoc networks (VANET) provide vehicle-to-vehicle (V2V) and vehicle-to-infrastructure (V2I) communication. These applications are foreseen to improve traffic safety considerably, and to enable innovative infotainment applications and business models.

Security is concerned with protection against malicious manipulation of IT systems and plays an important role when designing and implementing such applications. The difference between IT safety and IT security is depicted in Figure 9.1.

The majority of software and hardware systems in current vehicles are *not* protected against tampering. One reason for this is that past vehicle IT systems provided little incentive for malicious manipulation; there was little gain in compromising a

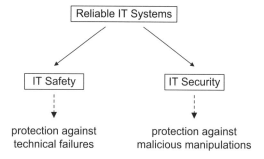

Figure 9.1 The relationship between IT safety and IT security.

single vehicle, and such a compromise did not affect other vehicles. Additionally, security tends to be an afterthought in any IT system as evidenced by several case studies including the development of the Internet. As can be seen in this example, implementing IT security afterwards is relatively inefficient.

VANET applications consist of safety and commercial applications. VANET applications need security functionality in order to protect the driver, the manufacturer, component suppliers, and service providers. Safety applications must be protected to avoid malicious manipulation, potentially causing harm to the vehicle driver, and commercial applications must be protected to prevent loss of revenue. The manufacturer and component suppliers need to protect the applications to ensure safety and reliability, and avoid a loss of image in case of a security breach. While security objectives and solutions are well researched on common PC-based environments, the approach for VANET is very different. Some obvious differences are listed below.

- Embedded computing platforms used in vehicles have low-cost processors which are limited with respect to computational capabilities and memory. Hence, the usage of cryptographic primitives and protocols is limited. Vehicles have a lifespan of at least a decade and it cannot be assumed that the built-in computing platforms will be upgraded during this time.

- The bandwidth for external communication is limited. Again, vehicles have a lifespan of at least a decade and it cannot be assumed that the bandwidth capabilities will be upgraded during a vehicle's lifetime.

- Attackers of vehicles often have physical access to the vehicle and the vehicle's electronic control units (ECU). Furthermore, remote attacks via communication channels might be mounted. In the traditional PC world, attackers are usually external entities that mount remote attacks via the Internet whereas in the vehicular world attackers might additionally include the owner or third parties that have physical access to the vehicle, e.g. mechanics, valets, etc.

- VANET will in most situations not have permanent connections to a fixed infrastructure such as the Internet. Therefore, critical issues such as privacy and control of the network need to be handled in a different way from PC-based networks. Furthermore, it might be costly to establish the necessary Internet-like organizational aspects in VANET since these traditional approaches cannot simply be reproduced.

9.1.1 Outline

This chapter is organized as follows. After describing the state of the art in the next section, we present challenges of providing data security in VANET in Section 9.2, followed by attacker and applications models in Section 9.3. We explore aspects regarding supporting infrastructure in Section 9.4. In particular, these aspects include the management and handling of a public key infrastructure (PKI). We then present protocols for providing secure communication, secure positioning, and identification of misbehaving nodes in Section 9.5. Furthermore, we consider privacy aspects and solution approaches in Section 9.6, and show implementation aspects including appropriate key lengths, physical security, organizational aspects, and software updates in the field in Section 9.7. Finally, an outlook and conclusions are given in Section 9.8.

9.1.2 State of the art

VANET are an emerging research area, both in academia and in industry. There are several completed and ongoing projects. While the early projects mainly considered the feasibility of VANET, recent projects include security aspects. The Vehicle Safety Communications Consortiums (VSCC) under a cooperative agreement with the US Department of Transportation (USDOT) worked on security solutions that strongly influenced the IEEE P1609.2 standard – Consortium (2006); IEEE (2006). Currently the main industrial projects in the USA are performed by the Vehicle Infrastructure Integration (VII) initiative (Initiative 2008) as well as by the Vehicle Safety Communications 2 (VSC2) Consortium in the Vehicle Safety Communications Application (VSC-A) project. Their results are expected to influence the IEEE P1609.2 standard. This standard defines the over-the-air message format for VANET and currently suggests attaching an ECDSA (elliptic curve digital signature algorithm) digital signature to each message. Furthermore, either a certificate or a certificate digest needs to be attached to each message. This standard also defines message content encryption as well as the format of certificate revocation lists (CRL). In Europe, security is considered within the Network on Wheels (NoW) and the SEVECOM (Secure Vehicular Communications) project. Most European automotive manufacturers are grouped in the Car2Car Communication Consortium (C2C-CC) that works closely with these projects. All of these industrial projects include security aspects, in particular message authentication as well as privacy. A common approach to security is to apply the IEEE P1609.2 standard regarding message authentication and assume a deployed public key infrastructure (PKI). Privacy protection is designed on top by equipping each vehicle with multiple certificates, and changing certificates regularly. These approaches are detailed and extended below.

There is also increasing academic research available. Raya et al. (2006b) provide an overview of crucial security aspects. Raya and Hubaux describe a security and privacy framework for VANET (Raya and Hubaux 2005). Hubaux et al. (2004) describe further security issues, specifically concerning privacy and secure positioning. Gerlach et al. (2007) describe a general security architecture for VANET. Finally, Parno and Perrig (2005) give an overview of challenges, adversaries, attacks, properties of VANET, and useful security primitives for design solutions. Research is, in particular, performed in privacy protection for VANET as well as in control mechanisms such as certificate issuance and revocation. We will mention further related work as it arises in the sections below.

9.2 Challenges of Data Security in Vehicular Networks

Data security in the PC and office world is well researched, although largescale devastating attacks still occur. Security in vehicular networks poses different security threats and also has different requirements. In the following we discuss challenges of data security in VANET from a holistic point of view.

Risk potential Due to the close coupling with the physical environment, the risk involved in vehicular networks can be much larger than the risk in conventional IT applications. The hacking of an automotive safety-critical application system can have far more immediate physical consequences than hard disk data destroyed by a computer virus. It is often argued that safety applications will never take over control of a vehicle but will support the driver by providing information. Even so, the compromise of such safety applications poses a threat to the driver. Imagine a safety application notifying a driver of an imminent impact – there is a distinct possibility that the warned driver's reaction to the warning will cause a collision. If a safety application can be successfully compromised, even a tiny fraction of drivers acting erroneously due to the attack represents a serious threat to safety due to the huge number of vehicles that could be attacked. Generally speaking, the familiar threats of conventional IT systems (hacking, phishing, pharming, etc.), which target mainly abstract digital data, are extended to VANET to create threats involving our safety-critical physical environment.

Financial assets There are a variety of promising applications based on vehicular communication that involve financial aspects, such as digital infotainment content (e.g. music and video files as well as navigation maps), locationbased services (e.g. special deals at the next motel), and built-in automotive payment functions (e.g. road tolling). Consequently there is a large incentive to manipulate such systems for financial gain. The system might be manipulated by the vehicle owner (e.g. in the case of digital content) or by third parties (e.g. by competing motels). Tampering with non-safety applications imposes far less risk of being prosecuted than does tampering with safety applications. Whereas in the second case, police authorities might heavily pursue any illegal modifications, in the first case, industry needs to defend itself.

New business models Closely related to financial assets are new business models. There will be many vehicular applications where the business model relies on strong security functionality. In such systems, manipulation may lead to a loss of revenue. Time-limited feature activation of services such as telematics services or satellite radio are examples. It is envisioned that selling software for vehicles will make for a completely new business model. For instance, software optimizations for a sporty or gas-saving engine adjustment might be sold in the same way as updates to the vehicle's entertainment system. VANET communication channels are a potential distribution channel for content deployed under new business models. At the same time, tamper-resistant or tamper-evident hardware might be required to protect the business model. While such hardware is not at the core of a secure VANET, it might be attractive to deploy a single secure hardware platform in each vehicle that is shared by VANET applications as well as applications from these new business models to reduce cost.

Cost The cost of security solutions is crucial in any IT system. There is an understanding of security in the PC world by end-users who are willing to invest in security solutions such as anti-virus software and firewalls. However, in the automotive domain there is little willingness by vehicle buyers to spend money for security. Therefore, security solutions need to be especially cost efficient.

Usability In the PC world, many users have gotten used to installing security software and configuring it to a certain degree. However, a vehicle driver expects not to deal with any electronic issues, and certainly not with a security configuration. Therefore, security should be configured and adjusted automatically. If adjustments by external entities are necessary, then they should only be implemented during a workshop visit or by automatic updates via VANET communication channels.

Mobility Vehicles are, by design, highly mobile. Furthermore, vehicles usually do not move in groups but will mainly encounter other vehicles they never have had contact with before. However, most vehicles do not randomly move but stay in a certain area most of the time, say 50 km around the owner's home, or even move in a repeating pattern, for example the route of a daily commute. Contact with other vehicles might be limited to only a few seconds such that establishing a secure channel cannot take too long. Furthermore, the communication quality might be affected by the velocity of vehicles, resulting in packet loss.

Privacy Privacy is already a concern in conventional IT systems. In VANET, there is a high correlation between user and vehicle. Privacy concerns include disclosure of a driver's location and behavior. Today, almost all movement patterns of an individual can be traced by tracking their vehicle. Further privacy concerns might be involved in financial transactions carried out on VANET. Privacy is both a *technical* and an *organizational* matter.

Reliability Malicious modifications can harm the reliability of vehicular networks. There is a trend in the automotive domain to allow (remote) software updates. Even though this function offers great opportunities to both users and manufacturers, unauthorized software updates can lead to serious safety and liability issues, and to financial loss.

Market penetration Vehicles are expected to be equipped with VANET radios over the next decade. However, it is expected to take a considerable time until all vehicles are VANET enabled (in the USA, around 7 million vehicles are sold per year whereas there are around 243 million vehicles altogether). Furthermore, it is unclear to what degree or when there will be supporting infrastructure available in the form of roadside units (RSUs). Therefore, potential security solutions should work with a low penetration rate of radio-enabled vehicles and small number of deployed RSUs. A means of accelerating this process might be introducing incentives to vehicle owners and automotive manufacturers.

Legislation Legislative requirement might force VANET to provide strong security and privacy solutions. Legislation might require both technical solutions and organizational mechanisms for the vehicles and for the supporting infrastructure. Given the complexity of a security certification process à la Common Criteria or FIPS 140-2, this is not an easy undertaking for industries without much prior experience in security. For instance, in Europe, trucks are required to use a digital tachograph that must be shown by security certification to be tamper resistant (Commission 2003).

9.3 Network, Applications, and Adversarial Model

In this section, we describe the network, applications, and adversarial models to complete the VANET setting outlined above. We do not consider a specific model instantiation here but describe common characteristics that hold for most VANET scenarios.

9.3.1 Network model

Network nodes include on-board units (OBUs) installed in vehicles and roadside units (RSUs) installed next to the road. It is possible that the network will be set up with only vehicles' OBUs and without any RSUs. Both OBUs and RSUs are expected to implement dedicated short range communication (DSRC) radios. DSRC is currently standardized; IEEE 802.11p addresses the physical (PHY) layer and medium access control (MAC) layer whereas the upper layers (network and above) of the communication stack are being developed within IEEE P1609 (channel management, resource manager, and security). The current DSRC standard suggests seven channels for the USA, each 10 MHz wide and in the 5.8 GHz frequency spectrum. Each channel is expected to allow a bandwidth of 3 to 6 Mbit/s and in the near future up to 27 Mbit/s. Safety-critical messages need to be sent on a dedicated channel, called the control channel. While in many scenarios a single-hop communication model suffices, particularly for safety applications, a multi-hop topology is desirable for other applications. The packet size varies between 50 and 200 bytes for safety applications whereas it might be far larger for non-safety applications. Each packet's header imposes an overhead of at least 46 bytes at 3 Mbit/s at the MAC and PHY levels, and another 11 bytes of overhead is introduced according to the over-the-air message format of IEEE P1609. The transmission range is approximately 300 meters in light traffic situations, but 800–1000 meters have been observed.

To support VANET operations, certificate authorities (CAs), governments, or network operators will deploy roadside units (RSUs), which are fixed-position infrastructure DSRC radios that the deploying entity can use to spread and gather information pertinent to the VANET (e.g. distribute traffic and road condition information or gather traffic data for traffic reports). RSUs are likely to be sparsely placed owing to the cost of deploying large numbers of RSUs (Resendes 2008). Even if RSUs are deployed more densely than anticipated, there will be a deployment period during which they are not all online. Thus, any VANET application or design must handle this case and provide a valid solution during incremental deployment.

RSUs might be connected to a fixed infrastructure (e.g., Internet via GSM) but they do not need to be. Both vehicles and RSUs might be privately owned (private vehicle, company owned RSU) or public (police vehicle, public transportation vehicle, government owned RSU). Vehicles are highly mobile but they do not form static groups. Vehicles might move very fast (with speeds of more than 200 km/h on a highway) or they might be packed very densely (e.g. on a 12-lane highway). The number of nodes in the network might reach several hundred million or even in the range of a billion worldwide, but at the time of initial deployment the market penetration will be very low. Organization of the network might be provided by local authorities. Furthermore, in most cases a node will only communicate with its close neighborhood such that complexity of the network stays manageable. OBUs and RSUs will probably be separate pieces of hardware and come with their own computing platforms. This computational platform might provide tamper evidence or resistance. However, because of the high cost, such resistance cannot be assumed. A GPS receiver is necessary for anticipated safety-related applications and are already built into most of today's vehicles. Therefore, it can be assumed that vehicles are aware of their location and the current time. A loss of the GPS signal is possible, e.g. in urban areas or in a tunnel. After a loss of the GPS signal, modern vehicles will use further vehicle sensor data including velocity and steering angle to estimate the current location. However, it can be assumed that after at most 5 minutes without GPS, the accuracy of the location will be degraded so much that safety applications will not work properly and safety systems will be turned off.[1]

9.3.2 Applications model

Applications for VANET can roughly be categorized into safety and non-safety applications. In safety applications, vehicles usually broadcast a *beacon* which

---

[1] Note that after a loss of GPS, the ability to maintain an accurate estimate of time is much greater (and cheaper) than maintaining an accurate estimate of location.

includes velocity, location, and further vehicle status data, and is broadcast frequently, e.g. 3–10 times per second. The intended receivers are vehicles in the direct communication neighborhood such that no routing is required. There are other safety applications where an RSU detects a dangerous situation (e.g. an icy road condition) and broadcasts this information to approaching vehicles. Non-safety applications are different in nature. In general, the focus is less on short latency but transmission of large data files, multi-hop communication, and service provision. In the following, we summarize typical requirements for safety and non-safety applications without having a particular application in mind. A more comprehensive requirements list has been developed as part of the SEVECOM project (Kroh 2006). Note that these requirements only include dedicated VANET applications. In this chapter we do not consider applications that depend on an external communication channel such as telemetrics via cell phone network or satellite radio. Some vehicles might come with special applications and security solutions, such as police vehicles. We do not consider such special solutions but assume that all vehicles use the same security protocols.

Safety applications

As already described, safety applications are mainly based on V2V communication but there are also V2I safety applications. The V2V safety applications include emergency breaking notification, warning of vehicles in the blind spot, and vehicles being on a collision course (e.g. at an intersection). On the other hand, red-light notification and icy-road warnings are safety applications based on RSUs and V2I communication. In the following, we summarize typical requirements that will cover the majority of safety applications.

Latency Since vehicles are highly mobile and might move at high velocity, the overall time delay between the time of an original event occurs (say, an emergency brake) to the reception of that information by the receiver's vehicle safety application should be small. Typically it is expressed that a delay of 100 ms is acceptable where the overhead due to security should be at most 20 ms.

Security objectives The correct operation of safety applications needs to be ensured even in the presence of attackers. Therefore, cryptographic protocols that provide message authenticity are mandatory. Confidentiality of transmitted information is usually not required in safety applications since the transmitted information is locally observable by anyone (such as location and velocity). However, this argument needs to be considered more carefully when considered through the lens of privacy which is discussed below. Finally, it might be required to be able to map a received message uniquely to a vehicle in such a way that there is no doubt about that vehicle having broadcast the given message. For instance, such a property can be used in case of a hit-and-run accident where the hit vehicle or neighboring vehicles store the messages received at the time of impact. Again, this requirement needs to be considered in a different perspective for privacy purposes. Finally, note that so far there are no active countermeasures to denial of service (DoS) attacks used. An adversary can always physically jam the communication channels.

Messages Safety applications involve beacon messages transmitted by vehicles as well as status information transmitted by RSUs. Furthermore, vehicles transmit warnings of imminently threatening situations such as emergency braking. While beacon messages are transmitted regularly 3–10 times per second, further messages are transmitted infrequently on an as-needed basis. Therefore we assume here that each vehicle broadcasts 10 message per second. On a highly populated highway there might easily be 100 or even more vehicles in the direct neighborhood of a vehicle. Therefore, it is estimated that vehicles need to be able to process at least 1000–2500 received messages per second depending on the available channel bandwidth. There are no sessions established but messages are to be understood on their own (stand-alone). It follows that communication is unilateral.

Topology In most cases, safety applications are based on a single-hop communication and notify the direct neighborhood. However, safety applications based on multi-hop communication might extend available reaction time to a threat and should not be ruled out.

Infrastructure A public key infrastructure (PKI) is assumed to be in place. The PKI is under the control of a single authority, or the power might be shared by several authorities and might be designed in a hierarchical fashion. All vehicles are registered and are issued at least one certificate. The role and organization of the PKI are described later in detail.

Privacy Privacy is a core requirement. Privacy is mainly demanded by vehicle buyers. DSRC-enabled vehicles regularly broadcast signed information such as location and speed. This information potentially allows tracking of vehicle movements over time, and might provide information to third parties that could be used in a manner not originally intended (e.g. police using velocity information embedded in beacon messages to automatically issue speeding tickets). Already today drivers sacrifice a part of their privacy every time they use their vehicle. Vehicles can easily be tracked via road toll information, and license plate information can be recovered by automatic processing of surveillance camera data. While it is desirable to keep the level of privacy after introducing VANET at the same level as before, it is unclear today if it is even possible to compare the level of privacy before and after introducing VANET owing to a lack of clear and transparent metrics. Privacy involves a variety of stake holders, including the vehicle owners and drivers, the vehicle manufacturers, law makers and law enforcement, and transportation authorities. Note that privacy protection is both an organizational and a technological issue. We believe that organizational consensus needs to be found before technological solutions can be fine-tuned and implemented. Privacy is discussed below as a major part of this chapter.

Detection and revocation For safety applications, an efficient mechanism for detecting malicious and malfunctioning vehicles is crucial as is the ability to revoke these vehicles quickly. The criteria when a vehicle is considered to be misbehaving are manifold and not fully defined yet, including inconsistent message content and external input from the physical world (e.g. vehicles that are known to be tampered with). In some cases, it might be even desirable to identify a particular vehicle, e.g., in a hit-and-run accident. Detection of (misbehaving) vehicles is required for revocation. Therefore there is a tension between revocation and the level of privacy; the higher the level of privacy the more inefficient is revocation and vice versa. We elaborate these properties later on.

Non-safety applications

Non-safety applications are based on V2V and V2I communication. In general, RSUs will be heavily involved in non-safety applications. Such applications include payment services, distribution of multimedia files and commercial information, traffic optimization, Internet access, location-based services, fleet management, and many more. The requirements for non-safety applications are less strict than for safety applications since they are not safety critical. At the same time, the requirements are more diverse and depend on the considered application. Nonetheless, we summarize typical requirements in the following.

Latency and messages In most cases, latency is not as important as it is for safety applications. Also, message frequency is lower. Messages are not regularly broadcast by vehicles but only by RSUs. There might be a requirement for a certain throughput in order to support the exchange of large data files, and several messages might be exchanged as part of a bilateral communication session.

Security objectives Message authenticity and integrity is usually required to avoid malicious manipulation. Confidentiality is required for all transactions in which a business model or personal data needs to be protected such as with financial transactions and the broadcast of digital content. However, commercial promotions or offers need to be protected from alternation but need not be encrypted.

Topology Non-safety applications are based on single-hop (e.g., RSU to OBU) and multi-hop communication. Applications might depend on either or both nodes.

Infrastructure A PKI is assumed to be in place.

Privacy Privacy is a core requirement in non-safety applications as well. In addition to location privacy, traditional privacy aspects arising from monetary transactions need to be considered. These aspects are considered in detail below.

Detection and revocation Non-safety applications usually involve intensive communication with RSUs. RSUs will be connected to central communication systems, maybe even to the Internet. Misbehaving vehicles performing transactions with central servers via RSUs can be detected and revoked by traditional approaches. The certificate revocation list (CRL) can be distributed to all central servers. It might also be necessary to distribute the CRL to vehicles in order to lock out misbehaving vehicles. While detection of misbehaving nodes and revocation in the setting of non-safety applications appear to be an easier task than in the setting of safety applications, the incentive to misbehave might be higher in non-safety applications, thus requiring stronger protection mechanisms. In particular, this holds whenever an adversary's gain is financial in nature.

9.3.3 Attacker model

We now give an overview of the attacker and the possible nature of the attacks. We start by describing common capabilities and classify attacker capabilities and properties. Then we classify attacker categories and finally describe attractive attacks. Neither our list of the attacker's capabilities nor our list of the possible attacks can be comprehensive. Attackers are led by their motives unless they are irrational. These motives might be of economic nature, curiosity, malignity, or a competitive spirit. While attacks of an economic nature can be assessed in an economic manner, the remaining attacks need to be assessed in an absolute manner by comparing required and available financial and computing resources as well as manpower.

Adversaries

We assume that an attacker has full access to a generic DSRC radio. Therefore adversaries are able to listen to the channel and read all messages in their reception range. Furthermore, attackers are able to actively broadcast new messages, replay messages, and tunnel messages over another channel to another location. An attacker might mount a Sybil attack by manipulating a DSRC radio to assume several identities or by deploying several DSRC radios in a single vehicle. An attacker might act rationally or purely maliciously but is bound by financial resources and available manpower. Furthermore, the attacker is computationally bound. While the attacker has access to powerful computing devices that are far more capable than the computational platforms deployed in vehicles, they are limited by the technology available at the time of the attack. We assume that there will be no known attack to standardized cryptographic primitives during the lifetime of a VANET and that the advancement in attacking cryptographic primitives is based on technology's advancement that roughly follows Moore's law. We also assume that there are no global devastating attacks possible such as an insider publishing crucial secret keys, an attacker compromising the infrastructure servers, etc. While we cannot rule out such attacks, we believe that they can be handled by traditional security policies and by implementing well-known organizational and technological security mechanisms. In the following, we describe the main properties and capabilities that makes an attacker:

Insider/outsider An attacker might have access to insider knowledge. Insider knowledge includes detailed knowledge about the VANET's design and about its configuration. The level of insider knowledge is fine-grained. It ranges from detailed knowledge about a subsystem to knowledge of secret keys. Note that insider knowledge might be distributed without having the attack in mind, e.g., by improper organization or by dissatisfied employers.

Road coverage The attacker is able to cover a certain percentage of the road network. While a basic attacker has control of a single DSRC radio and covers a range of at most 1000 meters, an organized attacker might deploy a grid of several dozens or hundreds of DSRC radios.

Technical expertise This describes the attacker's expertise in mounting attacks. As mentioned above, we assume that an adversary is not able to mount an attack on cryptographic algorithms. We also hold the chances low that an adversary is able to compromise the infrastructure network and capture restricted data. Therefore by technical expertise we mainly refer to the adversary's capability to mount physical attacks on the computing platforms of OBUs and RSUs in order to extract program code and secret keys.

Resources Attackers have a certain limited budget at their disposal. The budget is described in terms of financial and human resources. Furthermore, the adversary needs tools to mount attacks. The adversary might be an illegal organization that has access to large financial resources and aims to make a financial gain. Another adversary might be a group of people connected by the Internet but with few financial resources and that act out of curiosity.

We now list typical attacker classes and describe them by the above capabilities and properties.

Curious hacker The curious hacker has no insider knowledge, except what is available on the Internet. They have very limited resources and obtain their information via the Internet. Furthermore, they publish their findings on the Internet. Therefore, virtual groups of curious hackers work together connected by the Internet. These groups can become extremely large, as they are, for example, with popular objectives such as hacking the iPhone. The technical expertise of the curious hacker ranges from average to outstanding. In the extreme case, the curious hacker has cutting-edge technical expertise and access to necessary tools at their workplace. While a single curious hacker has very limited resources and might not have access to a DSRC radio, the virtual group of curious hackers might be able to cover a vast amount of the road network. We expect that curious hackers will focus their energy on consumer products rather than VANET security, and once they focus on VANET they will probably analyze non-safety applications more intensely than on safety applications.

Academic hacker Academic hackers are closely related to curious hackers. Academic hackers are professors and students at a university. They act for research purposes and out of curiosity. Academic hackers have cutting-edge expertise and technology, and follow the results of curious hackers on the Internet. They might exchange their findings with other academic hackers but also with curious hackers. Academic hackers usually provide their results to the system provider first to give them a chance to fix the security weaknesses before publishing results.

Malicious hacker A malicious hacker will deliberately cause harm. Malicious hackers will collect available information from the Internet and potentially work together with curious hackers without revealing their motives. They might act without deeper purposes, or they might have terroristic motivation. In the later case, these malicious hackers potentially have access to large monetary resources and they can buy expertise and manpower without revealing their true motivation. The combination of powerful resources and malignity makes malicious hackers a dangerous adversary.

Organizational hacker An organizational hacker is driven by a well-defined motive. In most cases, the motive is financial in nature. However, government organizations that use data in a manner contrary to that defined by policies or laws might also be considered as an organizational hacker. Organizational hackers act rationally to maximize their objective. They have powerful resources and technical expertise, and they can use their resources to cover a considerable part of the road networks. Organizational hackers might have access to direct insider knowledge, or they might be able to get insider knowledge by bribing insiders. Furthermore, organizational hackers follow the development of the other hacker categories as far as it is published on the Internet. Organizational hackers do not publish any of their findings, or they only make public the fact that they were able to compromise the system (e.g. in order to sell their hack). Organizational hackers are a considerable threat due to their powerful resources. However, they act rationally since they seek to optimize the financial gain.

End users The formerly described attacker *actively* mounts attacks. However, the largest attacker class encompasses 'end users' who apply the attacks. The end users are informed through the Internet. On the Internet, programs and descriptions are offered that explain all the steps of an attack in detail such that end users do not need great technical expertise. End users do not understand the attacks themselves, but they are able to exploit the attacks for their purposes. Their purposes are manifold, ranging from tracking down their neighbors, to a kid starting an attack out of curiosity, to controlling the traffic lights on their commute to work, to copying digital content from their vehicles to their home entertainment systems, to checking out how many vehicles are parked in a garage, etc. End users comprise vehicle drivers and owners as well as companies. Companies will be intimidated to apply such attacks if they are forbidden by law. Individuals are tempted to mount attacks on non-safety-critical applications but they will be reluctant to endanger their safety by mounting attacks on safety-critical applications.

Attacks

In this section we give an overview of attacks that are attractive to adversaries. We do not claim that this list of attacks is complete or that they are ordered by importance or severity. In general, attacks include eavesdropping, manipulation of messages (e.g. forging, injection, replay, and tunneling), and physical attacks to extract keys. As already mentioned, denial of service (DoS) attacks are out of scope here but will nonetheless be described.

Denial of service In a DoS attack, an attacker either disturbs the communication channel or overwhelms the vehicle's available resources to exclude a peculiar vehicle or all vehicles in transmission range. DoS attacks can be mounted by physically jamming the communication channel to block any communication, or by broadcasting message packets that take up a vehicle's computing resources. For instance, an adversary can flood the communication channel with flawed messages. While there are no comprehensive technologies known to encounter DoS attacks, we believe that they do not pose a considerable threat. In the end the safety level is reduced to today's state.

Message manipulation In a message manipulation attack, an attacker injects forged messages or suppresses transmitted messages. The motives of an attacker might be manifold. An attacker might want to introduce flawed information to convince other vehicle drivers to take an alternative route, giving themselves a clear path. Alternatively, an attacker might selectively suppress packets warning of a traffic jam to give themselves a clear path on the alternate route. Nodes acting as relays can easily mount such attacks. Furthermore, a driver might try to masquerade as a priority vehicle, such as an emergency vehicle, to reduce their commute time. In general, an attacker can modify packet information including location, vehicle status, identifier, and special events (e.g. emergency breaking).

Replay and tunnel attacks Closely related to the previous attack are replay and tunnel attacks. In these attacks, a message is replayed at a later time or in a different location. In the later case, the message is tunneled to another location by an external communication channel (e.g. GSM) and replayed.

Eavesdropping A basic attack is eavesdropping of messages. An attacker records received messages and analyzes them. Police could use this information to issue speeding tickets, or companies could use data mining to extract information about drivers' behavior. The latter aspect brings us to privacy.

Breach of privacy An attacker could try to track vehicles and their drivers in order to gain information about general behavior (e.g. a client profile) or specific behavior (e.g. to blackmail a specific driver). The attacker might be powerful and have a global observation network using RSUs. However, the adversary might also only deploy a single DSRC receiver in order to collect information about a single compromising location. Furthermore, the adversary might couple information gained by a DSRC receiver with other information such as digital camera pictures. Finally, the attacker might use RF fingerprinting to identify and recognize vehicles.

Masquerading and Sybil attacks In a Masquerading attack, the attacker pretends to be another vehicle by using false identification, whereas in a Sybil attack they pretend to be multiple vehicles concurrently. Especially if each vehicle holds multiple keys, a malicious vehicle could use multiple keys simultaneously, giving the attacker an advantage. This advantage could take many different forms. The attacker could use multiple identities to artificially congest a roadway and make other drivers think that they should take an alternative path to avoid the congestion. If revocation is based on voting for malfunctioning or malicious vehicles, then performing a Sybil attack gives the attacker a greater number of votes. In particular, this is true in cases of local revocation in a constrained area or because it is hard to vote one way or another because of incomplete information.

Extraction of secret keys An attacker might mount an attack to extract secret keys from an OBU or an RSU. Since we assume that only verified cryptographic mechanisms are implemented, the extraction of keys is only based on physical attacks. The attacks include reading out keys of the non-volatile or volatile memory as well as applying so-called side-channel attacks. We will describe further details and countermeasures below.

Financial exploitation All the attack examples above describe attacks aimed at personal comfort, prevention of prosecution, or rare disruption. We foresee that most attacks will actually be mounted for financial gain. Such attacks will be implemented as a combination of the above attacks. For instance, keys will be extracted to gain unrestricted access to digital content, or identities will be falsified to evade toll fees.

Having described application requirements and adversary aspects, we now approach security solutions. In short, cryptographic protocols are applied to counteract message forging, eavesdropping, replay, and tunnel attacks. Further protocols can be applied to protect privacy and avoid masquerading. Finally, secure hardware might be deployed to prevent the extraction of keys.

9.4 Security Infrastructure

A security infrastructure provides services that enable cryptographic and privacy-preserving protocols, in particular, key management in VANET. We start by introducing the basics of cryptography as the basis for the following sections.

9.4.1 Cryptography services

Even though security depends on much more than cryptographic algorithms – a robust overall security design including secure protocols and organizational measures is needed as well – cryptographic primitives and schemes are in most cases the atomic building blocks of a security solution. Properly combined cryptographic primitives and schemes are required to enable the following security services.

Confidentiality (sometimes misleadingly called *privacy*) is a service ensuring that information is kept secret from all but authorized parties.

Integrity is a service ensuring that system assets and transmitted information cannot be modified by unauthorized parties. Modification includes writing, changing, changing the status, deleting, and injecting transmitted messages. It is important to point out that integrity relates to active attacks as well as technical errors and therefore it is concerned with detection rather than prevention. Moreover, integrity can be provided with or without recovery.

Authentication (more precisely *message origin authentication*) is a service concerned with assuring that the origin of a message is correctly identified. Notice that origin authentication implies integrity.

Identification (more precisely *entity authentication*) is a service establishing the identity of an entity (e.g., a person, computer, credit card).

Non-repudiation is a service which prevents the sender of a message from denying having created that message.

Access control is a service restricting access to resources reserved for privileged entities.

Security services are provided by employing the two known cryptographic cipher categories: symmetric and asymmetric (public-key) cryptography.

Symmetric-key cryptography

Symmetric-key cryptographic algorithms are commonly the basic building blocks of any secure system which minimally requires confidentiality. These algorithms are used to encrypt messages in bulk and to provide secure storage of data. In this kind of cryptographic algorithm, the keys used for encryption and decryption are the same for both communicating entities and hence they are called *symmetric ciphers*. It can be considered as a locked box, with the messages inside, that is sent to the other party. If the other party has the right key to the lock, then that party can open the box and read all the messages. The security of a symmetric cipher depends on the key (the algorithm is assumed to be public). The exchange of these keys between the parties should be done using a secure channel, e.g., provided by a public-key cryptographic system.

Symmetric-key algorithms are mainly divided into two categories: *block ciphers* and *stream ciphers*. Block ciphers encrypt the messages in data blocks of fixed length. A very widely used block cipher is the Advanced Encryption Standard (AES) (US Department of Commerce/National Institute of Standard and Technology 2001). AES supports variable block and key sizes of 128, 192, and 256 bits giving a choice of different security levels based on its application. AES has been optimized for efficient software and hardware implementations. Block ciphers are often used as building blocks for further cryptographic methods such as message authentication codes (MACs). Unlike block ciphers, stream ciphers encrypt a plain text bit by bit. The most famous example is the one-time pad (OTP) encryption (also called Vernam cipher), which is the only known cipher that can be proven to be unbreakable. The OTP works by bitwise XOR of the plain-text with a one-time key which is of the same length. The problem of having a secret key of the same length as the message to be transmitted over a secure channel makes OTP encryption inconvenient to use in practice. This shortcoming is overcome by using a pseudo-random number generator as the source for the secret key (but the unconditional security holds no more). Today's stream ciphers operate on a single bit of plain text (or a few bytes of data), XORing with a pseudo-random key stream generated based on a master key and an initialization vector. Stream ciphers are especially useful in situations where transmission errors are highly probable because they do not have error propagation. Furthermore, stream ciphers mostly provide a higher throughput in comparison with block ciphers.

Public-key cryptography

The main function of symmetric algorithms is the encryption of information, often at high speeds. However, there are two problems with symmetric-key schemes:

1. They require secure transmission of a secret key before being able to exchange messages.

2. If in a network environment each pair of users shares a different key, this will result in many keys. For a network with n users, $\frac{n(n-1)}{2}$ individual keys have to be shared beforehand. Hence, this is not feasible in large-scale networks such as VANET.

The idea behind public-key (PK) cryptography can be visualized by making a slot into the locked box so that everyone can deposit a message (like a letter box). However, only the receiver can unlock the box and read the messages inside. This concept was first proposed by Diffie and Hellman in 1976. Public-key cryptography is based on the idea of separating the key used to encrypt a message from the one used to decrypt it. Anyone who wants to send a message to another party, e.g., to Bob, can encrypt that message using Bob's *public key*. However, only Bob can decrypt such messages using his *private key*. It is understood that the private key should be kept secret at all times whereas the public key is publicly available to anyone. Furthermore, it is computationally infeasible for anyone, except Bob, to derive the private key from the public key (or at least to do so in a reasonable amount of time). There are three basic mechanisms where public-key algorithms are used:

- key establishment and key exchange
- digital signatures
- data encryption.

These mechanisms can be implemented using one of the following public-key algorithm families:

- Algorithms based on the *integer factorization problem*: given a positive integer n, it is computationally hard to find its prime factorization, e.g., RSA.

- Algorithms based on the *discrete logarithm problem* (DLP): given α and β it is computationally hard to find x such that $\beta = \alpha^x \bmod p$, e.g., the Diffie–Hellman key exchange and the digital signature algorithm (DSA).

- Algorithms based on *elliptic curves* rest upon the DLP on the algebraic structure of elliptic curves over finite fields.

Despite the differences between their underlying mathematical problems, all three algorithm families perform complex operations on very large numbers, typically 1024–4096 bits in length for the integer factorization and discrete logarithm systems, and 160–256 bits in length for elliptic curve systems. This results in a poor throughput performance compared to symmetric ciphers. Nevertheless, PK algorithms solve the key distribution problem in an elegant way, since the public part of the key can be distributed via an unsecured channel. Hence, one can establish a secure link between two parties without the need for a previously exchanged secret. Thus PK encryption is normally used for transmitting only small amount of data, such as symmetric keys. PK algorithms are not only used for the exchange of secret keys but also for authentication by using digital signatures. Digital signatures are analogous to handwritten signatures. They enable communication parties to prove to a third party that one party has actually generated the message, also called non-repudiation. Since the digital signature is a function of the message content and the private key, only the holder of the private key could have produced the signature. In practical terms, we use the private key for signing (thus only the holder of the non-public private key can sign a document) and the public key for the verification (thus everyone can verify the signature using the openly available public key). For practical implementations, using the RSA algorithm for digital signatures, a significantly smaller public key can be chosen to make the verification of an RSA signature a very fast and facile operation. However, the private RSA key needs to be full length for security reasons. Hence, RSA should be used in applications where the verification is done on the embedded platform and the signing on a personal computer or server. Instead, elliptic curve cryptography (ECC) should be used for applications where the embedded device performs encryption or signature generation as well as decryption or signature verification. Therefore, ECC is often the choice for VANET.

9.4.2 Key management

Safety beacons will contain precise information about vehicle positions, velocities, and accelerations. Vehicles will present warnings to drivers based on information gathered by the vehicle from received safety beacons. These warnings will inform the driver about potentially dangerous situations, such as, hazardous road conditions, excessive speed approaching curves, and emergency braking behavior by other vehicles (Robinson et al. 2006). If these warnings are presented when there are no hazardous situations, then users may become desensitized to the warnings, or the warnings themselves may pose a safety threat. If an attacker can inject falsified packets into the VANET causing this desensitization or causing accidents because drivers do react to the falsified warnings, significant harm could be done as a result of the VANET instead of resulting in the VANET helping reduce vehicle crashes or the severity of crashes. These undesirable outcomes could also result from erroneous information generated by malfunctioning hardware. It is likely that the perverse attractiveness of these undesirable outcomes will cause some users to intentionally generate falsified safety beacons. It is also likely that hardware will fail. We state these assumptions in the following axiom.

Axiom 9.4.1 *Some users of a VANET will misbehave or will have equipment that malfunctions.*

Since vehicles may malfunction and users may misbehave, we want to limit the amount of damage such behavior can cause in a VANET. We state this formally in the following axiom.

Axiom 9.4.2 *Users that misbehave or vehicles that have malfunctioning equipment should be EXCLUDED from the network in order to limit the damage caused by these entities.*

Specifically, we would like to remove such vehicles and users from the network. The main purpose of the security infrastructure is providing key management. Vehicles' OBUs and RSUs need to be equipped with keys in order to perform cryptographic operations. At the same time, vehicles need to be able to securely obtain other vehicles' keys in order to securely communicate with these other vehicles. Furthermore, it is desirable to provide services to renew vehicle keys and revoke them, e.g., if a vehicle misbehaves. The main mechanisms provided by a key management scheme are the following:

1. vehicle registration and certificate issuance

2. key distribution

3. key renewal

4. vehicle revocation.

Key management might provide services using symmetric cryptography only or using PK cryptography. While a purely symmetric solution is theoretically plausible, it would require an online central server. Using an online central server, each communication would require the establishing of a session from sender through central server and to receiver. We rule out this approach in our given setting due to the lack of a reliable online connection and the stringent latency requirements of VANETs.

The correct and widely accepted approach is to implement a PKI. A PKI basically is a mechanism that allows binding public keys to corresponding identities by means of a certificate authority (CA). The CA is an entity that every vehicle will trust, and might be a government agency or a government certified organization. The identity is described by a unique string which might be the vehicle's identification number

(VIN), the license plate number, or a random but unique string. A certificate is similar to a passport being issued by a trusted authority, the government. The passport proves the holder's identification by including an ID and a picture (modeled as public and corresponding private key by a certificate), and certifies authenticity due to forgery-safe properties (modeled by the CA's signature). Each vehicle is deployed with at least one certificate and the corresponding private key. Once a vehicle desires to securely communicate with another vehicle, the communication partners exchange their certificates and verify each other's certificate, after which they can use PK cryptography mechanisms such as digital signatures or key establishment.

As a consequence of Axioms 9.4.1 and 9.4.2, we need to bind safety beacon information to a certificate. Binding this information to a certificate allows vehicles to differentiate between correctly behaving vehicles that should be trusted and malfunctioning or misbehaving vehicles that should not be trusted. This binding also provides a mechanism for excluding misbehaving vehicles from the VANET. We express this requirement in the following corollary.

Corollary 9.4.3 *In order to differentiate between trusted and untrusted vehicles and to exclude untrusted vehicles, safety beacon information should be bound to certificates, belonging to vehicles.*

To remove a vehicle from the network, all of a vehicle's certificates need to be invalidated. To invalidate a certificate on the network, the corresponding certificate must be revoked.

Corollary 9.4.4 *A VANET invalidates a certificate with the goal of removing an untrusted vehicle from the VANET. This process is called revocation.*

Note that if a vehicle has more than one certificate, it is possible that a CA will, at some time, only revoke some, but not all, of the untrusted vehicle's certificates. Only after the CA revokes the remaining certificates of the untrusted vehicle will revocation be complete. The reason this might occur is discussed below. However, even in this case, it remains the goal of the CA to remove the untrusted vehicle. Thus Corollary 9.4.4 remains in force.

Certificates

A certificate is a vehicle's public key and identifier signed by the CA. Let PK_A be the public key of vehicle A, and ID_A the identifier. Let $\|$ be the concatenation of two strings. Then a certificate is computed as

$$Cert(A) := Sig_{CA}(PK_A \| ID_A), PK_A, ID_A.$$

Note that certificates usually include additional data, e.g., expiry date and issuer. Also note that vehicles can be equipped with many certificates. There is a one-to-one relationship between public key PK_A and secret key SK_A. For instance, in the case of ECC the public key is computed as,

$$PK_A := SK_A \cdot Q,$$

where Q is a public base point. Certificates usually include additional data such as lifespan, supported algorithm, etc. In order to verify a certificate, a vehicle needs to obtain the CA's public key. The vehicle can then verify the certificate's signature in order to validate the public key and the identification. The CA's public key acts as root key and is the basis of trust. It will be installed in vehicles at the time of production. The size of certificates, signatures, and public keys is summarized in Table 9.1 according to the format and key sizes defined in IEEE P1609.2. The public key is in compressed format (Harper et al. 1993). The same table also provides run-time estimates for signature generation and verification based on our cryptographic implementation on an embedded CPU running at 400 MHz.

Table 9.1 ECDSA sizes and performance.

ECC-256	
Certificate size	≈120 bytes
Signature size	64 bytes
Public key size	33 bytes
Secret key size	32 bytes
Signature generation	6 ms
Signature verification	23 ms

Recently a new class of certificates was proposed called implicit certificates. One type is the elliptic curve Qu-Vanstone (ECQV) implicit certificates (Research 2006). Instead of explicitly stating an identifier, a public key, and a signature, ECQV certificates have the size of a public key, and the information needs to be reconstructed from the implicit certificate. Although an implicit certificate requires only 33 bytes instead of 97 bytes for a public key and CA signature, and also bandwidth overhead is a crucial performance indicator, we are not aware that implicit certificates have been suggested or applied to VANET applications yet.

PKI topology

There might be several certificate authorities (CAs), not just one. For instance, there might be a CA for each country, or there might be a CA for each state of a country. In the latter case, the PKI can be organized in a hierarchical manner. The topology usually reflects the organization's hierarchy. There is a root certificate (a self-signed certificate), issued by the root CA, say the US DOT. The root CA issues certificates for each sub-CA, e.g. each state. Each sub-CA then issues certificates for vehicles registered in that state. Each vehicle is equipped with the root certificate. If a vehicle needs to verify another vehicle's certificate, it performs the following computations:

1. Obtain the vehicle's certificate. If the vehicle is registered with the same sub-CA, continue to Step 3. Otherwise, obtain the vehicle's sub-CA certificate. The vehicle might broadcast both certificates.

2. Validate the sub-CA certificate using the root certificate.

3. Verify the vehicle's certificate using the sub-CA certificate.

4. Use the vehicle's public key, for instance, to verify messages originating from that vehicle.

This approach requires the verification of two certificates. In general, a PKI can be designed to involve any number of hierarchy levels. However, a PKI of n levels requires nodes to verify up to n certificates. However, these verifications only need to be performed once for each sub-CA certificate and they can be precomputed. Nonetheless, for efficiency reasons, especially in VANET, a flat hierarchy is preferred. One approach is to have a single CA but several registration authorities (RAs). For instance, each state of the USA operates an RA. The RA forwards certificate requests to the CA to issue a certificate. The CA might include a state identifier in the certificate if required.

Node registration and certificate issuance

The registration process might be performed by the OBU or RSU manufacturer, or by the vehicle OEM. The registration might be handled by the CA or a separate RA. The CA then generates a private/public key pair, issues a certificate for the public key, and returns the certificate and the secret key to the node in a secure manner. If a separate RA is involved, the CA returns the certificate to the RA, which forwards it to the node. This might improve the level of privacy if the CA does not know the node's ID and the RA cannot obtain the certificate because it is securely sent from CA to node. A more secure alternative is to have the node generate the public/private key pair, and provide only the public key to the CA for issuing a certificate.

Certificate distribution

There are two main approaches for distributing certificates: the sending node might provide its certificate, or the receiving node might look the certificate up in a public directory. In the latter case, RSUs connected to a central database are required. If one of the communicating nodes is an RSU, this approach is reasonable. In the extreme case, all certificates are stored in a node at production time. However, this approach does not scale well and does not allow new nodes to be added. Therefore, in most cases nodes will distribute their certificates. Before considering the distribution in more detail, it is useful to divide communication into two main categories:

1. *1–1*: Two nodes are communicating and possibly establish a session.

2. *1–n*: A single node broadcasts information to many nodes, probably without establishing a session.

In the first category, certificates are exchanged as part of the session setup. In most cases, the nodes will exchange certificates, verify the public keys, perform a key-agreement scheme based on the public keys, and finally use symmetric schemes for efficiency reasons. The first category is usually used in non-safety applications.

In the second category, messages are broadcast without establishing sessions first. The message is probably digitally signed but not encrypted. A receiver needs to obtain the sender's certificate before it is able to verify the message. A straightforward way to do this is to attach the sender's certificate to each broadcast message. However, there might be settings where one can do it more efficiently.

As already mentioned, a certificate might be attached to a message. Alternatively, the certificate digest can be attached. The certificate digest is a unique digest that can easily be computed, usually by applying a hash function to the certificate. Every time the receiver obtains a certificate, it checks whether it received the certificate in the past. If not, the receiver verifies the certificate and stores it together with the certificate digest. Based on this strategy, Hu and Laberteaux (2006) presented further approaches to certificate distribution in the context of broadcast message authentication of beacon packets:

1. *Periodic broadcast*: The certificate is broadcasted periodically, say every 500 ms. Usually the certificate is sent in a piggy-back fashion together with a message.

2. *Certificate on demand*: Whenever a vehicle A encounters a node B, for which it has no certificate stored, A concludes that B also does not have A's certificate yet. Therefore, vehicle A broadcasts its certificate in this situation. Node B will act in the same manner and broadcast its certificate upon encountering vehicle A. Nodes might even go a step further and actively request certificates by including in their messages a list of identifiers of other nodes, for which they do not have yet a certificate. In order to avoid channel congestion, a hold-off time might be introduced. Therefore, a node must wait a certain time, say 500 ms, before rebroadcasting its certificate.

3. *Hybrid broadcast*: The periodic and on-demand broadcast can be combined in such a manner that certificates are broadcast on demand by default. However, if no certificate is demanded for a certain period of time, the certificate is broadcast in a periodic fashion until a certificate is demanded.

It was shown above that certificates are twice as large as signatures. Therefore, optimizing the certificate distribution mechanism in a VANET is a proper means of considerably reducing the over-the-air bandwidth overhead due to security protocols.

Certificate revocation

Nodes are equipped with certificates and use these for secure communication. However, nodes might misbehave on purpose or by malfunction. Therefore it is desirable to introduce mechanisms to revoke the certificates of such nodes (cf. Corollary 9.4.4). The following steps are necessary for revoking a certificate:

1. detection of misbehaving nodes

2. revocation of detected nodes

3. distribution of the revocation information.

In the first step, misbehaving nodes need to be identified. The identification can be performed by a central agency that analyzes communication data received by RSUs. Additionally, cooperating vehicles may identify malicious nodes. Such approaches are presented in more detail below. In the next step, the corresponding certificates are identified and put on a certificate revocation list (CRL). If certificates include an expiration date, they are removed from the CRL once they expire. Typically, only the certificate digest of the certificate is put on the CRL, which is then signed by the CA. In the last step, the CRL information needs to be distributed to all nodes. The naive approach is to regularly broadcast the entire CRL in VANET by RSUs. RSUs with a constant connection to the CA gain immediate access to the CRL. Vehicles receive the CRL, verify it using the CA's public key, and store it. Whenever a vehicle communicates with another node, it first checks whether the other node's certificate is listed on the CRL. CRLs might grow large such that the distribution of CRLs might be a burden for the communication channel and might lead to congestion. Consequently there might be a delay between when the CRL is updated and when a vehicle receives it. In this time window, vehicles are vulnerable to communication with malicious nodes. There are several strategies for improvement. However, none of these approaches comes without drawbacks such that for a given setting a careful analysis is required to select an appropriate strategy. The following strategies can also be combined.

Differential CRLs The naive approach described above is to publish the entire CRL. CRLs are updated regularly, say whenever a new certificate is added, or periodically. Therefore, besides the full CRL, differential CRLs (also called delta-CRLs) are created. A differential CRL only lists those certificates that were added since the last update. Differential CRLs are numbered such that a node detects missing previous differential CRLs. Differential lists can reduce the over-the-air bandwidth overhead significantly. However, there are methods required to make sure that vehicles are able to obtain missing differential lists.

Partitioned CRLs Another approach is to partition CRLs in order to keep the size of individual CRLs smaller. Each certificate might include an indicator on which partition's CRL it would be put in case of revocation. CRLs can be partitioned once a certain threshold is reached (e.g. 10 000 entries), by geographic location, among all CAs, or certificates might be uniformly distributed to partitions. Therefore, the indicator might be an explicit certificate field stating, say, the state of registration or simply a partition number. In case of a hierarchical PKI, the corresponding CRL might be implicitly given by the sub-CA the certificate was issued by. CRLs are issued by the CA, and in case of a hierarchical PKI by the sub-CAs. In most cases, vehicles need only check certificates against a single partition CRL, in particular if partitions are implemented according to geographic location. However, mechanisms need to be in place such that vehicles obtain enough CRLs from other partitions in order to be able to immediately check a certificate with high probability without requesting a CRL first.

Certificates with short lifespans An alternative approach is to use certificates with short lifespans. For example, a vehicle may be deployed with a certificate with a one-day lifespan. The CA still maintains a CRL. However, the CRL does not need to be distributed at all, or it stays very short such that distribution becomes easier. If a vehicle is listed on the CRL, the CA does not issue a new certificate once the old one expires. Considerable bandwidth is saved since the CRL becomes small or does not need to be distributed at all. However, vehicles need to be able to connect to the RSU network regularly in order to automatically receive new certificates. The more often that vehicles are guaranteed to have such a connection, the shorter the lifespan that can be chosen, and the smaller the maximum possible time delay between detection of a malicious vehicle and its actual revocation (the time when the certificate expires).

Most industrial VANET projects plan to use explicit CRLs. The IEEE P1609.2 standard defines a certificate and CRL format including limited lifespan and explicitly supports partitioned CRLs. Since the standard defines message formats but no mechanisms, it is up to the VANET operator to define and implement proper certificate revocation mechanisms. Further details can be found in the VSC report (Consortium 2006). Also, there is literature available on CRLs used for Internet nodes, e.g., Housley et al. (2002).

Raya et al. (2006a) present VANET-specific methods for revocation. They suggest three methods. The RC^2RL (revocation using compressed certificate revocation lists) protocol uses Bloom filters to compress CRLs. The size savings are provided at the cost of additional computational burden on the nodes' side. The RTPD (revocation of the tamper-proof device) mechanism assumes that all nodes are equipped with a tamper-proof device (sometimes also called a trusted platform module – TPM – or trusted computing platform). Instead of broadcasting a CRL, the CA sends a command to a vehicle's tamper-proof device to turn off. In particular, if a vehicle holds more than one certificate, this mechanism is very efficient. However, an implementation needs to make sure that an adversary cannot simply detect such disable command messages and filter them out. Therefore, messages need to be encrypted between the CA and the device in such a way that an attacker cannot distinguish commands to turn off from other messages, e.g., based on size. This protocol is envisioned to support a scheme based on CRLs and not take the only line of defense. The third protocol, Distributed Revocation Protocol (DRP), is a mechanism for nodes to detect misbehaving nodes and warn their neighborhood. Therefore, it can be seen as a temporary revocation from the neighborhood with the possibility of forwarding this information to the CA for further action. Protocols for detecting misbehaving nodes will be discussed in more detail below.

Certificate renewal

We described above the way nodes are equipped with certificates at production time in order to enable secure communication later. Furthermore, we explained that certificates should have a limited lifespan to reduce the CRL size or even make a CRL obsolete. Therefore, there needs to be a method of receiving new certificates in a secure manner. Certificate renewal might be provided in certified workshops only, or it might be performed automatically in the field, e.g., via RSU. In any case, there should be mechanisms in place such that dishonest workshop employees cannot manipulate the renewal process. We suggest a renewal process in the following discussion. A change of the root certificate or addition of another root certificate can also be done in a similar fashion.

A vehicle A holds a private/public key pair SK_A/PK_A as well as the current certificate $Cert(A) := Sig_{CA}(PK_A \| ID_A)$. The vehicle now generates a new private/public key pair SK'_A/PK'_A. It signs the new public key including some further information for the CA by computing $S := Sig_{SK_A}(PK'_A)$ and sends $(PK'_A, S, Cert(A))$ to the CA. The CA first verifies the certificate by checking whether $Ver_{CA}(Cert(A)) \stackrel{?}{=}$ 'valid', i.e., by checking whether $Cert(A)$ was signed with CA's secret key. If so, the CA verifies whether $Ver_{PK_A}(PK'_A, S) \stackrel{?}{=}$ 'valid' where PK'_A is the message to verify against the signature S. If the verification is successful, the CA issues a certificate $Cert'(A) := Sig_{CA}(PK'_A \| ID_A)$ and sends it to A, for example, via an RSU.

The process needs to be slightly changed if A wants to select a new identity ID'_A without revealing old and new identity on the transmission channel in plain text. Again, A generates a new key pair SK'_A/PK'_A as well as a new identity ID'_A. A then assembles a message $M := (PK'_A, ID'_A)$, signs the message $S := Sig_{SK_A}(M)$, encrypts it using the CA's public key as $E := Enc_{CA}(M \| S \| Cert(A))$, and finally sends E to the CA. The CA decrypts E as $(M \| S \| Cert(A)) = Dec_{SK_{CA}}(E)$, verifies the certificate $Cert(A)$ and the signature S as described above. The CA now issues a new certificate $Cert'(A) := Sig_{CA}(PK'_A \| ID'_A)$, encrypts it for A as $E := Enc_{PK'_A}(Cert(A'))$ and sends it to A. Finally A decrypts E to obtain the new certificate.

Interoperability

We showed that there might be several CAs. Even if a country uses a single CA, it is highly probable that each country will operate its own CA. Therefore, there need to be mechanisms in place such that vehicles registered in one country can cross the border and participate in the VANET of another country. It should even be possible that two vehicles of different countries are able to communicate with each other while in a third country. The following approaches enable such interoperation:

Cross certification CAs accept each other. Therefore, they sign the certificate of another CA to indicate that vehicles registered with the other CA are allowed to participate in VANET communication. If a vehicle A now encounters a node B from another country, it needs to obtain B's cross-signed root certificate, verify the cross-certification, and then verify B's certificate using the cross-certificate's public key. In this instance, it does not matter where the vehicles meet. Nonetheless, vehicle A registered with CA_1 entering another territory under the authority of CA_2 should automatically be equipped at the border with cross-certificates. The vehicle needs the cross-certificate of CA_1 signed by CA_2, but also the cross-certificate of CA_2 signed by CA_1. A uses the latter to verify the certificate of node B which is registered with CA_2, and A provides the former to B in order to allow it to verify A's certificate.

Note that while the additional computational overhead for A is a single certificate verification, each node B needs to perform an additional certificate verification upon encountering A. However, B only needs to compute the verification once for each cross-certificate.

Temporary certificates Another method is to provide temporary certificates to vehicles entering the territory of an alien CA. There needs to be a cross-certification of CAs beforehand. Once vehicle A passes the border, it requests a temporary certificate. The process is very similar to a certificate renewal. A generates a temporary key pair SK_T/PK_T, signs the temporary public key $S := Sig_{SK_A}(PK_T)$, and sends $(PK_T, S, Cert(A))$ to the alien CA_1. The alien CA_1 verifies the cross-certificate, checks whether the certificate $Cert(A)$ is valid, verifies the signature S, issues a temporary certificate $Cert(T) := Sig_{CA_1}(PK_T \| ID_A)$, and sends it including the cross-certificate $Sig_{CA_2}(Cert(CA_1))$ to A. A now verifies the cross-certificate, then verifies the certificate $Cert(T)$, and finally uses it. This method is elegant since after this time, A does not impose in other nodes additional computational overhead for certificate verification.

9.5 Cryptographic Protocols

In this section, we present the main cryptographic protocols applied to VANET. First, we consider authentication protocols. Authentication – knowing to whom you are talking – is the core requirement for establishing a trust relationship. Another core requirement is confidentiality, which is provided by encryption algorithms. Other protocols include key agreement, secure positioning schemes, and protocols for identifying misbehaving nodes. We discuss privacy in detail in a separate section later on. In the following we assume our VANET to use ECC with a key length of 256 bits according to IEEE P1609.2 since ECC shows the best overall performance of established cryptographic systems on computationally resource constrained platforms deployed in vehicles.

9.5.1 Certificate verification

Vehicles and RSUs exchange certificates before starting secure communication. All protocols require the verification of certificates before continuing. Therefore, we define a general method CERT_CHECK to verify certificates as described in Algorithm 1. We assume a flat PKI for efficiency reasons. The input to the algorithm, $C(A)$ is either a certificate $Cert(A)$ or a certificate digest, which is usually the truncated output of $H(Cert(A))$ where H is a hash-function.

The CERT_CHECK algorithm requires a single signature verification or none at all. We assume that browsing the CRL and storing the certificate requires negligible time. Therefore, CERT_CHECK runs in 11 ms when verifying a certificate (cf. Table 9.1), and in negligible time when checking a certificate digest. In the following discussions, we describe the time required for CERT_CHECK by $time_{CC}$.

Algorithm 1 Certificate verification *CERT_CHECK*.
1: **if** C(A) is a certificate **then**
2:    Verify whether $Ver_{CA}(C(A)) \stackrel{?}{=} \text{'valid'}$.
3:    Store $C(A)$ together with a validity flag and its digest.
4: **else if** C(A) is a certificate digest **then**
5:    Verify whether $C(A)$ is stored and valid.
6: **end if**
7: Verify the CRL for $C(A)$.

9.5.2 Encryption

Encryption is one of the core security services to provide confidentiality and avoid eavesdropping. Encryption can be provided by both symmetric and asymmetric cryptography. In the case of symmetric cryptography, it is essential that the parties share a secret key, which might be agreed upon by a key agreement scheme as described in the following section. Encryption and decryption are performed using the same key, and are denoted by

$$E := Enc(m, K)$$

and

$$m := Dec(E, K)$$

where *m* is an arbitrary message string. A popular encryption algorithm is the standardized block cipher AES (US Department of Commerce/National Institute of Standard and Technology (2001). Block ciphers can operate in several modes Menezes et al. 1997). Every party that holds the secret key K is able to decrypt such messages. Therefore, symmetric encryption can be used in a pairwise (*1:1* relationship) or groupwise (*1:n* relationship) fashion.

Asymmetric encryption is usually performed using RSA or ECC. Since we favor ECC for VANET, we focus on ECC encryption here. The elliptic curve integrated encryption scheme (IEEE 2003) (ECIES; sometimes also called elliptic curve augmented encryption scheme or elliptic curve encryption scheme) basically combines a key agreement scheme followed by symmetric encryption into a single scheme. While it is possible to separately run a key agreement scheme based on public-key cryptography, as described above, and then execute a symmetric scheme, as just described, ECIES does not require interaction and is more efficient. We denote encryption as

$$E := Enc_{PK}(m),$$

and decryption as

$$m := Dec_{SK}(E)$$

where *PK* and *SK* are the public and private keys, respectively. An ECIES encryption takes roughly two point multiplications, whereas a decryption takes only a single point multiplication to determine the key for efficient symmetric encryption. The run-time performance for an embedded CPU running at 400 MHz platform is summarized in Table 9.2 and assumes that only small payloads are encrypted such that the computational demand for the public-key operation outweighs the actual symmetric encryption.

Table 9.2 ECIES performance.

Encryption	23 ms
Decryption	17 ms

9.5.3 Key agreement

Key agreement allows two parties to agree on a symmetric key by interacting over an insecure channel in such a way that an eavesdropper is not able to derive the key. Therefore, public-key methods are used. The computed shared key is then used for symmetric cryptography. It is usually only useful to execute a key agreement scheme in a VANET when there is a long-term or intensive relationship between two nodes, not when two nodes meet only once during their lifetime. Examples where key agreement is useful include the communication between an OBU and an RSU if the vehicle passes that RSU daily during its commute, or between OBU and a service provider. There are two main strategies:

1. Two parties perform a protocol in such a way that both parties influence the outcome. A standard example is the Diffie–Hellman key exchange protocol based on ECC.

2. Party A randomly chooses a key, encrypts it using B's public key, and sends it to B. This approach is often based on RSA and is usually referred to as key transport. In this strategy, A has full control over the key selection process.

The key agreement protocol of choice for ECC is the elliptic curve Diffie–Hellman (ECDH) key exchange protocol (ANSI X9.63-1998 1998), described by Algorithm 2. Note that the public key is derived from the secret key in ECC as $PK := SK \cdot Q$.

Algorithm 2 Diffie–Hellman key agreement.
1: A sends B its certificate $Cert(A)$.
2: B sends A its certificate $Cert(B)$.
3: A and B run CERT_CHECK.
4: A computes $K := SK_A \cdot PK_B = SK_A \cdot SK_B \cdot Q$.
5: B computes $K := SK_B \cdot PK_A = SK_B \cdot SK_A \cdot Q$.

At the end, both A and B share a common key K. The performance of ECDH is summarized in Table 9.3. The point multiplication $k \cdot PK$ takes more time than a signature generation but less time than a signature verification. Indeed, owing to the design of ECDSA signature verification, the time for signature verification equals the time for a point multiplication plus the time for a signature generation, resulting in $23 - 6 = 17$ milliseconds for the point multiplication running on an embedded CPU at 400 MHz. The over-the-air (OTA) bandwidth overhead is the sum of two certificates.

Table 9.3 ECDH performance.

Run-time (A & B)	$time_{CC} + 17$ ms
OTA bandwidth	240 bytes

An actual implementation of the ECDH protocol might be refined to include an ephemeral value such that the computed shared secret between the same parties differs each time they run the protocol. Performance improvements were introduced by the ECMQV protocol. Further details can be found in the IEEE P1363 standard (IEEE 1999, 2003).

The described EDCH protocol provides pairwise key agreement. There are also schemes to provide a groupwise key agreement. After running such a protocol each member of the group shares a common secret key. Again, there are two flavors; in the first one there is a dedicated node that selects a random secret key and distributes it to the other nodes, whereas in the second one each node contributes a share and is then able to compute the common secret. The latter approach has been intensively researched for ad-hoc networks (Augot et al. 2007). However, this approach requires two or more rounds and is computationally demanding. A naive method for key transport is to let the dedicated node randomly select a secret and then encrypt it for all nodes successively using each node's public key. A more sophisticated approach shifts a part of the computational burden from the dedicated vehicle to vehicles that have already received the key. Again, this approach is resource-demanding. The underlying assumption is that vehicles trust each other for the messages protected by the group key. We expect that such an approach can only be applied to vehicles that travel together over a long period of time but not to groups of vehicles that randomly meet for only a few seconds. Therefore, expensive group key agreement schemes are possible for such platooning scenarios. We expect this to be an unusual case and do not go into further detail here.

We suggest the following approach for the case where a set of vehicles is grouped around an RSU. Here, the RSU selects a random key, and sends the encrypted key to a vehicle once it enters the RSU's transmission coverage. This approach is reasonable if the RSU is computationally powerful, and it might be applied to a variety of applications including privacy protection, and reducing the computational burden in congested areas by using symmetric cryptography rather than asymmetric cryptography. This mechanism is described in Algorithm 3. New vehicles are easily added to the group, and leaving vehicles do not need to be especially considered.

This approach is vulnerable to adversaries that are able to extract keys from an OBU. These adversaries gain K and are then able to manipulate the communication. Therefore, vehicles need to restrict the type of messages encrypted with K, and the

Algorithm 3 RSU initiated group key agreement.

1: RSU randomly selects a key K.
2: Once RSU receives a new certificate, $Cert(A)$, it runs CERT_CHECK and extracts PK_A.
3: RSU computes $E := Enc_{PK_A}(K)$ and sends E to A.
4: A decrypts $K := Dec_{SK_A}(E)$.

RSU needs to actively scan for attacks and react properly. Such reaction might be troublesome since a simple change of keys will not suffice. In the worst case, the RSU needs to stop the service and alert law enforcement.

9.5.4 Authentication

Authentication is the core security requirement in VANET. Authentication provides message integration in order to avoid message manipulation. Basically, all applications in a VANET require authentication. Authentication comes in different flavors: message authentication versus entity authentication (identification), or broadcast, pairwise, and groupwise authentication. We require that, by default, each message in a VANET includes authenticated position information and a timestamp such that replay and tunnel attacks are avoided.

Digital signature broadcast message authentication

Broadcast message authentication provides the security service where a single node authenticates a message and n receiver nodes are able to verify the message. This authentication mechanism is the default for VANET and it is standardized in IEEE P1609.2. No session is established, but messages are sent in a stand-alone manner. For instance, broadcast message authentication is used for authenticating vehicles' beacon messages and safety messages as well as RSUs' safety messages and information messages. The basic broadcast message authentication protocol is described in Algorithm 4. The nodes might be either OBUs or RSUs.

Algorithm 4 Broadcast message authentication.

1: Node A signs a message m as $S := Sig_{SK_A}(m)$ and broadcasts $(m, S, C(A))$ where $C(A)$ describes either A's certificate or its certificate digest.
2: Receiver B runs CERT_CHECK and extracts A's public key.
3: B verifies the CRL for $C(A)$.
4: B checks whether $Ver_{PK_A}(m, S) \stackrel{?}{=} \text{'valid'}$.

Table 9.4 provides an overview of ECDSA broadcast message authentication protocol performance running on a 400 MHz CPU. An ECDSA certificate has a size of 120 bytes, whereas a certificate digest of 8 bytes is sufficient and is denoted as $|C(A)|$.

Table 9.4 ECDSA broadcast message authentication performance.

Run-time (sender)	6 ms
Run-time (receiver)	$time_{CC} + 23$ ms
OTA bandwidth	64 bytes + $|C(A)|$

An attractive signature scheme based on ECC was proposed by Pintsov and Vanstone (2001). The elliptic curve Pintsov–Vanstone signature (ECPVS) provides message recovery, and more importantly for VANET, comes with a small signature size. Instead of ECDSA's 64 bytes it only requires 32 bytes for a signature. However, to the best of our knowledge such signatures have not been proposed or applied to VANET so far.

Pairwise and groupwise authentication

Previously we described broadcast authentication where there is a 1:n relationship between sender and receivers. However, there are also cases in VANET where there is a 1:1 relationship. For instance, a vehicle might daily approach the same RSU during its commute, or repeatedly demand the same service from a provider. We call this a pairwise relationship and apply different authentication protocols in this case. Establishing pairwise authentication is only useful if there are repeated interactions. Otherwise, the above mechanisms for broadcast authentication can also be used for a single receiver.

The authentication mechanism of choice for pairwise authentication is a symmetric message authentication code (MAC). Computing a MAC is several orders of magnitude faster than generating a digital signature. MAC algorithms are often based on hash algorithms, and a typical MAC family is the HMAC-SHA algorithm. In our case, using HMAC-SHA-256 is reasonable, and it outputs 32-byte tags. Note that in many settings truncating the output tag is reasonable, especially in systems where manipulation is detected after several invalid MAC verifications. Algorithm 5 shows how MACs are used.

Algorithm 5 Message authentication code.

Require: Nodes A and B share a common secret key K.
1: Node A computes the message authentication code $M := MAC(m, K)$ over message m using shared key K, and sends (m, M) to B.
2: B receives (m', M') and computes $\bar{M} := MAC(m', K)$.
3: B accepts the message if and only if $\bar{M} = M'$.

Using a MAC requires that A and B share a common secret key. In a VANET, we observed that predistributed keys are not feasible but that such a shared secret needs to be derived by means of a key agreement scheme as was described above. A correct design executes a key agreement scheme, and then uses a key derivation function

(KDF) for the agreed session key to derive two separate keys used for encryption and authentication, thus avoiding excessive exposure of the agreed master key. We measured the average time of an HMAC-SHA-256 operation using 512-bit blocks to be 18 μs whereas using a single 512-bit block takes 36 μs running on a 400 MHz CPU. The performance is summarized in Table 9.5. However, the time to perform a key agreement needs to be added for completeness.

Table 9.5 MAC performance.

Run-time (sender)	$\lvert m \rvert /512 \times 14\mu s$
Run-time (receiver)	$\lvert m \rvert /512 \times 14\mu s$
OTA bandwidth	32 bytes

A groupwise authentication can be performed in exactly the same manner. First, a key agreement scheme needs to be executed such that a group of nodes obtains a common shared key K. Then, Algorithm 5 is performed with a sender A and several receivers B. Also in the case of groupwise authentication an execution is only useful and efficient if the group exchanges several messages before breaking up again, or if the nodes will regroup. Otherwise, using a broadcast authentication scheme provides superior performance.

TESLA broadcast message authentication

TESLA is another broadcast message authentication algorithm, which was introduced by Perrig et al. (2000, 2001, 2002). TESLA provides run-time efficient authentication based on a mixture of digital signatures (in our case ECDSA) and MACs generated using symmetric cryptography at the cost of authentication delay at the receivers' side. Therefore, it combines the above-described ECDSA broadcast authentication and MAC authentication. TESLA was originally intended for authenticating broadcast streams where a delay in authentication is negligible, such as authentication of a multimedia file stream or a stock market ticker. Hu and Laberteaux (2006) applied TESLA to VANET, and conclude that, despite some obstacles, TESLA is an appropriate authentication mechanism for VANET.

TESLA uses time to provide asymmetric signature properties with symmetric functions. The sender Alice first generates a hash-chain with temporary keys $k_i = h(k_{i+1})$ for $i = 0 \ldots n$.[2] First, the final element k_0 is broadcast to all receivers in an authenticated manner, e.g., by digitally signing it. Then Alice sends messages m_i authenticated by k_i in time interval t_i. TESLA is based on the security condition that such a message is only accepted during the time interval t_i but not later. In the next time interval, Alice discloses k_i and the receivers verify m_i. The protocol works as presented in Algorithm 6. The receiver must buffer messages before they can be verified. Furthermore, there needs to be time synchronization between sender and receiver. Otherwise, after a key was opened an attacker could use that key to forge messages. We observed above that vehicles in VANET come with GPS units and thus

---

[2]Remember that hash functions are one-way functions such that deriving k_{i+1} from k_i is computationally infeasible.

are constantly synchronized to a single global time signal such that accurate time synchronization can be assumed. When GPS is lost, time can be estimated much more accurately (or cheaply) than position estimates. That is, by the time the clock skew becomes so large as to become a problem, the location estimate is useless, and need to be broadcast.

Algorithm 6 TESLA broadcast authentication.

1: Initially, A signs $S := Sig_{SK_A}(k_0)$ and broadcasts $S, C(A)$.
2: Each receiver B runs CERT_CHECK and verifies S.
3: **for** message m_i in time interval t_i, $i = 1$ to n **do**
4:     A computes $M_i := MAC(m_i, k_i)$ and broadcasts M_i, m_i.
5:     B checks whether it received M_i, m_i in time interval t_i and buffers it.
6: **end for**
7: **for** message m_i in time interval t_{i+1}, $i = 1$ to n **do**
8:     A broadcasts k_i.
9:     B checks whether $M_i \stackrel{?}{=} MAC(m_i, k_i)$.
10: **end for**

Table 9.6 summarizes TESLA performance for a 400 MHz CPU. We assume that there are n messages broadcast per protocol run. A digitally signed key needs to be sent out regularly to allow newly encountered vehicles to verify messages. Therefore, it appears reasonable to choose n in the range of 3 to 10 assuming that messages are broadcast every 100 ms, i.e., signed keys are sent out every 300 to 1000 ms. The time for computing a symmetric message authentication code (MAC) is negligible in this setting. Hu and Laberteaux (2006) argue that a 10-byte key and 10-byte MAC output are sufficient for a VANET setting since they have a very short lifespan of only a few milliseconds.

Table 9.6 TESLA broadcast authentication performance.

Run-time (sender)	$6/n$ ms
Run-time (receiver)	$time_{CC} + 23/n$ ms
OTA bandwidth	$64/n + 20 + |C(A)|$ bytes

Identification

Unlike message authentication, identification (or entity authentication) enables a claimer to prove knowledge of a secret that only the claimer knows, thus proving its identity. An identification process needs to include timeliness in order to prove that the claimer definitely has knowledge of the secret. Timeliness might be proven by an interactive challenge–response protocol or by authenticating a timestamp (Menezes et al. 1997). Since there is a globally accurate time source available in VANET, we suggest using the timestamp method described in Algorithm 7. There are several variations of this scheme. In particular, the digital signature can be replaced by a symmetric MAC or by encryption, and mutual entity authentication can be performed more efficiently than running the scheme twice. We expect that identification will be applied to identify a vehicle to an RSU or to a service provider, e.g., for tolling, payment services, and subscription services.

Algorithm 7 Entity authentication with timestamp.

1: A computes $S := Sig_{SK_A}(t \| B)$ and sends $S, t, Cert(A)$ to B.
2: B runs $CERT\_CHECK$, verifies that the timestamp t is acceptable, and checks whether $Ver_{PK_A}(S, t \| B) \stackrel{?}{=} 'valid'$.

Other authentication mechanisms

There is a wide variety of authentication schemes available in the literature. However, the requirements of VANET are highly specific such that, to the authors' best knowledge, no such protocol has been suggested for VANET. The asymmetric MAC broadcast authentication scheme composes multiple symmetric MACs to reflect an asymmetric digital signature (Canetti et al. 1999). However, the scheme does not scale well for the number of vehicles anticipated. One-time and k-time signature schemes, such as Perrig's BiBa broadcast authentication (Perrig 2001), require very large public keys in the range of 10 Kbyte. Therefore, they are not well suited for today's envisioned VANET having low bandwidth communication channels. Related to one-time and k-time signature schemes is so-called signature propagation and traversal that authenticate several packets with a single signature. For instance, the hash-value of the first packet is attached to the second, and so on. The last packet of such a chain is finally signed. The receiver needs to buffer all messages until it finally authenticates all packets of the chain by verifying the digital signature. The computational burden, as well as the authentication tag size, is reduced at the cost of time delay. Applying such an approach conflicts with the stringent time delay requirements in VANET. However, for non-safety VANET applications it might be appropriate.

Recently, ID-based signature schemes have had a revival, based on pairing-based cryptography. It has even been shown that ID-based schemes are efficient enough to run on sensor network devices (Oliveira et al. 2007). ID-based signature schemes have an advantage in VANET since they do not require distributing certificates, saving considerable bandwidth. However, ID-based signatures are still an order of magnitude slower than conventional ECDSA signatures.

Additionally, group signatures may be used. In a group signature scheme, each member of a group holds a private key, and there is a single public key for the group. Each group member is able to create a signature in such a way that the message can be verified using the group's public key. A verifier is not able to determine which group member actually signed the message. The verifier is also not able to determine whether two messages were signed by the same group member.

A simple setup might define all vehicles as a single group. Unfortunately, the group signature schemes known so far are around two orders of magnitude slower than ECDSA. Therefore at this point, group signatures seem to be an incorrect approach for computationally constrained platforms in vehicles.

9.5.5 Secure positioning

Most VANET projects assume that secure positioning will be provided by GPS, which provides both time and location for all vehicles worldwide. Today's vehicles usually have a GPS receiver built in for navigation systems such that it can be concurrently used by VANET at little additional cost. GPS depends on line-of-sight communication with satellites and therefore does not work properly in certain settings such as in tunnels or in an urban canyon like Manhattan. Modern GPS devices in vehicles use a vehicle's speed and direction in order to determine the vehicle's position after loss of the GPS signal until the GPS signal is recovered. Clearly, without GPS the accuracy of the extrapolated location degrades over time. Another weakness of GPS is its vulnerability to attacks. An attacker can simply tamper with the GPS sensor to make the vehicle think that it is at a falsified location. For instance, an attacker can either override the GPS signal or disconnect the GPS antenna and insert a forged GPS signal in the vehicle's communication bus. Both attacks could be mounted relatively easily. These attacks are limited to a single vehicle or a small area. An attack that forges the global GPS signal appears to be unrealistic and is usually not considered to be a vulnerability.

Based on these observations, Hubaux et al. (2004) suggest an approach based on RSUs that applies distance bounding. Here the RSUs are assumed to be trustworthy and they determine a vehicle's location. Distance bounding determines an upper bound for the distance between verifier and claimant. Algorithm 8 describes the scheme suggested in Hubaux et al. (2004). First, the claimant vehicle A and the verifier RSU B verify identification in a mutual manner. The claimant A then commits to two nonces, n_A, n'_A. Such a commitment is usually provided by applying a hash function H to the committed value and publishing the hashed value. The commitment avoids the situation where the claimant alters its choices after publishing the commitment. The verifier B sends out a challenge c to the claimant A and measures the time until the response is received. The claimant A is not able to compute the response before receiving the nonce. Obviously, the claimant can pretend to be further away than it actually is. However, the claimant cannot pretend to be closer.[3] In order to decouple such schemes from the time required to compute the response, very efficient operations such as XOR are used to calculate the response. The time for actually computing the response and other time overheads (e.g. network processing delay) are deducted. Based on the measured time and the speed of radio transmissions (the speed of light), the verifier can then calculate the upper distance bound. Finally, A opens the second nonce such that B can verify whether the original commitment holds. A sends its certificate or certificate

---

[3]We assume here that attackers do not collude. In particular, we assume that there are no colluding attackers that have access to the same valid public/private key pair, and that only a single vehicle is able to respond to the verifier's requests.

DATA SECURITY IN VEHICULAR COMMUNICATION NETWORKS

digest, depending whether B obtained A's certificate beforehand during another communication. As a final step, RSU B might send the determined location to A such that A can compare it to its GPS location, and use it for its safety applications.

Algorithm 8 Distance bounding in VANET.

1: Vehicle A generates random nonces n_A, n'_A and computes the commitment $c := H(n_A, n'_A)$.
2: A sends c to the verifier B.
3: B generates a random nonce n_B and sends it to A.
4: A computes $n := n_A \oplus n_B$ and sends the result to B.
5: The verifier B measures the time t between sending n_B and receiving the result n and determines the distance based on t.
6: A computes the signature $S := Sig_{SK_A}(A \parallel n'_A)$ and sends $S, n'_A, C(A)$ to B.
7: B runs CHECK_CERT, verifies whether $Ver_{PK_A}(S, A \parallel n'_A) \stackrel{?}{=} \text{'valid'}$, calculates $n_A := n \oplus n'_A$, and checks whether $C \stackrel{?}{=} H(n_A, n'_A)$.

If distance bounding is used by a vehicle in a VANET and by at least four RSUs (or three RSUs to determine the location in two dimensions) with known location, at the end of running Algorithm 8 each RSU determines an upper bound on the vehicle's distance. If these distances are interpreted as a sphere around each RSU, and if there is an intersection of all spheres that lies in between the four RSUs, the vehicle's position is determined. A cheating vehicle that enlarges its alleged distance by delaying a response will be detected. Certainly, real-time submission of information via DSRC is crucial in this situation, and even small time delays distort the results. The MAC and network layer might introduce unpredictable delays before a packet is actually sent out which might prevent precise timing. Switching communication channels might introduce additional delays. The delays in Algorithm 8's distance bounding algorithm mainly come from the delay of a digital signature generation. We assume 16-byte nonces, and we truncate the hash function output to 16 bytes as well. Table 9.7 summarizes the performance for a 400 MHz CPU.

Table 9.7 Distance bounding.

Run-time (vehicle)	6 ms
Run-time (RSU)	$time_{CC} + 23$ ms
OTA bandwidth	$128 + |C(A)|$ bytes

9.5.6 Identification of misbehaving nodes

Identification of misbehaving nodes and revocation of those nodes is a crucial issue in VANET. Even if there are effective mechanisms in place to revoke vehicles, misbehaving vehicles first need to be identified. Vehicles can misbehave in a variety of ways. They might malfunction, their sensor input might be manipulated, or their cryptographic keys might be extracted in order to forge messages. Vehicles may also be stolen or vehicles with expired registration or insurance policies might not necessarily be evicted from the VANET. Alternatively, drivers of such vehicles might take more risks than average drivers such that safety applications are more useful for these drivers.

Raya et al. (2006a, 2007) present a scheme to detect misbehaving nodes and then temporarily evict them. In the misbehavior detection system (MDS), a node uses its own sensory and GPS input, received messages, and a set of evaluation rules to detect faulty received messages. MDS needs a rule basis to detect both the manipulation of protocol execution and the manipulation of transferred data. MDS then evaluates the actual behavior of a node in the direct one-hop neighborhood based on the expected average behavior. If this evaluation is above a well-defined threshold, a vehicle is identified as misbehaving. Note that MDS does not distinguish between data sources and data relays in a multi-hop scenario. Therefore, a vehicle that forwards faulty messages will be detected as misbehaving and will be contained. After MDS identifies a misbehaving node, its identity is passed to the local eviction of attackers by voting evaluators (LEAVE) protocol. LEAVE provides a service such that an honest majority can temporarily evict a misbehaving node. Consider vehicles running the LEAVE protocol broadcasting accusation messages to the shared neighborhood warning of a misbehaving node. Each vehicle supporting the accusation then signs this message until a well-defined threshold t is reached. If $n \geq t$ vehicles support the accusation evidenced by their signatures, a disregard message is broadcast which instructs all neighbors of the misbehaving node to ignore its messages. At the same time, the disregard message is forwarded to the CA so that it can take further action, such as permanently revoking the identified vehicle. While these works describe mechanisms supporting such a scheme, it is up to the implementer to define a set of evaluation rules. The scheme is based on the assumption that there is an honest majority of vehicles available in the neighborhood. In most scenarios, it can be expected that there is a huge majority of honest vehicles but only a small fraction of misbehaving ones. However, attackers might collude or a single attacker might control several manipulated DSRC radios in his vehicle, using multiple radios (also known as the Sybil attack). The adversary is then able to maliciously incriminate behaving nodes. Therefore, there needs to be a mechanism to include appropriate evaluation rules in MDS and to uncover such misbehaving on the CA's side.

Golle et al. (2004) present a more general approach, where received data is compared to a physical model of the VANET that dictates certain rules and statistical properties of events (e.g. two vehicles cannot be at the same location, and vehicles usually do not move at 200 mph). A vehicle accepts data if it conforms with its VANET model with a high probability and otherwise the data is rejected. This approach considers colluding attackers and Sybil attacks. However, the authors do not describe details of an actual algorithm. There is a wide variety of further work available in this area. However, most of this work is set in the Internet, static sensor networks, or other scenarios that have very different assumptions.

9.5.7 Summary

We reviewed protocols for VANET that provide authentication, confidentiality, secure positioning, and identification of misbehaving nodes. In our view, authentication is the basic core requirement for implementing more complex protocols. Broadcast message authentication is especially useful in safety applications where short latency is required and heavy load due to beacon messages may be observed. Broadcasting nodes might be vehicles or RSUs. We propose one of two approaches, namely ECDSA digital signatures, standardized for VANET in IEEE P1609.2, and TESLA. While ECDSA comes with a heavy computational demand, TESLA requires timeliness. Pairwise authentication schemes between two nodes appear to be especially useful in relationships between vehicle and RSU or service provider that span over several communication sessions or involve a single message-intensive communication session. For pairwise authentication, first a key agreement scheme is executed, e.g. ECDH, followed by a symmetric pairwise MAC. In such a pairwise relationship, proof of identity might also be required. We expect that groupwise authentication and key agreement mechanisms will in particular be applied in settings where an RSU sets up a group for reducing the burden on channel due to security overhead. Moreover, it will probably only be applied to a few dedicated applications such as permanent platooning. Group signatures will, if at all, mainly be used in applications for the sake of privacy where there are no latency requirements. Table 9.8 summarizes these authentication schemes. Finally, we suggested considering implicit certificates as well as short digital signatures such as ECPVS for VANET.

We also considered secure positioning, and detection of misbehaving nodes. We believe that the detection of misbehaving nodes is a core requirement for the successful deployment of VANET. We are not aware of any actual implementations of such schemes and encourage further research and development in this area. We considered positioning based on GPS as sufficient for most applications, and the application of secure positioning protocols in areas without GPS coverage and for applications that are highly dependent on a reliable position.

9.6 Privacy Protection Mechanisms

Privacy is a central requirement for VANET. We are mainly concerned with how specific key assignment methods affect vehicular privacy in VANET, and we define privacy in this context as the inability to link a broadcast signature to a vehicle or a usefully small group of vehicles. Though operating a vehicle already involves significant privacy risks from technologies such as automated toll collection and automatic license plate recognition (Chang et al. 2004; Naito et al. 2000), VANETs are unique in that vehicles use relatively long-range, non-line-of-sight radio communications to very accurately advertise their positions in safety beacon messages. Furthermore, these advertisements are signed using some key that has been assigned to that vehicle, and, according to the IEEE 1609.2 draft standard, are accompanied by a certificate, certificate digest, or certificate chain that attests to the validity of the key (IEEE 2006). Most of the contemplated VANET designs

Table 9.8 Authentication protocols.

Application type	Authentication type	Application comment	Protocol comments
Safety applications	ECDSA broadcast authentication	Sign beacons and warnings	High computational burden
	TESLA broadcast authentication	Sign beacons and warnings	Introduces timeliness
Non-safety applications	ECDSA broadcast authentication	Sign periodical messages (commercials, etc.)	Only few signatures required
	TESLA broadcast authentication	Sign periodical messages (commercials, etc.)	Only few signatures required
	Pairwise ECDH + MAC + encryption	Communication session (tolling, payment, etc.)	Efficient for repeating/intensive sessions
	Groupwise key transport + MAC (+ encryption)	RSU centered grouping Permanent vehicle grouping (platooning)	Vulnerable to attacks Initially high computational burden
	Group signatures	Privacy preserving	Very high computational burden

require VANET vehicles to regularly broadcast their certificate(s). Even if a certificate contains only the vehicle's pseudonym, if care is not taken in how the VANET is designed, and specifically, in how keys are assigned to vehicles, it may be possible to remotely track individual vehicles in a VANET using their certificates and signatures. That said, Axiom 9.4.2 and Corollary 9.4.4 require an efficient revocation process. Given this requirement, it may not be possible to provide the same level of privacy with a VANET as compared to what privacy would exist without the VANET. We evaluate the privacy provided by key assignment methods, the robustness of those methods as defined by the properties we describe in Section 9.6.1, and the ability of those methods to maintain VANET security goals.

We can divide attempts to preserve privacy into: 1. preserving privacy from the CA, or 2. preserving privacy from non-CA entities (e.g., other vehicles). There are many reasons to be interested in providing privacy to vehicles, including protection from big-brother behavior of governments and corporations, and auto manufacturers' concerns of acceptability of VANET to consumers. We will discuss the motivation for providing privacy when we discuss the details of CA privacy and non-CA privacy in Section 9.6.1. However, our primary concern is to analyze the viability of various key assignment methods for a VANET with respect to the privacy and security these methods provide.

Vehicles may be assigned multiple certificates so that long-term vehicle behavior, e.g., positions, cannot be correlated to a single vehicle. By changing signing keys, and correspondingly certificates, a vehicle may achieve greater privacy. Additionally, vehicles may share certificates so that certificates do not correspond to vehicles in a one-to-one manner. In other words, observing the use of one certificate multiple times does not equate to observing a vehicle multiple times. Similarly, observing a vehicle multiple times does not equate to observing the use of the same certificate multiple times.

Others have proposed the use of group signatures for obtaining privacy while maintaining the goal of binding safety beacon information to vehicles, as stated in Corollary 9.4.3 (Parno and Perrig 2005; Raya and Hubaux 2007). RSUs may be required to generate group signatures, which means that many roadways will not support group signatures during incremental deployment. Parno and Perrig have noted that using group signatures comes at the cost of not being able to attribute misbehavior to a vehicle, thus failing to support Axiom 9.4.2 (Parno and Perrig 2005). Raya and Hubaux have also noted that group signatures are computationally expensive and therefore may not be suitable for VANET where vehicles have insufficient computing power (Raya and Hubaux 2007). Consequently, we will not consider further the use of group signatures in our discussion of key assignment and privacy in VANET.

Privacy is significantly affected by how keys are assigned in a VANET. However, privacy may also be impacted by other factors outside of key assignment that will affect the level of privacy a vehicle can maintain. Specifically, it is possible to correlate broadcast VANET data with information obtained through other methods, such as cameras, and it may be impossible to defend against determined attackers who use both sources of information. We will discuss the problem of tracking in more detail in Section 9.6.3.

9.6.1 Properties

We assume that vehicles are assigned a number of keys, d. This number, d, may be larger than 1, allowing vehicles to change or rotate the keys they use in order to prevent long-term tracking from correlating safety beacons signed with the same key. Each key may be shared among a number of vehicles, g, which likewise may be larger than 1. Allowing g to be larger than 1 has been proposed as another mechanism for increasing vehicular privacy, making vehicles indistinguishable from other vehicles that have been assigned that same key (Xi et al. 2007). We will also use the logical result from Axiom 9.4.1 that there is a fraction of revoked vehicles, f. Generally, we will apply these properties to key assignment methods, which we will discuss in Section 9.6.2.

CA privacy

A CA is a centralized organization that signs vehicles' keys for the purpose of generating certificates. Since a vehicle's key information must pass through some CA, the CA often has privileged information about the identity of a key owner. Because the CA has this information, the CA may become a useful tool for law enforcement. One specific concern with the CA having this information is that government law enforcement may subpoena the CA to enforce the law, including for driving violations (e.g., speeding). Auto manufacturers are concerned that without CA privacy, a VANET system will not gain acceptance among buyers for this reason. Other reasons to retain privacy from a CA include possible misuse by an observing government agency or employee for political or personal reasons, and the possibility for unintentional leakage. For the latter reason, consider the following scenario. Privacy from CAs is not maintained, and law enforcement uses VANET information to track vehicles to gather evidence for prosecution. This information will need to be retained to possibly be presented in a court of law. A side-effect of having to store this information is that the government agency that stores this information will be a centralized target for hackers who want the VANET information. User privacy can be reduced through another mechanism if this information is stored. There have been many cases of government agencies or their employees leaking privacy-sensitive information in unintended ways, e.g., a lost USB flash drive (BBC News 2008a,b), a misconfigured web server, or a misplaced organizational laptop (TSA Public Affairs 2008). Thus, it might be better to provide privacy from a CA when designing the VANET. It may be possible for the CA to perform its duties (e.g., revoking vehicles and assigning keys to new vehicles) without retaining sufficient information for compromising privacy; however, the CA initially has access to this information since the CA assigns keys and therefore, we must consider that the CA has access to this privacy-sensitive information.

One way to describe privacy in this context is whether evidence gathered in the form of vehicle position or speed data signed by a valid key would be definitive evidence in a court of law. If it can be demonstrated that this evidence suffers from an unacceptably high *false positive rate*, such VANET evidence should not be definitive in a court of law, no more than establishing the guilt of the driver because he owns a 'blue car.' A false positive rate is an unacceptably high probability that one vehicle could be mistaken for another if the distinction between the two is based only on which keys the vehicles hold. For a VANET, having privacy from a CA means that a CA, even if subpoenaed, would only know imprecise or unreliable information about the owner of any given certificate. We describe the condition of a CA having poor ability to link a certificate to a VANET vehicle as having 'privacy from a CA.' If the VANET designer's goal is to provide maximum privacy to VANET vehicles (above all other considerations), then privacy from a CA is an attractive attribute. As we show below, maintaining privacy from the CA comes with significant compromises to other design goals. Increasing g leads to increasing the number of vehicles among which vehicles are indistinguishable to the CA. Thus, vehicle privacy is increased by increasing g. This is true for both CA and non-CA privacy, the latter of which we will discuss next.

Non-CA privacy

Maintaining privacy from a non-CA entity may be important for other reasons. If it is possible to remotely track vehicles through their VANET messages, a corporation (a non-CA entity) might be able to track a user's shopping habits and correlate that to the user's home address. Thus, the corporation might specifically target VANET users with advertisements. Additionally, VANET messages could be used by private investigators to track the people they are observing. In fact, private investigations have already made use of electronic tolling information from the E-ZPass[4] and Fast-Lane[5] toll plaza systems (Hager 2007).

Generally, by increasing d, we increase privacy from non-CA entities. By assigning a large number of keys to vehicles, non-CA entities will not know if broadcasts signed with two different keys came from the same or two different vehicles, based on key information alone. Care must be taken in how certificates are constructed so that vehicles cannot be identified by information included in their certificates. Clearly, the use of other information, even the information signed in the broadcast, can be used to reduce a vehicle's privacy. We will leave the discussion along this path for Section 9.6.3. Increasing d, however, increases the cost of revocation, either in the size of the certificate revocation list (CRL) or in the computational cost of processing the certificate revocation. We will see below that each of the considered methods of key assignment provides non-CA privacy.

Revocation

Correct and innocent vehicles may be harmed by malfunctioning or malicious vehicles' false information. For example, if a malfunctioning vehicle broadcasts an incorrect position, another vehicle may display a false warning to its driver. A malicious vehicle might also generate incorrect roadway congestion information, causing other vehicles to believe that the roadway that the malicious vehicle is on is more congested. Thus, the deceived vehicles may take a different route, leaving the malicious vehicle's roadway less congested. In our following discussions, we

---

[4] see http://www.ezpass.com/
[5] see http://www.masspike.com/travel/fastlane/tollplaza.html

make no distinction between malfunctioning and malicious vehicles, describing them collectively as 'untrusted' vehicles.

Our discussion of the basic principles which we assume hold for a VANET, which we gave in Section 9.4.2, resulted in our deducing that a VANET should use a PKI structure and asymmetric cryptography to sign packets in order to protect users from unlimited damage from vehicles that are untrusted. The vehicle is assigned keys from a CA and uses them to sign messages. Receiving vehicles may simply assume that this information is valid once they verify the correctness of the message signature and the validity of the associated key. The information may also be subject to other tests for validity, including claimed acceleration, velocity, and position (Golle et al. 2004; Studer et al. 2007).

Revocation is a mechanism for protecting correct and innocent vehicles from the effects of untrusted vehicles. Stated more concretely, a CA revokes a key by publicly announcing that the key is no longer valid. Receivers thus distrust any information signed by a key once it learns that a CA has revoked that key. As we will discuss in detail below, it is possible for a vehicle to have more than one key. Therefore, it is also possible that a specific vehicle will have some, but not all, of its keys revoked. An untrusted vehicle, if left with at least one unrevoked key, can continue to operate with a valid key. Thus we stress that a vehicle is not revoked until all d of its keys are revoked.

We define a 'revocation event' as the following sequence of events:

1. One or more entities observe and report to the CA that a vehicle, using a specific key, acted in an untrustworthy way.

2. The CA revokes the reported key and (perhaps) other keys assumed to be associated with the same reported vehicle (details discussed below). The CA creates an updated CRL containing these newly revoked keys.

3. The CA uses some method to disseminate this new CRL to all vehicles in its area of responsibility.

At a high level, a revocation event occurs when a vehicle 'is caught' acting in an untrustworthy manner while using one of its keys, causing the CA to revoke one or more of the keys.

We now consider the speed of a revocation process. We will show in Section 9.6.2 that there is a trade-off between revocation speed and privacy. While all three steps above contribute to the speed of a revocation process, the first and third steps have previously received consideration in the literature (Golle et al. 2004; Laberteaux et al. 2008; Papadimitratos et al. 2008; Studer et al. 2007). However, to our knowledge, the impact of the second step on privacy and revocation speed has not received its due consideration. Thus, the second step will be the focus in what follows. Since we focus on the second step, instead of specifying revocation speed in minutes, we define a 'faster' revocation process as one that requires fewer revocation events to revoke all of a vehicle's keys.

Intuitively, when a CA receives a report of a vehicle's use of a specific key linked to untrustworthy behavior, depending on the information the CA has, the CA can respond by revoking only one of the vehicle's keys (i.e., the reported key), revoking all of the vehicle's keys, or revoking some fraction of the vehicle's keys. If the mechanism of revocation is to be useful for protecting correct and innocent vehicles, revocation should be *fast*. If an untrusted vehicle's keys cannot be revoked quickly, vehicles cannot fully trust the PKI to perform its core mission, i.e., to identify an untrusted vehicle at the time of contact. The slower the revocation, the larger the window of opportunity for untrusted vehicles to do damage to the VANET. Since the creation and maintenance of any PKI is non-trivial and often expensive, it would be unwise to create a PKI with known slow revocation properties. We restate our assumption that the number of untrusted vehicles is proportional to n, the population of vehicles. Therefore, after a settling time, it can be assumed that $f \cdot n$ cars are fully revoked.

Brittleness

If vehicles share keys, i.e., $g > 1$, then each revocation event affects innocent vehicles as well as the target vehicle. *Brittleness* describes how large of an effect revocation of an untrusted vehicle has on the privacy provided by the key assignment method to these innocent vehicles. If we assume that g is large, that is, a large number of vehicles share any single key, then revoking all the keys of an untrusted vehicle impacts a large number of users that shared keys with the now-revoked vehicle. As a result, the privacy retained by non-targeted vehicles is reduced, i.e., the number of pseudonyms for these non-targeted vehicles is reduced.

For example, suppose there is a VANET consisting of five vehicles ($n = 5$), each of which are assigned two keys ($d = 2$). Each key is shared by two vehicles ($g = 2$). Now suppose that one of the vehicles is revoked; that is, all of its keys are revoked. Assuming that a single vehicle does not share both keys with the revoked vehicle, then there are two vehicles that have only a single valid key remaining after the revocation event. Before the revocation, for each key, each vehicle could hide among a group of two vehicles, itself and the other vehicle that shares the key with it. However, after the revocation, the size of this group is reduced to one. Thus, the privacy of the vehicles who share keys with the revoked vehicle is reduced because after revocation no other vehicle shares the keys held by the revoked vehicle. If d were increased to three, then the size of the group among which a vehicle is indistinguishable would also be larger, thus providing more privacy to vehicles. Increasing g to three would increase privacy by increasing the size of the group that a vehicle hides among, but it would also increase the loss of privacy experienced by innocent vehicles.

Again, we describe the effect of innocent vehicles losing privacy when an untrusted vehicle is revoked as *brittleness*. In comparing various key assignment methods, brittleness describes how large an effect a single revocation event has on vehicle privacy. Increasing g increases the brittleness of the network. Brittleness can be decreased by increasing d. If a large number of vehicles lose a key owing to a revocation, holding more keys reduces the adverse affect to innocent vehicles.

Collapsibility

The property of collapsibility reflects how resilient a key assignment method's security properties are to key compromise. If a vehicle's hardware and correspondingly its keys are compromised, a significant number of vehicles may be unable to use their keys to sign safety beacon messages with the security properties intended by Axiom 9.4.1 and Corollary 9.4.3.

Collapsibility can be decreased by decreasing g. Intuitively, the smaller the number of vehicles that share a key, the smaller the number of vehicles that will be affected by a hardware and associated key compromise.

Key collisions

A key collision occurs when two or more vehicles use a key simultaneously and are within a two-hop radio range of each other. When a key collision occurs, a non-colliding vehicle that overhears both colliding vehicles' transmissions may think that a single vehicle is claiming multiple locations and is either malfunctioning or malicious. By increasing g, the probability of having key collisions increases. Key collisions will increase the number of vehicles wrongfully revoked, and present an opportunity for malicious vehicles to use a shared key to revoke another vehicle's keys. When $g = n$, key collisions may be expected and therefore ignored.

9.6.2 Key assignment

In this section, we consider what performance various key assignment methods can provide in terms of the properties given in Section 9.6.1. The CA is responsible for assigning keys to vehicles, therefore the CA will be the holder of privacy-sensitive vehicle data, such as keys and relationships between keys and vehicles. The CA keeps these keys so that the CA can revoke untrusted vehicles. How keys are assigned affects vehicle privacy and the usability of a VANET.

Key assignment design space For our discussion in this section, a key assignment method describes how many keys each vehicle will own (d), as well as how many vehicles are assigned the same key (g). To fully analyze the g–d design space, we divide the space up into four different regions, which are illustrated in Figure 9.2. We will explore the properties of each of these regions below. As we will show, the choice of g and d will determine the amount of privacy available from the CA and from non-CA entities. We specifically investigate the impact of each key assignment region on the level of privacy that a vehicle can maintain with respect to the CA. We will also discuss vehicle privacy from non-CAs. In this section, we will show that there is a trade-off between how much privacy a vehicle can maintain from the CA and how quickly a vehicle can be revoked from the network.

There are three main approaches for assigning a vehicle's signing keys which cover the g–d design space, as shown in Figure 9.2:

1. (a) There is only one key in the network. Each vehicle is supplied with this same key. ($g = n, d = 1$).

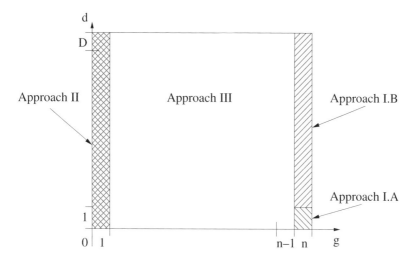

Figure 9.2 The g versus d design space explored in relation to CA privacy.

   (b) There are many keys in the network, and each vehicle has every key. ($g = n, D \geq d > 1$).

2. Each vehicle is loaded with a set of keys. Keys are not shared between vehicles. ($g = 1$).

3. Each key is shared among a group of vehicles. ($n > g > 1, D \geq d \geq 1$).

We denote the maximum number of keys possible in a VANET as D, and the total number of vehicles as n. We assume a fixed D, though D may be very large. Without making this assumption, the problem of choosing a key assignment method becomes much harder due to trying to assign keys to vehicles when $g \neq 1$ and still being able to provide privacy to vehicles using newly added keys. We evaluate a static n as an assessment of VANET privacy properties at any given instance. The result of our analysis below will lead us to the conclusion that Approach II is the best of the three key assignment methods.

Key assignment Approach I

We divide Approach I ($g = n$) into two subapproaches: Approach I.A ($d = 1$) and Approach I.B ($d > 1$). In Approach I.A all vehicles are provided with a copy of the same key. Under this approach, signing a message can be a simple, symmetric cryptographic operation since the key distribution problem is trivially solved, and non-repudiation is not possible. Since all vehicles have the same key, the CA cannot identify which vehicle signed an individual message based on the message's signing information. Similarly, no non-CA entity can tell two signatures apart based on key information alone. Thus, CA and non-CA privacy is complete with regard to key information. Unfortunately, a single hardware failure or successful attack would result in the complete compromise of such a solution, since no message could be trusted following the compromise of a single shared secret key. Thus, Approach I.A is extremely collapsible.

One difference between Approach I.A and Approach I.B is that instead of having only a single key for the entire network, there are many keys, though each vehicle still shares all of the keys. Approach I.B suffers from the same problem of collapsibility as Approach I.A: a single hardware compromise results in all of the keys being compromised, which means that no message could be trusted. Similarly, the CA still cannot identify the transmitter of a message for revocation purposes. Approach I.B can also suffer from key collisions, where Approach I.A did not have this problem. However, Approach I.B possesses the same level of privacy from CAs and a non-CAs as Approach I.A did.

Approach I.A and Approach I.B share some common performance in terms of brittleness and Sybil attack resilience. Both Approach I.A and Approach I.B are extremely non-brittle, in that the privacy of systems using Approach I.A or Approach I.B is not reduced by the revocation of malfunctioning or misbehaving vehicles, but the revocation of a single vehicle would result in the collapse of the VANET. Both Approach I.A and Approach I.B are highly susceptible to Sybil attacks since correct vehicle behavior and a vehicle using the common key(s) is indistinguishable from safety beacon information alone.

Key assignment Approach II

Approach II provides each key to only one vehicle ($g = 1$), i.e., no single key is held by more than one vehicle. This type of distribution solves Approach I's problem with key compromise and collapsibility. The primary advantage of the second approach, where each key is known only to one vehicle, is that revocation is efficient: once a single misbehavior is matched to a key, all the keys belonging to that vehicle can be revoked in a single revocation event, thus resulting in the exclusion of that vehicle from the VANET. In this approach, fast revocation can be the same as complete revocation, since each key is held by a single vehicle. A second advantage of this approach is that it can provide the property of non-repudiation, if public keys are used for signing. The disadvantage of this approach is that each key uniquely identifies a node, raising privacy concerns.

The CA will need to be able to correlate a vehicle's keys in order to enable fast revocation. For example, the CA may keep a list of all keys for each vehicle. These lists may be subpoenaed by a law enforcement agency that wishes to track certain VANET users. More generally, consider a design that attempts to protect users' privacy from the CA. Suppose the CA intentionally keeps incomplete information about users' keys. In such a case, the CA may not be able to perform a full vehicle revocation in a single revocation event. However, as discussed in Section 9.6.1, if revocation is not fast (i.e., achieved in a small number of revocation events), the PKI has diminished value. Thus, even in such a case, there must be a process to keep revocations fast. Consequently, in this approach, the same mechanism that is used to enable fast revocation can also be used for privacy compromise. In other words, if there is a mechanism that is useful for quickly revoking a vehicle's certificates, the same mechanism can be used to compromise a vehicle's privacy since the CA must be able to revoke all of the keys of that vehicle and therefore all the keys must be known to the CA. This linkage between quick revocation and privacy compromise holds whether a single CA has all of the information necessary for revocation or the information is dispersed among several CAs. Thus, no system based on having a single vehicle per key ($g = 1$) can provide both fast revocation and privacy from the CA. If a VANET designer is willing to sacrifice maintaining complete privacy from the CA, then this assignment method is tenable.

Since vehicles can be loaded with a large number of keys, d, and keys may never need to be reused, the non-CA privacy provided by Approach II is high. Only the CA can know the keys assigned to a vehicle, not non-CA entities, if $d \neq 1$. Additionally, if $d \neq 1$, then Sybil attacks are possible. However, since all keys are unique to their vehicles, there are no key collisions.

Similarly, since vehicles do not share keys with other vehicles (i.e., $g = 1$), then Approach II is not brittle. When a vehicle's key is revoked, the privacies of other vehicles in the VANET are not reduced because no other vehicle has been assigned the revoked key.

Key assignment Approach III

Approach III provides each key to several vehicles. This approach solves the problem of a CA definitively knowing which vehicle possesses each key, since each key is shared among a group of vehicles. However, as shown below, this improved CA-privacy comes at the expense of slower revocation. We mathematically explore this trade-off between revocation speed and CA-privacy below.

Approach III solves the issue of key compromise and collapsibility that Approach I had since not all vehicles are assigned all the same keys in the VANET. If $d = 1$, then every vehicle shares its only key with $g - 1$ other vehicles. In this case, Approach III is extremely brittle since a single revocation results in $g - 1$ other vehicles also being revoked. Therefore, we do not consider $d = 1$ further for Approach III. Since $d \neq 1$ and $g \neq 1$, Approach III provides non-CA privacy because signatures are not attributable to individual vehicles ($g \neq 1$) and non-CA entities do not know which keys have been assigned to individual vehicles. Sybil attacks are again possible in Approach III because vehicles are assigned multiple keys. We will include a discussion of the brittleness of Approach III in our mathematical assessment of Approach III below.

Approach III attempts to avoid the trade-off of maintaining privacy from the CA and enabling fast revocation that Approach II had to make. Again, one goal of sharing keys is that, when law enforcement detects that a vehicle is using a certain key, they cannot affirmatively link that key back to a single vehicle. In particular, because other vehicles share the same key, law enforcement cannot affirmatively prove that a single vehicle was the violator. It may be desirable to set an even tougher goal for vehicle privacy in this scenario. Since evidence in a court of law builds a case, it may be desirable that the information obtained from a VANET be even less probative such that it cannot be efficiently used to build a case. A casual inspection of Figure 9.2 may lead the reader to think that the design space of Approach III is large.

However, practical considerations greatly constrain the design of key management of Approach III. We now list four primary constraints.

Constraint 1 Assuming the CA can only revoke a single key for a single reported infraction (single reported key), revoking a vehicle requires that each of its keys be individually revoked, which upper bounds the number of keys per vehicle (d), if the efficacy of revocation is to be preserved. Generally this constraint is applicable, but we will discuss below the case when this constraint does not hold. (At the time a key is revoked, it may be possible to infer which other keys a vehicle holds besides the key being revoked. We will discuss this possibility in greater depth below.)

Constraint 2 If a constant fraction f of all nodes have had all of their keys revoked as results from Axiom 9.4.1, then excessive sharing will result in all keys within the system being revoked. (Intuitively, if each key is shared by g nodes, and $g \approx 1/f$, then all of a vehicle's keys will be revoked with high probability. We will investigate this outcome in mathematical detail below.)

Constraint 3 A privacy compromiser might observe *transitions* between keys. For example, a law enforcement officer may observe a single vehicle speeding, and during this observation, the speeding vehicle transitions from one key to another. With each additional key observed by the law enforcement officer, the pool of suspects shrinks. If the number of keys that the law enforcement officer observes in this way (ρ) is sufficiently large, but the degree to which each key is shared (g) is sufficiently small, then the probability that more than one vehicle has all such keys approaches zero, contravening the objective of providing anonymity from the CA.

Constraint 4 Key collisions will cause additional revocations in a VANET. A designer must mitigate these additional unnecessary revocations by either increasing d or decreasing g.

Considering Constraint 1, it may be possible for the CA to infer which additional keys are held by a vehicle given a single key from the vehicle. This situation might arise and be of significance during revocation. Using the single key reported for revocation, a CA may be able to infer additional keys held by the offending vehicle, and the CA may use this knowledge to revoke more than one key at a time. However, this option is not considered by current approaches. We will show below that for other reasons, mathematically described, Approach III does not allow for both privacy and fast revocation. If the CA can infer more than one key given a single key, Approach III simply provides even less privacy. Additionally, any inference method useful to the CA for revocation purposes will be useful to other agencies for their various purposes (e.g., law enforcement using inference to more completely track and ticket speeding vehicles). Consequently, we will not consider the use of inference by the CA in our discussion below, and we will consider Constraint 1 to apply.

Designers of a VANET must be careful in how they choose the parameters mentioned above. Failing to do so can result in consequences that may not be immediately apparent. Table 9.9 shows the notation we will use in the following

Table 9.9 Privacy design parameters.

g	Number of vehicles sharing each key (cars/key)
d	Number of keys held by each vehicle (keys/car)
n	Number of vehicles in the VANET
ε	Probability of false positive
f	Fraction of vehicles with all certificates revoked
ρ	Number of keys from one vehicle required by a single observer to break privacy
w	Wrongful revocation rate
σ	Vehicle encounter rate
C	Key collision rate

discussion. To illustrate these design decisions and the constraints discussed above, we consider two bounding cases under two opposite assumptions: complete independence in terms of key assignment, that is, keys are assigned completely at random, and complete dependence, that is, any two cars that share a single key in common will also share all of their keys in common. In the following discussion, we will assume g and d to be constants, that is, keys are shared among groups of equal size and each vehicle is loaded with the same number of keys. A vehicle that has had some keys revoked will have less than d keys remaining that it can still use. We will justify these assumptions after presenting our mathematical analysis of privacy for Approach III.

Independent distribution

Design In Approach III, under the independent key distribution assumption, each key is held by multiple vehicles. Therefore, the CA does not know which vehicle among the group that holds a certain key is the vehicle being reported for revocation. In other words, the CA cannot know from a single revocation report more than the single associated key, and only one key can be revoked at a time. Thus, the number of keys per vehicle d must be limited so that revocation is still an effective mechanism.

Let us assume that all keys are distributed independently in a VANET that has n vehicles. Let us also only concern ourselves for now with revocation effects due to the fraction of vehicles, f, that have had all of their keys revoked. The probability that an arbitrary key is not revoked is equal to the probability that no vehicle in the group of revoked vehicles holds that key. Quantitatively, the probability that an arbitrary key is not revoked is $(1-f)^g$. We call w the fraction of vehicles wrongfully revoked.[6] We define an innocent vehicle as a vehicle outside the group of justifiably revoked (untrusted) $f \cdot n$ vehicles. Under the independent key distribution assumption, each innocent vehicle may or may not share a key with one of the $f \cdot n$ untrusted vehicles. Consequently, the probability that an innocent

---

[6]Wrongful revocation may be highly unacceptable to users. The average consumer may not accept or understand when a vehicle service provider (e.g., repair shop) explains that their vehicle or VANET safety enhancements are not functioning because some other vehicles are misbehaving and the network is designed knowing that this could happen.

vehicle has all of its keys revoked is,

$$w = (1 - (1-f)^g)^d. \tag{9.1}$$

Equation (9.1) states that for each of a vehicle's d keys, at least one of the $g - 1$ other holders of those keys are in the completely revoked, untrusted vehicle group.

Above, we mentioned that one metric of privacy is the likelihood of being correctly or incorrectly identified after an observation. If a vehicle blends into its surroundings and enjoys high privacy, then the likelihood of it being misidentified is high. Consider, for example, identification based on hair color (high likelihood of misidentification, since many people share the same hair color), DNA matching (low likelihood of misidentification) and blood type.[7] To aid in the intuition, imagine cases where a court of law tries to identify a defendant based on some identifier, e.g., hair color, DNA matching, blood type. Essentially, the higher the likelihood for misidentification, the less likely the court will treat the evidence as definitive. Thus, a person who is identified by some highly shared characteristic, e.g., brown hair, retains a higher level of privacy than the person who is identified with some unique characteristic, e.g., certain combinations of DNA markers. This method of measuring privacy is apt for a discussion of the privacy that VANET users maintain from the CA and those (such as law enforcement officers) that can subpoena the CA. Over the past few years, protecting privacy has become a more pervasive issue in society. Thus, VANET may be unacceptable to users if governments or police can employ VANET data to issue driving violations, such as speeding.

Continuing this line of thought, a vehicle that broadcasts identifiers that are widely shared maintains more privacy from the CA than if it broadcasts identifiers shared by a small group of vehicles. To illustrate, consider a scenario where a law-enforcement-controlled listening station receives a VANET packet from a speeding vehicle. If that packet contains a non-reputable signature created by a key held by only one vehicle ($g = 1$), then that vehicle has essentially no privacy. Via a subpoena to the CA, law enforcement could use the unique identifier to determine the identity of the vehicle.

If, on the other hand, the identifying key is held by exactly three vehicles ($g = 3$), then the vehicle has more privacy than before. Then consider that law enforcement finds a vehicle shown to hold the offending key and arrest the vehicle owner. In this case, if law enforcement has no more evidence, they must admit that there is only a 1/3 chance that they have the correct driver, i.e., the misidentification rate (given that the defendant is shown to have the key) is still 2/3. Courts should not convict based solely on such evidence since it has such a high misidentification rate. Of course, if the key is held by even more vehicles, the misidentification rate, and thus privacy, increases.

If instead the law-enforcement-controlled listening station is able to observe the same vehicle using two different identifiers, e.g., the speeding vehicle switches from one signing key to a different signing key while being observed by the listening station, then the situation changes: Assuming as we do throughout this subsection

---

[7]Some current resources list blood type as having at least 0.6% likelihood of misidentification. The least common blood type in the United States is AB−, which is present in 0.6% of the population.

that the keys were assigned independently, the pool of possible vehicles (i.e., vehicles which hold both observed keys) shrinks significantly. We now develop a general mathematical analysis of privacy when an arbitrary number of certificates are known to come from a single vehicle.

Suppose a vehicle is observed to be behaving in a untrusted way by a law enforcement officer, and the officer observes the vehicle using ρ keys. The probability that a second arbitrary vehicle other than the observed untrustworthy vehicle shares ρ keys[8] with the observed vehicle is

$$\varepsilon = \left(\frac{g-1}{n-1}\right)^\rho \quad (9.2)$$

where ε is the probability of *mistaken identity* or of a *false positive*. We will refer to ε as the false positive rate below. Again, the false positive rate is the probability that one vehicle is mistaken for another vehicle based on key information. Here, the two vehicles can be mistaken for each other because they have at least ρ keys in common. Consider the following scenario. One vehicle is observed using ρ specific keys. A second vehicle is compelled to admit that it also possesses the same ρ keys. In this scenario, the second vehicle has a probability of being mistakenly identified as the first with probability ε. One analog of this false positive rate is the false positive rate for DNA matching between two random people. Under the assumption of independent key distribution, Equations (9.1) and (9.2) result in a trade-off between the false positive rate and the wrongful revocation rate, which we will illustrate below.

Initially, one might think that the false positive rate should be minimized. However, a high enough false positive rate (e.g., from using DSRC keys to identify rule-breakers) implies that such evidence would not be definitive. In other words, those that would like to discourage law enforcement from using DSRC keys for identifying suspects would want ε to be so large that such use would be widely discredited. This argument is similar to the idea that law enforcement cannot convict a driver of a 'green car' simply because they observed a green car breaking the law; the likelihood of mistaken identification is too high. Essentially, the larger ε, the less privacy is sacrificed by DSRC users to the CA.

A false positive may occur if the innocent vehicle and the observed vehicle share at least ρ keys. If we specify a lower bound, ε'_ρ, which provides an acceptable amount of privacy for a given ρ, then, solving Equation (9.2) for g, we get,

$$g(\varepsilon'_\rho) \geq 1 + (n-1)\varepsilon'_\rho{}^{\frac{1}{\rho}}. \quad (9.3)$$

Thus, $g()$ is $\Theta(n)$, that is, given a number of observed keys ρ and a privacy bound, ε'_ρ as the number of cars n increases, the number of keys held by each car, $g()$, would also need to increase linearly with n. Additionally, since $g()$ is $\Theta(n)$, $(1-f)^g$ goes to 0 and $w = (1 - (1-f)^g(\varepsilon'_\rho))^d$ goes to 1 as n goes to ∞. w going to 1 implies that once all the f vehicles are completely revoked, the remaining $1 - f$ innocent

---
[8]In a privacy compromising situation ρ would need to be the number of keys observed by the privacy attacker.

vehicles are also revoked. Thus, Approach III with independent key distribution does not scale with increasing network size. When we fix the false positive rate (ε), the maximum fraction of cars wrongfully revoked (w), number of cars (n), and number of connected keys (ρ), we find a bound on d that satisfies Constraints 2 and 3 by combining Equations (9.1) and (9.2):

$$d \geq \frac{\log(w)}{\log(1 - (1-f)^{1+(n-1)(\varepsilon'_\rho)^{1/\rho}})}. \tag{9.4}$$

Example Consider the situation in the USA where $n \approx 200$ million. We will look at the situation where there is only a small amount of privacy provided by the VANET; the design problems only become more acute by increasing the provided privacy. For these examples, we will set $w = 10^{-4}$, $f = 10^{-5}$, and $g = 2 \cdot 10^6$. If we specify $\varepsilon'_\rho = 10^{-2}$ (a false positive rate of 1 in 100), and $\rho = 1$, then from Equation (9.4), $d \approx 4.5$ billion. Thus, in order to maintain a false positive rate of 1 in 100 (a reasonable amount of privacy), vehicles will require ≈ 4.5 billion keys. As ρ increases, for constant d, ε'_ρ must decrease rapidly, that is, privacy rapidly decreases. If we keep d set at ≈ 4.5 billion, set $\rho = 2$, ε'_ρ decreases to 1 in 10 000 (smaller privacy). For $\rho = 3$, ε'_ρ decreases by another two orders of magnitude to 1 in 1 million. Thus, even for a small amount of privacy (small ε), each vehicle must have a very large number of keys (large d), and privacy rapidly decreases with increasing numbers of observed keys (increasing ρ), even for small number of observed keys. If we maintain the assumption that the CA does not use any inference to link multiple certificates to a vehicle (recall our discussion of when Constraint 1 holds), then $d = 4.5$ billion requires 4.5 billion revocation events to revoke a single vehicle, which could take a long time.

Dependent distribution

Design Now, let us assume keys that are no longer independently distributed but are distributed in a completely dependent manner, that is, if any two vehicles share a single key, then they share each of their keys. At a high level, this type of distribution will suffer increasingly from wrongful revocation as f and g increase. The larger the groups of vehicles are, that is, g, the more wrongfully revoked vehicles there will be. Also, the severity of this effect is intensified by increasing f.

Consider again Constraint 1 from above in this new manner of distribution. When a vehicle is reported for revocation purposes, the CA can do one of two things: revoke a number δ of a vehicle's keys or revoke all of a vehicle's keys. A dependent distribution is essentially Approach II where the CA could keep a list of unique keys held by each vehicle, except now the CA can keep a list of unique keys held by a single group of vehicles. Thus, revocation can be fast for this manner of distribution for the same reasons as revocation could be fast for Approach II above. This results in d not needing to be bounded for fast revocation purposes.

Since certificates are no longer independently distributed, the false positive rate is, $\varepsilon = \frac{g-1}{n-1}$, and after fixing ε, $g \geq 1 + (n-1)\varepsilon$. With a dependent distribution, ε is no longer a function of ρ. In this design, the false positive rate cannot be decreased by increasing the number of keys a vehicle holds. The probability that a vehicle is wrongfully revoked has become $w = 1 - (1-f)^g$. Again, the wrongful revocation rate cannot be reduced by increasing the number of certificates held by a vehicle.

Example Consider again the scenario we had in our independent distribution example ($f = 10^{-5}$, $n = 2 \cdot 10^8$). If we want even only a small amount of privacy, $\varepsilon = 10^{-4}$, then $g \geq 20{,}000$ and $w \approx 0.18$, that is, approximately 18 in every 100 vehicles will be wrongfully revoked. To preserve a wrongful revocation rate of $w = 10^{-4}$, as used in our independent distribution example, $g = 10$ and $\varepsilon \approx 4.5 \cdot 10^{-8}$, that is almost no privacy is provided. The dependent distribution example illustrates that a completely dependent distribution is unrealistic. However, in an actual deployment, the distribution would also not be completely independent. The actual key distribution would suffer from a higher wrongful revocation rate than the independent distribution due to key sharing, and the effect of increasing d is reduced.

Key collisions

Because more than one vehicle can be using the same key at the same time in the same location, under either distribution, independent or dependent, key collisions will occur. When an observer hears two other 'vehicles' claiming two separate locations, each using the same key, the observer has no way to differentiate between the case of two different innocent vehicles using the same key by coincidence, or one untrustworthy 'vehicle' pretending to be multiple vehicles. In such cases, it seems plausible that the observer will report the common key as untrusted, with the goal of triggering its revocation. There is a rate at which keys will be removed from the VANET due to unnecessary revocations from key collisions. These revocations will come from vehicles who overhear other well-behaved vehicles that happen to be using the same key within the radio range of the overhearing vehicle (the vehicles whose keys are colliding need not be within radio range of each other for this to occur). We define σ to be the rate at which a vehicle encounters other vehicles (e.g., 200 vehicles per day).

Intuitively, the expected number of revocations due to key collisions is proportional to the rate at which a vehicle encounters other vehicles, σ, and the number of vehicles that share each key, g. The expected number of revocations due to key collisions is inversely proportional to the number of keys held by each vehicle, d, and the total number of vehicles in the VANET, n. That is, the more vehicles another vehicle encounters, or the more vehicles that share keys with that vehicle, the higher the probability that key sharing vehicles will encounter each other and a key collision revocation will occur. Conversely, the more keys that cars are loaded with, the less likely it is that vehicles that share keys will use their shared keys concurrently. A more detailed and rigorous mathematical analysis of key collisions is beyond the scope of this text but may be found along with a more detailed analysis of key assignment in the literature (Haas et al. 2009).

Discussion

In reality, designers of a VANET may choose a system with the key distribution lying somewhere between complete independence and complete dependence. The above discussion illustrates that privacy from the CA may be unattainable while maintaining reasonable network performance (e.g., more malicious vehicles are revoked than legitimate vehicles, i.e., $w < f$) using Approach III.

Let us reconsider our assumptions that g and d are constants, and allow g to be a function of the individual key under the mathematical framework developed above.[9] Suppose g is chosen such that $g_{lo} \leq g \leq g_{hi}$. For the case of independent key distribution, the keys that are shared by g_{lo} vehicles would be considered *low-privacy keys* since there is a smaller population of vehicles for an individual vehicle holding one of these keys to hide among. The low-privacy keys provide a smaller ε to vehicles than choosing the constant g would have provided. Consider also that keys shared among g_{hi} vehicles result in a larger w, that is, more vehicles are revoked wrongfully. Thus, these keys have *higher risk* for revocation than keys distributed with constant g. The same outcome is obtained if keys are distributed in a completely independent or in a completely dependent manner.

Consider allowing d to vary from vehicle to vehicle. Let $d_{lo} \leq d \leq d_{hi}$. If keys are distributed completely independently, vehicles with d_{lo} keys will have a smaller false positive rate, and thus are *low-privacy vehicles*. Similarly, vehicles with d_{hi} certificates have a higher false positive rate and are *high-privacy vehicles*. This type of distribution implies different classes of service in terms of privacy for different vehicles. When keys are distributed in a completely dependent manner, having a non-constant d does not affect either the false positive rate or the wrongful revocation rate.

Thus, allowing g and d to vary results in both poorer privacy and more wrongful revocations. Considering the above arguments for how to assign keys to vehicles, we can eliminate designs that are infeasible, given our goals of maintaining privacy, enabling fast revocation, and building a robust system. Key assignment Approach I is infeasible because a single CA or vehicle compromise results in complete key compromise for the VANET. Approach III is infeasible because it cannot provide both CA privacy (large ε) and fast revocation (small d). As a result, we are left with Approach II, where limited CA privacy is possible, but is not able to be retained for vehicles that have keys revoked. We believe Approach II to be superior for key assignment because the safety properties of the VANET are preserved (e.g., vehicles are removed quickly for infractions through revocation), it provides a highly non-brittle VANET, and it results in no wrongful revocation, unlike Approach III.

9.6.3 Tracking vehicles

Because VANETs are used for safety applications including collision avoidance, each vehicle broadcasts accurate position information. Even if each advertisement from a vehicle was signed using a different public key, these keys can be correlated based

---

[9]The mathematics of a scheme where g and d vary are not actually the same. In fact, having non-constant g and d may make the mathematics of an analysis, such as we have carried out above, intractable.

on, for example, known limits of acceleration, deceleration, and car spacing (Golle et al. 2004). In particular, given two advertisements of location and velocity, a vehicle can calculate the acceleration needed in order for the two advertisements to have come from the same vehicle. A vehicle can use the acceleration information in a filter to decide which certificates are correlated. In general, this disqualification process suffices to uniquely identify the vehicle that is changing certificates.

Making the tracking of vehicles through VANET messages hard may be important for privacy and personal safety issues. In a more sinister scenario, a sophisticated group of attackers could use vehicle tracking to tell when a user is away from his residence in order to steal things from the user's residence.

The problem of location privacy in wireless networks is not new. In previous literature, the use of *silent periods* (Hu and Wang 2005) and *mix zones* (Beresford and Stajano 2004) allow a transmitter to increase privacy across different sessions. In VANET, however, turning off a vehicle's transmitter for a long period of time can be contrary to the safety objective. Safe strategies for occasional silence and determining if it is possible to use such silence while maintaining safety goals are important areas for future research.

Another method for tracking vehicles is to use *RF fingerprinting* (Brik et al. 2008; Hall et al. 2003). Due to imperfections in the electronics of RF transmitters, each transmitter has a unique *fingerprint*. An RF fingerprint is observed in the unique imperfections in the waveform produced by a set of radio equipment. When two certificates are transmitted using the same RF fingerprint, an adversary can quickly determine that the two certificates belong to the same sender. Recent work (Brik et al. 2008) has shown that fingerprinting in this manner is useful even for distinguishing among 802.11 devices from the same manufacturer and of the same model.

Previous analysis of actual vehicle mobility shows substantial periodic and habitual behavior (Jetcheva et al. 2003).[10] When a vehicle is in certain locations, such as the owner's driveway, any VANET transmissions are extremely privacy compromising because the number of cars that may be in that driveway is very small. Since location is transmitted together with the public key, an adversary can correlate keys simply by observing transmissions from a person's driveway.

An adversary or an authority can also correlate VANET transmissions with real-world observations. For example, the police could couple VANET-based speed violation detection with on-road cameras and automated license plate recognition to issue tickets automatically. Parking lot security cameras can also be used to correlate the keys that belong to a single vehicle. Thus, there may not be a mechanism for preventing a determined attacker from compromising privacy in a VANET or on a roadway in general.

9.6.4 Evaluation

Approach II is the only viable method for key assignment that will preserve the principles laid out in Section 9.4.2. Choosing Approach II may come with the

---

[10]This paper uses the movements of buses, which are highly periodic. The reason the bus schedules and movements are highly periodic is that peoples' schedules are highly periodic (e.g., the times they go to and return from work).

sacrifice of CA privacy, but it is the only approach that results in a viable VANET and provides fast revocation. We have shown that, owing to collapsibility, Approach I is not viable. Similarly, we have shown that owing to the inherent trade-off between fast revocation and the false positive rate (i.e., privacy), plus the brittleness and potential evaporation of legitimate keys, Approach III is not viable.

Approach III is not totally dismissible, though from our analysis, Approach II provides a superior VANET design. We have investigated two methods for distributing keys under Approach III, completely independent and completely dependent distributions. There may be other methods for distributing keys under Approach III. For example, a geographic-based distribution, where certain keys are assigned to vehicles only within a certain geographic area, may provide superior performance to the distributions we have discussed. Geographic distribution may suffer from increased key collisions, due to higher densities of vehicles having been assigned a specific key, and may be restrictive in the sense that vehicles' privacy can only be maintained while they are within the geographic area in which the vehicle is designed to operate. This latter problem may be important to, for example, college students or military families who move large distances and take their vehicles with them. Whether a type of distribution exists that makes Approach III viable is an open problem for research.

The loss of CA privacy from Approach II may not be a significant loss in the big-picture of VANET privacy. Since the CA is a trusted entity, in that vehicles trust the CA to create valid certificates and revoke vehicles, lacking CA privacy may not be a significant loss. The CA would need to be trusted to discard information linking vehicle certificates to real-world vehicle user or owner identities. Tracking vehicles through the use of information broadcast in safety messages can allow both CA and non-CA entities to reduce vehicular privacy without the need to use keying information. This tracking problem is fundamental to the long-range, non-line-of-sight nature of radio communications. It may be impossible to prevent a determined attacker of privacy from reducing vehicles' privacies because such an attacker can combine safety beacon broadcasts with information from other sources such as cameras or automated tolling equipment to reduce or eliminate vehicles' privacies.

9.7 Implementation Aspects

In the previous sections, we mainly considered protocols for protecting VANET communication. When implementing a VANET there are a variety of further aspects to consider. We sketch the most important aspects in the following, including key length, physical security, organizational security, and update of software in the field.

9.7.1 Cryptographic schemes and key length

We mentioned above that for most VANET applications, elliptic curve cryptography (ECC) is the natural choice. ECC provides small certificates and efficient run-time performance. For these reasons, ECC is also the cryptographic system of choice in IEEE 1609.2. Based on ECC, ECDSA provides digital signatures whereas ECIES provides public-key encryption. IEEE 1609.2 implements point compression of public keys to reduce the size of certificates (Harper et al. 1993). Further potential performance improvements might be possible by using implicit certificates which further reduce certificate size, e.g., ECQV implicit certificates (Research 2006). Furthermore, using very small signatures based on elliptic curves might be applied to VANET (e.g. ECPVS – Pintsov and Vanstone 2001).

In some instances, RSA might be considered. RSA provides extremely efficient signature verification but it comes at the cost of expensive signature generation, and large signature and certificate size. Therefore, if many signatures are to be verified in a setting where there is sufficient bandwidth available, RSA might be appropriate. Other cryptographic schemes should be based on standard symmetric algorithms such as AES, the SHA family, and HMAC-SHA. Table 9.10 describes NIST recommended key sizes (Barker et al. 2005) for ECC, RSA, and symmetric key algorithms. Following these recommendations, a VANET being deployed today should use a key size for symmetric algorithms of at least 112 bits. However, there might be cases (e.g. broadcast authentication) where the information needs to be protected for less than a second such that using shorter key sizes might be reasonable.

Table 9.10 NIST recommended key sizes.

Security lifetime	ECC	RSA	Symmetric key algorithms
Through 2010	160	1024	80
Through 2030	224	2048	112
Beyond 2030	256	3096	128

9.7.2 Physical security

It is wise to base the security of a system on secret keys rather than on obscurity. An attacker that has full knowledge of a design must be prevented from compromising the system. Today, such a view is commonly accepted. However, almost all successful attacks on security protocols are based on exploiting implementation flaws and/or are attacks on the underlying computing platform. Today's technology simply does not allow implementing complex systems such as a VANET without any security weakness at acceptable cost. Therefore we suggest developing a VANET security system that conforms to and is evaluated based on a security certification standard such as the Common Criteria (CC) (Criteria 2007). While a CC evaluation process will not find all security weaknesses, it will eliminate many relevant vulnerabilities with a high probability. Interestingly, CC suggests keeping the design and the implementation of a security system confidential while the evaluation assumes that attackers have full knowledge of the design and the search for weaknesses may include examining the source code and hardware architecture.

Attacks often target the underlying computing platforms. Such physical attacks include side-channel attacks, probing attacks, hardware manipulation, and re-engineering. The following list gives an overview of these attacks:

Side-channel attacks A computing device leaks information via so-called side channels such as power consumption, electro magnetic radiation, or time behavior. This side-channel information can be measured and statistically analyzed to derive cryptographic keys.

Hardware manipulation Hardware components can be tampered with to gain information. For instance, the bus connection between a CPU and its volatile memory might be probed to extract secret keys, the program code might be extracted, or so-called *mod* chips might be soldered in to manipulate the hardware platform's behavior.

Re-engineering In a re-engineering attack, the program code is extracted, e.g. by using hardware probing. Then the binary program code is evaluated and disassembled to understand the program execution. The re-engineered software code is then scanned for security weaknesses. Finally, exploits based on identified weaknesses are developed.

Tamper-resistant hardware provides the means to avoid such attacks. Such hardware is especially designed to withstand tampering and side-channel attacks. Tamper-resistant hardware is designed to securely store secret keys and securely run cryptographic algorithms. Tamper-resistant hardware is typically used in monetary-sensitive applications including, for example, contactless and contact smart-cards used for credit cards. Such hardware platforms are typically security certified. However, the security certification needs to be renewed regularly (3–5 years) to meet the requirements due to advancing attacking technology. Therefore, it can be assumed that nodes deployed in the field for a lifespan of 10 or 15 years become vulnerable to physical attacks at some point, if their security platform is not updated once in a while. Such an update might conflict with low cost requirements and user comfort.

For many applications, a *tamper-evident* hardware platform is sufficient. These platforms do not provide protection from tampering but provide evidence of tampering. Such an approach is often employed by means of a seal. In vehicles, the status of the hardware integrity could be displayed by physical means, say a seal, or electronically, say an indicator in the main electronic control unit (ECU). However, the second approach would be especially vulnerable to tampering.

We are not aware of a thorough analysis of the extent to which tamper-evident or tamper-resistant hardware is necessary for VANET nodes. However, as a rule of thumb, it appears reasonable to provide tamper-evident hardware platforms for safety applications and tamper-resistent platforms for applications from which attackers may obtain a financial gain. In the case of safety applications this seems to be counter-intuitive. However, tampering with safety applications will be actively pursued by law enforcement such that preserving evidence of tampering is crucial. Furthermore, we expect that the number of adversaries tampering with safety applications will be rather small. We described above how misbehaving nodes can be identified and evicted in a VANET comprised of an honest majority of nodes such that the overall expected impact is low.

9.7.3 Organizational aspects

In this chapter we focused on the technical aspects of security in VANET. A wide variety of organizational aspects also need to be considered before deployment. Therefore, we briefly list the most important concerns:

Sharing of power Extensive power held by a single entity always poses a single point of failure and an attractive target for adversaries. Therefore, no single entity should have extensive power over the VANET but power should be shared between entities. For instance, the CA issues a vehicle's certificate and an RA registers the vehicle. Then the CA forwards the certificate to the RA which injects it to the vehicle. The CA and the RA need to collude in order to be able to map the vehicle's identity to the issued certificate.

Reflection of today's organization Today, organization of vehicle logistics is often shared between several departments. For instance, in the USA there is the USDOT, and states' departments of motor vehicles, each of which issues license plates. In other countries, there are similar regulations. It can be expected that organization of VANET in general, and with regards to security, has to reflect existing structures, especially for issuing and managing nodes' certificates. It is essential that keeping these structures does not endanger security but improves it, e.g., by sharing power.

Security policies Security policies are required for all aspects of VANET security. These policies have to cover CA procedures, as registration of vehicles, identification of misbehaving nodes, and revocation. Furthermore, there needs to be clear governance and regulations determining in which manner VANET data can be used. For instance, it should be defined before deployment if and to what extent data can be used from a hit-and-run accident or for catching speeding drivers. A disaster recovery policy that defines procedures for recovering from a security breach is essential.

Organizational security is a major part of a holistic security architecture. However, we expect that VANET-specific technical aspects will be used as input for designing proper organizational security procedures based on best practice in traditional large-scale applications.

9.7.4 Update of software and renewal of certificates

Experience shows us that software is usually not free of flaws, and it is desirable to be able to update software in order to introduce new features. At the same time, it is necessary to enable vehicles to be able to update key material. On the other hand, enabling a vehicle in such a way makes it vulnerable to attack. Many vehicles in the field already implement features to update software in ECU. For instance, most German vehicles implement the HIS security standard that provides an interface for secure software updates (Hersteller Initiative Software HIS 2006). The interface does not depend on the actual communication channel and allows for updating vehicle software by a wired connection, e.g. via a diagnosis device, or wireless connection, e.g. WiFi or VANET. As mentioned above, there should be policies about quickly updating security software in the field as well as in production if a security weakness becomes known.

At the same time, renewing certificates is an important issue. Certificates might be revoked (for valid reasons or by mistake) or certificates' lifetime might expire. In general, it is possible to renew certificates in the field using the VANET communication channel. However, for reasons of reliability and control, it might be preferred to renew certificates in certified workshops only. Section 9.4.2 described further details.

9.8 Outlook and Conclusions

In this chapter we have given an overview of challenges for VANET security, and described potential security protocols for a variety of security objectives including authentication, encryption, privacy preservation, secure positioning, and identification of misbehaving nodes. We expect that there will be only a few VANET worldwide, but each will be country- or even continent-wide and will comprise several hundred million nodes. VANET do not have a single purpose like mobile phone networks but come with a wide variety of applications that in turn come with a wide range of requirements. Therefore, a holistic security architecture needs to be defined that comes with a suite of security protocols. While the suite cannot be reduced to a single protocol for each area (e.g. a single authentication protocol) algorithms should be carefully chosen to keep the security module small, both in terms of complexity and code size. The security suite needs to include protocols for broadcast and pairwise authentication, encryption, privacy preservation, and identification of misbehaving nodes. Additional protocols might be added after careful consideration. We believe it is infeasible to design, implement, and deploy a security and application system in vehicles that will run for the entire vehicle lifetime without adaption. Therefore, secure updating of application and security software should be included from initial deployment.

There are many open research problems in VANET, ranging from more efficient authentication schemes specially suited to VANET requirements to a careful consideration of applying secure hardware to VANET applications. However, we see the biggest demand for efficient solutions in the area of privacy protection as well as identification of misbehaving nodes.

References

ANSI X9.63-1998 1998 Elliptic curve key agreement and key transport protocols. Technical report, ANSI.

Augot D, Bhaskar R, Issarny V and Sacchetti D 2007 A three round authenticated group key agreement protocol for ad hoc networks. *Pervasive Mob. Comput.* **3**(1), 36–52.

Barker E, Barker W, Burr W, Polk W and Smid M 2005 Recommendation for key management – part 1: General. *NIST Special Publication 800-57*, August 2005, National Institute of Standards and Technology. Available at http://csrc.nist.gov/publications/nistpubs/800-57/sp800-57-Part1-revised2_Mar08-2007.pdf (revised March 2007).

BBC News 2008a Firm 'broke rules' over data loss. Available at http://news.bbc.co.uk/2/hi/uk_news/politics/7575989.stm.

BBC News 2008b Probe into data left in car park. Available at http://news.bbc.co.uk/2/hi/uk_news/7704611.stm.

Beresford A and Stajano F 2004 Mix zones: user privacy in location-aware services. *Pervasive Computing and Communications Workshops, 2004. Proceedings of the Second IEEE Annual Conference on.* pp. 127–131.

Brik V, Banerjee S, Gruteser M and Oh S 2008 Wireless device identification with radiometric signatures. *MobiCom '08: Proceedings of the 14th ACM International Conference on Mobile Computing and Networking*, pp. 116–127. ACM, New York, NY, USA.

Canetti R, Garay J, Itkis G, Micciancio D, Naor M and Pinkas B 1999 Multicast security: a taxonomy and some efficient constructions. *INFOCOM '99. Eighteenth Annual Joint Conference of the IEEE Computer and Communications Societies. Proceedings. IEEE*, vol. 2, pp. 708–716.

Chang SL, Chen LS, Chung YC and Chen SW 2004 Automatic license plate recognition. *Intelligent Transportation Systems, IEEE Transactions on* **5**(1), 42–53.

Commission E 2003 Joint Interpretation Library (JIL): Security Evaluation and Certification of Digital Tachographs, JIL Interpretation of the Security Certification according to Commission Regulation (EC) 1360/2002, Annex 1B, Version 1.12. Technical report, European Commission.

Consortium VSC 2006 Vehicle safety communications project-final report, usdot hs 810 591. Available at http://www-nrd.nhtsa.dot.gov/departments/nrd-12/pubs_rev.html. Technical report, Vehicle Safety Communications Consortium.

Criteria C 2007 *Common Criteria for Information Technology Security Evaluation v3.1.*

Gerlach M, Festag A, Leinmüller T, Goldacker G and Harsch C 2007 Security architecture for vehicular communication. *4th International Workshop on Intelligent Transportation (WIT) 2007.*

Golle P, Greene D and Staddon J 2004 Detecting and correcting malicious data in vanets. *VANET '04: Proceedings of the 1st ACM International Workshop on Vehicular Ad Hoc Networks*, pp. 29–37. ACM, New York, NY, USA.

Haas JJ, Hu YC and Laberteaux KP 2009 The impact of key assignment on VANET privacy. *Security and Communication Networks*, 2009, Wiley Interscience, to appear.

Hager C 2007 Divorce lawyers using fast lane to track cheaters. *WBZ-TV – Breaking News, (originally appearing at http://wbztv.com/bios/local_bio_280170605) (accessed from http://msl1.mit.edu/furdlog/docs/2007-08-10_wbz_fastlane_tracking.pdf on April 8, 2009).*

Hall J, Barbeau M and Kranakis E 2003 Detection of transient in radio frequency fingerprinting using signal phase. *Proceeding (383) Wireless and Optical Communications.*

Harper G, Menezes A and Vanstone SA 1993 Public-key cryptosystems with very small key lengths. *Lecture Notes in Computer Science* **658**, 163–173.

Hersteller Initiative Software (HIS) 2006 HIS Security Module Specification, Version 1.1. Available at http://www.automotive-his.de/download/HIS%20Security%20Module%20Specification%20V1.1.pdf.

Housley R, Polk W, Ford W and Solo D 2002 (RFC 3280) Internet X.509 Public Key Infrastructure Certificate and Certificate Revocation List (CRL) Profile.

Hu YC and Laberteaux KP 2006 Strong VANET security on a budget. *Proceedings of the 4th Annual Conference on Embedded Security in Cars (escar 2006).*

Hu YC and Wang HJ 2005 Location privacy in wireless networks. *The ACM SIGCOMM Asia Workshop 2005* ACM.

Hubaux J, Capkun S and Luo J 2004 The security and privacy of smart vehicles. *Security & Privacy, IEEE* **2**(3), 49–55.

IEEE 1999 IEEE P1363 / D13 (Draft Version 13): Standard Specifications for Public Key Cryptography. Technical report, IEEE.

IEEE 2003 IEEE P1363a / D12 (Draft Version 12): Standard Specifications For Public Key Cryptography – Amendment 1: Additional Techniques. Technical report, IEEE.

IEEE 2006 *IEEE 1609.2 – Standard for Wireless Access in Vehicular Environments (WAVE) – Security Services for Applications and Management Messages*. Available from ITS Standards Program.

Initiative VII 2008 http://www.vehicle-infrastructure.org.

Jetcheva JG, Hu YC, PalChaudhuri S, Saha AK and Johnson DB 2003 Design and evaluation of a metropolitan area multitier wireless ad hoc network architecture. *The Fifth IEEE Workshop on Mobile Computing Systems & Applications (WMCSA 2003)* IEEE.

Kroh R 2006 VANETs Security Requirements.

Laberteaux KP, Haas JJ and Hu YC 2008 Security certificate revocation list distribution for vanet. *VANET '08: Proceedings of the fifth ACM international workshop on VehiculAr Inter-NETworking*, pp. 88–89. ACM, New York, NY, USA.

Menezes AJ, Oorschot PCV, Vanstone SA and Rivest RL 1997 *Handbook of Applied Cryptography*. CRC Press.

Naito T, Tsukada T, Yamada K, Kozuka K and Yamamoto S 2000 Robust license-plate recognition method for passing vehicles under outside environment. *Vehicular Technology, IEEE Transactions on* **49**(6), 2309–2319.

Oliveira LB, Aranha D, Morais E, Daguano F, Lo'pez J and Dahab R 2007 Tinytate: Identity-based encryption for sensor networks. Cryptology ePrint Archive, Report 2007/020. http://eprint.iacr.org/.

Papadimitratos PP, Mezzour G and Hubaux JP 2008 Certificate revocation list distribution in vehicular communication systems. *VANET '08: Proceedings of the Fifth ACM International Workshop on VehiculAr Inter-NETworking*, pp. 86–87. ACM, New York, NY, USA.

Parno B and Perrig A 2005 Challenges in securing vehicular networks. *Workshop on Hot Topics in Networks (HotNets-IV)*.

Perrig A 2001 The biba one-time signature and broadcast authentication protocol. *CCS '01: Proceedings of the 8th ACM Conference on Computer and Communications Security*, pp. 28–37. ACM, New York, NY, USA.

Perrig A, Canetti R, Tygar J and Song D 2000 Efficient authentication and signing of multicast streams over lossy channels. *Security and Privacy, 2000. S&P 2000. Proceedings. 2000 IEEE Symposium*, pp. 56–73.

Perrig A, Canetti R, Song D and Tygar JD 2001 Efficient and secure source authentication for multicast. *In Network and Distributed System Security Symposium, NDSS 2001*, pp. 35–46.

Perrig A, Canetti R, Tygar JD and Song D 2002 The tesla broadcast authentication protocol. *RSA CryptoBytes* **5**, 2002.

Pintsov LA and Vanstone SA 2001 *Postal Revenue Collection in the Digital Age*. Springer Berlin/Heidelberg, pp. 105–120.

Raya M and Hubaux JP 2005 The security of vehicular ad hoc networks. *SASN '05: Proceedings of the 3rd ACM Workshop on Security of Ad Hoc and Sensor Networks*, pp. 11–21. ACM, New York, NY, USA.

Raya M and Hubaux JP 2007 Securing vehicular ad hoc networks. *J. Comput. Secur.* **15**(1), 39–68.

Raya M, Jungels D, Papadimitratos P, Aad I and Hubaux JP 2006a Certificate revocation in vehicular networks. Technical report.

Raya M, Papadimitratos P and Hubaux JP 2006b Securing vehicular communications. *Wireless Communications, IEEE* **13**(5), 8–15.

Raya M, Papadimitratos P, Aad I, Jungels D and Hubaux JP 2007 Eviction of misbehaving and faulty nodes in vehicular networks. *IEEE Journal on Selected Areas in Communications* **25**(8), 1557–1568.

Research C 2006 Standards for efficient cryptography, SEC 4: Elliptic Curve Cryptography. Technical report, Certicom Research.

Resendes R 2008 The New 'Grand Challenge' – Deploying Vehicle Communications, Keynote Address – *The Fifth ACM International Workshop on VehicluAr InterNETworking (VANET 2008)*.

Robinson CL, Caminiti L, Caveney D and Laberteaux K 2006 Efficient coordination and transmission of data for cooperative vehicular safety applications. *VANET '06: Proceedings of the 3rd International Workshop on Vehicular Ad Hoc Networks*, pp. 10–19. ACM, New York, NY, USA.

Studer A, Luk M and Perrig A 2007 Efficient mechanisms to provide convoy member and vehicle sequence authentication in vanets. *Security and Privacy in Communications Networks and the Workshops, 2007. SecureComm 2007. Third International Conference*, pp. 422–432.

TSA Public Affairs 2008 TSA Suspends Verified Identity Pass, Inc. Clear® Registered Traveler Enrollment. Available at http://www.tsa.gov/press/releases/2008/0804.shtm.

US Department of Commerce/National Institute of Standard and Technology 2001 *FIPS PUB 197, Specification for the Advanced Encryption Standard (AES)*. Available at http://csrc.nist.gov/encryption/aes.

Xi Y, Sha K, Shi W, Schwiebert L and Zhang T 2007 Enforcing privacy using symmetric random key-set in vehicular networks. *ISADS '07: Proceedings of the Eighth International Symposium on Autonomous Decentralized Systems*, pp. 344–351. IEEE Computer Society, Washington, DC, USA.

10

Standards and Regulations

John B Kenney

Kenney Consulting

10.1 Introduction

Digital communication requires that the transmitter and receiver have a common understanding of how the information will be conveyed. At its base level this means a definition for the signals representing a logical 0 and 1. Multi-feature networks rely in addition on the definition of complex rules and packet structures. In many cases a large number of individuals and organizations have an interest in these definitions. Two primary means are used to create the rules that govern modern data networks: government regulations and cooperative standards. This chapter describes the current state of regulations and standards for VANETs.

For the purposes of this discussion a regulation is a rule put forward by a governmental agency, and has the force of law. An example regulatory agency is the US Federal Communications Commission (FCC). Typically, an organization with an interest in the outcome of regulatory rule-making has an opportunity to provide input and comments on a proposed regulation, but does not participate in the decision process. By contrast, organizations (or at least their employees) frequently play a direct role in forming cooperative standards. A standard is a set of rules that promote interoperability between equipment developed by distinct groups. In the context of data networks a standard usually defines a protocol or an interface. Standards do not usually have the force of law, though it is possible that a regulation will require conformance to a standard. A standards process is often sponsored by a neutral professional organization, for example the IEEE or the Society of Automotive Engineers (SAE).

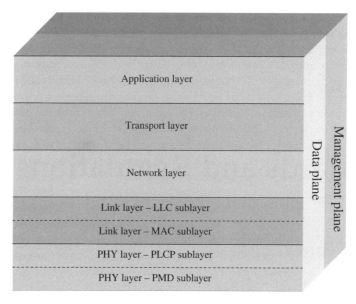

Figure 10.1 Open Systems Interconnection reference architecture.

This book is being written at a time when the regulations and standards governing VANETs are in flux. Some rules are relatively stable while others are subject to significant change. The discussion in this chapter attempts to provide guidance about the stability of the content it covers, but the reader is advised in every case to consult the source documents for updates.

10.2 Layered Architecture for VANETs

10.2.1 General concepts and definitions

Many readers will be familiar with the concept of a layered architecture for data networks. The well-known *Open Systems Interconnection (OSI)* (ITU-T X.200 1994) reference architecture includes five[1] layers, as illustrated in Figure 10.1. Dividing the entire problem space into layers, with well-defined interfaces between layers, promotes scalability, structured design, and technical evolution.

In several cases a given layer is further partitioned into sublayers. For example, the *physical (PHY)* layer is often divided into a *physical medium dependent (PMD)* sublayer and a *physical layer convergence procedure (PLCP)* sublayer. Similarly, in cases where devices share a physical medium, which is usually the case for a wireless VANET, the *data link* layer is often divided into a *medium access control (MAC)* sublayer and a *logical link control (LLC)* sublayer. These divisions are illustrated in Figure 10.1.

---

[1] The OSI model consists of seven layers, but layers 5 and 6 of that model are commonly omitted. To simplify the presentation, their functions will be considered as part of the application layer in this chapter.

In addition to the division of layers into horizontal sublayers, it is also common to divide a layered architecture into vertical *planes* representing the different actors in a network. The most common vertical division, illustrated in Figure 10.1, is between the *data plane* (also called the user plane) and the *management plane*. Actions in the data plane are initiated by applications and directly support the communication of data from one user application to another. The management plane is the locus of actions taken to manage and maintain the network, and so only indirectly supports user communication. Management plane actions include diagnostics, synchronization, and discovery and association of neighboring devices.

The terms *protocol* and *interface* generally carry the following meanings in the context of a data network layered architecture:

Protocol: The rules governing communication between layer n on one device and layer n **on another device**.

Interface: The rules governing information exchange between adjacent layers (layer n and layer $n-1$) or between planes **within a device**, where a 'device' could be a collection of physical units that collectively function as a single system.

10.2.2 A protocol stack for DSRC

Different types of VANETs will define the protocols and interfaces in Figure 10.1 differently. In order to illustrate important concepts, this chapter discusses regulations and standards associated with a specific use case: dedicated short-range communications (DSRC) in the US. DSRC is the focus of great attention in the US automotive industry, including government regulators, automobile manufacturers, communication equipment suppliers, and academic researchers. DSRC is also of interest in other countries (e.g. in Europe and Japan), though both the specific meaning of the term and the implementation model vary from place to place. The chapter covers some non-US DSRC details, but unless otherwise stated the discussion refers to the US model.

DSRC envisions two primary types of equipment: an 'on-board unit' (OBU) is a DSRC system within a vehicle (and hence highly mobile), and a 'roadside unit' (RSU) is a DSRC system that is used at a fixed point near a road. An RSU could be operated by a governmental agency or a private enterprise. Note that some writers use the terms on-board equipment (OBE) and roadside equipment (RSE) instead of OBU and RSU.

Figure 10.2 illustrates the protocol stack for US DSRC communication, including shorthand names of protocols intended for use at the various layers. For simplicity of presentation, the figure does not distinguish between the data and management planes. As noted above, as this book goes to press many of these protocols are under active development. Descriptions below provide an accurate snapshot of the status of each standard and regulation at this time.

The remainder of this chapter is organized according to Figure 10.2, working from the physical layer up to the application layer. Here is a brief summary of the topics covered:

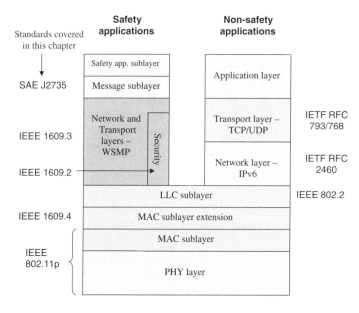

Figure 10.2 Layered architecture for DSRC communication in the US (IETF RFC 768 1980; IETF RFC 793 1981; IETF RFC 2460 1998).

Regulatory environment (Section 10.3): The US FCC has allocated 75 MHz of spectrum for DSRC communication, from 5.850 GHz to 5.925 GHz. This spectrum consists of a 5 MHz guard band and seven 10 MHz channels to support both safety applications and non-safety applications (see Section 10.3.1). For the purposes of this chapter, a safety application is one whose purpose is to directly prevent vehicle collisions, for example by providing a driver with a timely warning that braking or evasive action is required. Examples of safety applications are those aimed at preventing rear-end collisions, left-turn collisions, lane change collisions, and head-on collisions. A non-safety application is any other application. Non-safety applications include those related to automobiles and travel (e.g. tolling, navigation, traffic information) and more traditional applications such as Internet access. The separation is not always clear. For example, an application that provides information about a road construction zone ahead may impact safety indirectly, but will generally be classified as a non-safety application in this discussion.

Physical layer (Section 10.4): The PHY protocol for DSRC is being standardized under the IEEE Standards Association. More specifically, the popular IEEE 802.11 wireless local area network (LAN) standard is being amended to support DSRC communication. The amendment will initially be referred to as *802.11p Wireless Access in Vehicular Environments (WAVE)*, and will eventually be incorporated into a future release of the baseline 802.11 standard (IEEE 802.2 1998).

Link layer (Section 10.5): The MAC sublayer of the link layer will also be defined by the IEEE 802.11p WAVE amendment. The LLC sublayer of the link layer will use the existing, stable IEEE 802.2 standard. In the management plane, a MAC extension protocol defines how a device switches among the various DSRC channels allocated by the FCC. This protocol is being developed by a separate IEEE standards working group called 1609. The IEEE 1609 working group has authority for a set of related standards referred to here jointly as 1609.x. The MAC extension protocol is specifically designated IEEE 1609.4 *Multi-channel Operation* (IEEE 1609.4 2006).

Network and Transport layers (Section 10.6): Above the link layer the protocol stack separates. Safety applications are supported through one set of protocols while non-safety applications are supported through another set. The *IEEE 1609.3 Networking Services* standard defines a message, the *WAVE short message (WSM)*, and a protocol, the *WAVE short message protocol (WSMP)*, to support network and transport layer functions for DSRC safety applications. The 1609.3 standard also defines a message called the *WAVE service advertisement (WSA)*, which is used to advertise the availability of one or more DSRC services at a given location. A WSA might be used, for example, to advertise a traffic information service offered by an RSU. The *IEEE 1609.2 Security Services* standard defines mechanisms for encrypting and authenticating data plane WSMs and management plane WSAs.

Non-safety applications can also use WSMP, but in most cases are supported through a conventional Internet stack above layer 2. In particular, network layer services are provided by the *Internet Protocol version 6 (IPv6)*. Transport layer services for non-safety applications utilize the familiar *Transmission Control Protocol (TCP)* or *User Datagram Protocol (UDP)*. All three of these protocols: IPv6, TCP, and UDP are quite stable, and are defined by the Internet Engineering Task Force (IETF).

Application layer (Section 10.7): Application layer protocols could be almost anything. In particular, most non-safety applications that can run over the Internet can also run over a DSRC communication system (which may indeed have a backhaul link into the Internet). Some of these are covered by well-established standards. Others will be the subject of future standardization efforts or will be proprietary. *Safety* applications will likely have both a standard portion and a proprietary portion. The standard portion includes a common message format for conveying the state of a vehicle, an intersection, or other relevant information. The SAE DSRC technical committee is developing the *J2735 Message Set Dictionary* standard to define message formats.

10.3 DSRC Regulations

DSRC operations are subject to both cooperative standards and government regulation. This section looks at the regulatory side of DSRC. The emphasis is on DSRC regulations in the US, which are more stable and mature than in Europe. The status of regulations in Europe is also discussed briefly. Technologies similar to DSRC have been regulated and deployed in Japan as well.

10.3.1 DSRC in the United States

In the United States DSRC regulations are determined by the *Federal Communications Commission (FCC)*. FCC Regulations are published in the *Code of Federal Regulations, Title 47 (abbreviated CFR 47)* (CFR 47 Part 90; CFR 47 Part 95). This section is divided into discussions of general spectrum issues, regulations on RSUs, and regulations on OBUs.[2]

DSRC spectrum in the US

The most significant step taken by the US FCC was the allocation of 75 MHz of spectrum for DSRC services, i.e. for 'the improvement of traffic flow, traffic safety, and other intelligent transportation service applications' (CFR 47 §90.7). This allocation was made in October 1999, following passage of the *Transportation Equity Act for the 21st Century (TEA-21)* in 1998. In December 2003 the FCC updated its rules for licensing and operation in the DSRC band. Spectrum is scarce and valuable, so this allocation is a measure of the importance that the US government places on those improvements.

The spectrum is from 5.850 GHz to 5.925 GHz, which is commonly referred to as the 5.9 GHz band. The lower 5 MHz is reserved as a guard band. The remaining 70 MHz is divided into seven 10 MHz channels (see Figure 10.9 in Section 10.6.1). Channel numbers in this band are determined by the offset of the center frequency relative to 5.000 GHz, in units of 5 MHz. For example, the lowest 10 MHz DSRC channel occupies the range 5.855–5.865 GHz. Its center frequency, 5.860 GHz, is 860 MHz above the baseline, i.e. it is at an offset of 172 units of 5 MHz. So, this channel is designated Channel 172. Each subsequent 10 MHz channel is two more 5 MHz units offset from the baseline, so the DSRC channel numbers occupy the even integers in the range 172–184. Channel 184 occupies the range 5.915–5.925 GHz at the top end of the DSRC spectrum.

The FCC also permits the spectrum occupied by Channels 174 and 176 to be combined into a single 20 MHz channel, designated Channel 175, and similarly for the combination of Channels 180 and 182 into a 20 MHz Channel 181. These are of less immediate interest, however, so the focus of the discussion in this chapter is on the 10 MHz channels.

Special channel designations:
It is not expected that a DSRC device, whether an RSU or an OBU, will be equipped with separate radios for all of the seven DSRC channels; some will have only one radio and will access one channel at a time. The IEEE 1609 WG has devised a protocol that allows a device to switch channel operation over time (see Section 10.6.1) so that it can efficiently utilize the multi-channel spectrum. This protocol relies on the designation of one channel as the *control channel (CCH)*. Other

---
[2] The author wishes to thank Alastair Malarky for helping to explain the FCC DSRC regulations.

channels are referred to as *service channels (SCHs)*. In the US, Channel 178 in the middle of the DSRC spectrum is the control channel and the other six channels are service channels. The functions associated with these designations are explained in later sections.

On July 20, 2006 the FCC further designated service channels 172 and 184, at opposite ends of the spectrum, for specific purposes. Channel 172 is designated 'exclusively for vehicle-to-vehicle safety communications for accident avoidance and mitigation, and safety of life and property applications.' Channel 184 is designated 'exclusively for high-power, longer-distance communications to be used for public safety applications involving safety of life and property, including road intersection collision mitigation' (FCC 06-110 2006). RSUs licensed to non-governmental bodies are not allowed to operate on Channel 184. The designation of Channel 172 for vehicle-to-vehicle safety is somewhat incongruous with the IEEE 1609.4 channel switching protocol. The primary means for achieving 'accident avoidance and mitigation' is for each vehicle to frequently broadcast its state in so-called *basic safety messages (BSMs, see Section 10.7.2)*. Under the IEEE 1609.4 standard, it is contemplated (though not explicitly stated) that BSMs will be sent on the control channel (Channel 178 in the US). The FCC designation of Channel 172 does not require that BSMs be moved to that channel, but it would be consistent with the spirit of the designation. It would also create problems for single-radio devices that want to support both safety and non-safety applications. Reconciling the apparent contradiction between the IEEE 1609.4 standard and the 2006 FCC designation is an ongoing research question.

Power limits:
DSRC spectrum is shared among the OBUs and RSUs in a given area. With shared spectrum there is the potential that a signal from one transmitter will interfere with the ability of another transmitter to be heard. If the interferer is in the same channel as the target transmitter this is known as co-channel interference. If the interferer is in a different (spectrally near) channel this is known as cross-channel interference. The FCC regulates the transmit power of DSRC devices as a means of controlling both co-channel and cross-channel interference.

In DSRC, co-channel interference manifests itself in frame 'collisions', defined as the overlap in time and space of two or more transmit signals. The *medium access control (MAC)* protocol described in Section 10.5.1 attempts to avoid collisions, but it cannot eliminate them. The probability that a given transmission suffers a collision in a given region is proportional to the number of potential interferers in that region, which is itself proportional to the transmit power employed by each device.

Cross-channel interference is more difficult to avoid, and indeed it would be hard to justify allocation of spectrum for multiple channels if they could not be used at the same time and place. The principle means of controlling cross-channel interference is by defining limits on the power of a transmitter's signal at frequencies outside the assigned channel. In DSRC this is a two-step process: the absolute power that a transmitter may use in its target channel is constrained, and then the power that can be emitted in other channels is limited relative to the target-channel power. This latter limit is defined via a transmit spectral mask, which is further discussed in Section 10.4.1.

Table 10.1 FCC device classification.

Device class	Max. output power (dBm)	Communication zone (meters)
A	0	15
B	10	100
C	20	400
D	28.8	1000

The FCC defines two types of power limit: the maximum transmit power from the 'device' and the maximum transmit power out of the antenna. Device limits are defined according to one of four classes, which are also associated with desired transmission range. These are shown in Table 10.1 (CFR 47 §90.375). Antenna output power is specified in terms of the equivalent isotropically radiated power (EIRP). More specifically, the FCC regulations specify that the EIRP is to be measured 'as the maximum EIRP toward the horizon or horizontal, whichever is greater, of the gain associated with the main or center of the transmission beam' (CFR 47 §90.377).

The FCC regulations contain some explicit power limits, but also incorporate by reference others defined in *ASTM standard E2213-03* (ASTM E2213-03 2003). This standard defines a MAC and PHY protocol for DSRC, and is a precursor to the work in the IEEE 802.11 Working Group to define the 802.11p amendment. But, the standard also defines some things that will not be part of the 802.11p amendment. An example is that ASTM E2213 defines a detailed set of power limits according to channel and type of device (Section 8.9.1 of ASTM E2213-03 2003). Table 10.2 shows these limits, specified both for maximum output power and maximum EIRP, for each of the seven 10 MHz DSRC channels. The table also shows the frequency range and channel type of each channel (CFR 47 §90.377).

A 'public' RSU or OBU is one that is licensed to a governmental unit, e.g. a city police department or a county highway department. There are a few things to notice in this table. First, public devices are permitted to use higher EIRP than private devices on Channel 178 (the control channel) and Channel 184 (designated for long-distance public safety applications). On the other channels there are generally no distinctions between public and private devices. Second, devices operating on Channels 180 and 182 have lower power constraints than on the other channels. These channels are intended for shorter-range communication. Finally, on Channels 180 and 182 private OBUs are permitted higher output power (but not EIRP) than RSUs, and no limits are defined for public OBUs. OBUs are permitted higher output power than RSUs because their antennas generally have less gain. It is not clear whether the omission of public OBU limits for Channels 180 and 182 is intentional or not. This will likely be clarified when the FCC regulations are eventually updated to reflect the fact that IEEE 802.11p has superseded ASTM E2213 as the relevant MAC and PHY layer standard for DSRC.

The limits in Table 10.2 may be further reduced in a specific situation, for example as part of a site license for an RSU. Furthermore, both RSUs and OBUs are expected to use the minimum power necessary to support the application using the message.

Table 10.2 Maximum output power (dBm) and EIRP (dBm) by channel and device type.

Chan #	Public RSU Max. output	Public RSU Max. EIRP	Private RSU Max. output	Private RSU Max. EIRP	Public OBU Max. output	Public OBU Max. EIRP
172	28.8	33	28.8	33	28.8	33
174	28.8	33	28.8	33	28.8	33
176	28.8	33	28.8	33	28.8	33
178	28.8	44.8	28.8	33	28.8	44.8
180	10	23	10	23	n/a	n/a
182	10	23	10	23	n/a	n/a
184	28.8	40	28.8	33	28.8	40

Chan #	Private OBU Max. output	Private OBU Max. EIRP	Freq. range (MHz)	Channel type
172	28.8	33	5855–5865	Service
174	28.8	33	5865–5875	Service
176	28.8	33	5875–5885	Service
178	28.8	33	5885–5895	Control
180	20	23	5895–5905	Service
182	20	23	5905–5915	Service
184	28.8	33	5915–5925	Service

Note: 'n/a' indicates not applicable (not defined in ASTM E2213 or FCC regulations).

FCC regulations regarding RSUs

DSRC RSUs are authorized under Subpart M of Part 90 of CFR 47 (CFR 47 Part 90), which also incorporates by reference the ASTM E2213 standard. Beyond the power constraints discussed in the previous section, ASTM E2213 includes a comprehensive definition of a MAC and PHY protocol for DSRC. These protocols are based on IEEE 802.11. After ASTM E2213 was published in 2004, interested parties began working on an update to those MAC and PHY protocols, within the IEEE 802.11 Working Group. The result is the IEEE 802.11p amendment, which is nearing completion as this book is written (see Section 10.1). It is anticipated that after IEEE 802.11p is completed, the FCC will update its DSRC regulations to incorporate a reference to that standard, and will no longer refer to ASTM E2213 for the MAC and PHY protocols. However, since 802.11 is an international standard, the 802.11p amendment will not include some of the US-specific regulatory specifications that are in ASTM E2213, for example the detailed combination of power limitation by device type and channel number shown in Table 10.2. It is unclear whether the FCC regulations will continue to refer to those portions of ASTM E2213, or will extract those portions for inclusion directly in CFR 47.

In addition to the spectrum issues discussed above, the regulations address the following:

Licensing:
A license is required to deploy and operate an RSU. A single license can cover multiple RSUs. Each RSU must be individually registered before it can operate, and registrations are usually specific to a given site. Portable (non-site-specific) RSUs are allowed to operate from fixed points so long as they do not interfere with site-licensed RSUs. An RSU cannot operate while mobile. In addition to location, an RSU registration also includes other limitations, for example with regard to channels of operation, transmission power, direction of transmission, etc.

Co-existence with other services:
The DSRC frequency band is also in use by other services. Among these are Government Radiolocation services operated at military installations within the US. A DSRC RSU that is within 75 km of such an installation must coordinate its operation through the National Telecommunications and Information Administration. Section 90.371 of CFR 47 lists approximately 60 such locations.

Antenna height:
An RSU antenna is limited to 15 meters above the roadway bed surface. If the antenna is 8 or less meters, then the EIRP restrictions in Table 10.2 apply. If the antenna is between 8 and 15 meters the allowed EIRP is reduced.

Access priority:
The FCC assigns highest access priority, for example access to the wireless channel in the MAC protocol, to communications involving 'safety of life.' The next highest priority is assigned to communications 'involving public safety,' which includes all communication involving public RSUs by definition. All other communication is placed at a lower access priority.

FCC regulations regarding OBUs

DSRC OBUs are authorized under Subpart L of Part 95 of CFR 47 (CFR 47 Part 95). As with the RSU regulation, this subpart also incorporates by reference the ASTM E2213 standard. There are few additional regulations in Subpart L beyond those represented by ASTM E2213. One item of note is that OBUs are licensed 'by rule,' which means that any OBU that conforms to the regulations (principally the ASTM E2213 standard) is permitted to operate in the DSRC band. OBUs are also not restricted to specific locations, of course; they are permitted 'wherever vehicle operation or human passage is permitted' (ASTM E2213-03 2003). The OBU regulations also reflect the same three-level priority of access specified for RSUs above.

10.3.2 DSRC in Europe

The regulatory landscape in Europe for communication services is more complex and has more players than in the US. At the summit is the European Union (EU), which is a confederacy of 27 European nations. In the realm of radio spectrum, the EU has the authority to make allocations that the member nations are obliged to recognize, though the implementation of the allocation (e.g. time frame, legal mechanism) will vary from country to country.

In Europe the technology that is referred to in the US as DSRC is more commonly called *intelligent transport systems (ITS)*. The term DSRC in Europe is often more narrowly applied to a set of applications that includes electronic tolling. On August 5, 2008, the EU issued a decision to allocate 30 MHz of spectrum, 5.875–5.905 GHz, for ITS communication (EU Decision 2008). This 30 MHz is the center of the 70 MHz occupied by Channels 172–184 in the US DSRC spectrum. The brief decision report gave member nations six months to designate this spectrum for ITS use. The spectrum is to be licensed on a non-exclusive basis, as in the US. This decision was a very important step in harmonizing this technology between Europe and the US.

The decision report assumes a maximum EIRP of 33 dBm for ITS transmitters. This is one of the few ITS-operation technical details included in the decision.

It is not yet clear how the 30 MHz will be used, and a number of possibilities are being discussed. Some of the scenarios that have been discussed, at least informally, are:

- Define two channels, a 20 MHz channel from 5.885–5.905 GHz and a 10 MHz channel from 5.875–5.885 GHz. The 20 MHz channel would focus on vehicle-to-vehicle communication, and would thus carry the most critical safety messages. The 10 MHz channel would focus on vehicle-to/from-roadside infrastructure communication. In that sense the upper 20 MHz would resemble the US control channel (Channel 178) and the lower 10 MHz would be used similarly to one of the service channels. The European ITS community is generally more inclined than their US counterparts to require each vehicle to have at least two radios, which reduces the need for a channel switching protocol such as that discussed in Section 10.6.1.

- Define two 10 MHz channels, at the upper and lower ends of the spectrum, and leave the middle 10 MHz unused as a guard band to reduce cross-channel interference.

- Define a single 30 MHz channel, which clearly would eliminate the need to designate different channels for different types of communication.

There are also discussions about possibly augmenting this spectrum with additional bandwidth either above or below or both.

In 2008 the *European Telecommunications Standards Institute (ETSI)* published (ETSI EN 302 571 2008), a standard on ITS in the 5.850–5.925 GHz band (i.e. the same 75 MHz allocated for DSRC by the US FCC). Its scope is to cover the 'essential requirements of article 3.2 of the R&TTE directive.' The Radio & Telecommunications Terminal Equipment (R&TTE) directive was issued by the European Parliament in April 1999, and it relates to regulation of equipment intended to be connected to public networks. Article 3.2 of the R&TTE addresses 'effective use of the radio spectrum/orbital resource so as to avoid harmful interference.' Thus, this standard is primarily concerned with interference that an ITS device might cause. The bulk of the standard addresses testing methodology. The technical requirements portion of the standard limits transmissions in the band to 33 dBm EIRP, and 23 dBm/MHz spectral density assuming a 10 MHz channel. The standard specifies a transmit spectral mask that matches the Class C mask in the US FCC regulations.

Standardization efforts addressing the ITS problem in Europe more comprehensively have been undertaken by a variety of organizations. ETSI formed an ITS technical committee in October 2007 to progress standards for ITS deployment. A more ambitious standardization effort has been undertaken by an ITS committee within the *International Standards Organization (ISO)*. This group, ISO Technical Committee 204, Working Group 16, has developed an architecture for mobile communications called CALM (the acronym has been quoted as standing for a variety of phrases, but 'continuous air interface for long and medium distance' seems the most generally accepted). CALM will use, but is not limited to, wireless communication in the 5.9 GHz band. It does expect critical safety communication to use that band.

A common thread in the European regulatory and standards efforts is to harmonize with and build on developments in the US to the extent possible. So, for example, the IEEE 802.11p amendment for the MAC and PHY protocols (see Sections 10.4 and 10.5) will be an essential part of ITS lower layer communications in Europe. The IEEE 1609.x family of standards (see Section 10.6) is also seen as a core technology, albeit only one of a variety of approaches that have a home in the flexible CALM architecture. As many of the US-centric standards activities reach a level of maturity in the next few years, the parallel European efforts can be expected to remain in flux for a bit longer.

10.4 DSRC Physical Layer Standard

The next several sections examine the various layers of the DSRC protocol stack (Figure 10.2) in detail, working from bottom to top, and starting with the physical layer.

The physical (PHY) layer is the lowest layer in a protocol stack. A PHY standard defines a *protocol* between devices, and an *interface* to the next higher (data link) layer within a device. At the most general level, the task of the PHY protocol is to define the rules and formats for conveying digital 1s and 0s from a sender to one or more receivers. The inter-layer interface defines how the PHY layer supports services to the data link layer.

The PHY layer is often divided into two sublayers: a *physical medium dependent (PMD)* sublayer and a *physical layer convergence procedure (PLCP)* sublayer. The PLCP sublayer defines the mapping between a data link layer frame (set of bits) and the basic PHY layer data unit, the symbol. The PMD sublayer defines how the symbol is conveyed over the medium, including electrical, mechanical, and other physical characteristics; as the name implies the PMD depends on the specific medium over which communication takes place. In the case of VANETs, the medium is assumed to be wireless.

Initial efforts to define a PHY layer standard for DSRC were carried out under the auspices of ASTM International, a standards organization specializing in testing

Table 10.3 802.11 10 MHz OFDM channel basic parameters.

Parameter	Value
Number of data subcarriers	48
Number of pilot subcarriers	4
Total number of subcarriers	52
Subcarrier frequency spacing	156.25 kHz
Guard interval (GI)	1.6 μsec
Symbol interval (including GI)	8 μsec

and materials. The resulting standard, ASTM E2213-03 (ASTM E2213-03 2003), was largely based on IEEE 802.11 (IEEE 802.11 2007). In 2004 interested parties obtained approval to create a DSRC amendment to 802.11 within the IEEE 802.11 Working Group. This amendment, called wireless access in vehicular environments (WAVE) remains a work-in-progress as this book goes to press;[3] it will be designated IEEE 802.11p when it is released. The draft amendment is relatively stable. The discussion of the PHY layer and MAC sublayer in this chapter focuses on those aspects most likely to be included in the final WAVE 802.11p amendment.

IEEE 802.11 defines several PHY layer protocols. The WAVE 802.11p amendment is based on one of those, *orthogonal frequency division multiplexing (OFDM)* for the 5 GHz Band (Clause 17 of IEEE 802.11 2007). This OFDM PHY was first defined as the IEEE 802.11a amendment, and is commonly still referred to as 802.11a. The following subsections explain the basic 802.11 OFDM PHY and the modifications to this PHY proposed to enable WAVE operation.

The concept and theoretical basis of OFDM are covered in a separate chapter of this book. The present discussion focuses on the specific OFDM protocol defined in 802.11, including the modifications in the WAVE 802.11p amendment.

10.4.1 OFDM physical medium dependent (PMD) function

The OFDM protocol used in 802.11 is defined for three channel widths: 20 MHz, 10 MHz, and 5 MHz. Whereas most 802.11a implementations use the 20 MHz channel, DSRC will more commonly use the 10 MHz channel. The basic parameters of the 802.11 10 MHz OFDM channel are shown in Table 10.3. Four modulation techniques are available, each of which corresponds to a different number of bits encoded per subcarrier symbol. These are shown in Table 10.4. *Forward error correction (FEC)* coding is applied to the user bits, which has the effect of reducing the effective user bit rate but also of improving the probability of successful decoding. Three FEC coding rates may be employed: 1/2, 2/3, and 3/4.

There are twelve combinations of the four modulation rates and the three FEC coding options. Of these, eight combinations are allowed in 802.11 OFDM, as shown in Table 10.5. For example, BPSK uses one bit per subcarrier symbol and thus 48 bits

---
[3]Updates and pointers to evolving DSRC standards will occasionally be made available at http://www.vanetbook.com/.

Table 10.4 Modulation techniques used in 802.11 OFDM.

Modulation technique	Bits per subcarrier symbol
Binary phase shift keying (BPSK)	1
Quadrature phase shift keying (QPSK)	2
16-point quadrature amplitude modulation (16-QAM)	4
64-point quadrature amplitude modulation (64-QAM)	6

Table 10.5 Data rate options in a DSRC 10 MHz OFDM channel.

Modulation technique	Coded bit rate (Mbps)	Coding rate	Data rate (Mbps)	Data bits per OFDM symbol
BPSK	6	1/2	3	24
BPSK	6	3/4	4.5	36
QPSK	12	1/2	6	48
QPSK	12	3/4	9	72
16-QAM	24	1/2	12	96
16-QAM	24	3/4	18	144
64-QAM	36	2/3	24	192
64-QAM	36	3/4	27	216

per OFDM symbol. With 1/2 rate coding, there are 24 data bits and 24 coding bits per OFDM symbol. With 24 data bits per OFDM symbol and an 8 μsec symbol period, the resulting data rate is 3 Mbps.

PMD transmitter:
When the PLCP sublayer requests the PMD sublayer to transmit a packet, it supplies the coded bits that make up each OFDM symbol (the contents of which include MAC data and PLCP overhead described below). It also provides the data rate and the transmit power. The PMD sublayer performs the OFDM modulation, as described in more detail in the OFDM chapter of this book, including inverse FFT calculation, guard interval (cyclic prefix) insertion, wave shape filtering, RF modulation, and power amplification.

An 802.11 device (a.k.a. 'station' or STA) implementing the OFDM 10 MHz PHY must support the 3, 6, and 12 Mbps data rates. The other rates are optional.

Transmit power out of the 802.11 device (which is distinct from power out of the antenna) is represented internally as an integer number of milliwatts, up to 10 000, providing a range from 0 dBm to 40 dBm. As discussed in Section 10.3.1, the FCC defines four device classes (Class A, B, C, and D), each of which has a maximum transmit power far less than 40 dBm.

Each device class is also associated with a transmit spectral mask, which limits the out-of-band energy of a transmitter. These masks are defined in the IEEE 802.11p amendment. A given mask specifies a frequency-dependent upper bound on the permitted power spectral density (PSD) of the transmitted signal. The PSD is specified in terms of a 100 kHz resolution bandwidth. Both the frequency and

Table 10.6 Power spectral density limits for 10 MHz DSRC channels in the US (from IEEE 802.11p).

Freq. offset	±4.5 MHz	±5.0 MHz	±5.5 MHz	±10.0 MHz	±15.0 MHz
Class A	0 dBr	−10 dBr	−20 dBr	−28 dBr	−40 dBr
Class B	0 dBr	−16 dBr	−20 dBr	−28 dBr	−40 dBr
Class C	0 dBr	−26 dBr	−32 dBr	−40 dBr	−50 dBr
Class D	0 dBr	−35 dBr	−45 dBr	−55 dBr	−65 dBr

the PSD in the mask function are expressed in relative terms: frequencies are offsets from the center frequency of the signal, and PSD is relative to the peak PSD of the signal (i.e. in dBr). Each mask is a piecewise linear function of frequency offset, and thus can be defined by specifying the PSD limits at the breakpoints between the linear segments. The transmit spectral mask values at the breakpoints are shown for each device class in Table 10.6. Progressing from Class A to Class D, each class allows a higher maximum transmit power and enforces a tighter spectral mask. For a frequency that is between two breakpoints, the mask limit is formed by the line segment connecting the breakpoint values. For example, between 5.5 MHz and 10 MHz offset, the Class A mask has slope: (8.0/4.5) dBr/MHz. So, at an offset of 7.5 MHz the Class A mask limit is: -20 dBr $- (2.0 \cdot 8.0/4.5)$ dBr $= -23.6$ dBr. The mask is flat at offsets greater than 15 MHz.

Of the four transmit spectral masks, the one most likely to apply to OBU transmissions is Mask C, shown in Figure 10.3 for a 10 MHz channel.

PMD receiver:
The PMD receiver performs the demodulation steps described in more detail in the OFDM chapter of this book, including automatic gain control (AGC), clock recovery, RF demodulation, guard interval removal, and FFT. When the PMD sublayer wishes to pass a received packet up to the PLCP sublayer, it also makes available the received signal strength indication (RSSI).

Receiver performance is specified in IEEE 802.11 in terms of minimum sensitivity and channel rejection. Minimum sensitivity is defined as the minimum absolute signal energy for which a reference 1000 byte packet must be correctly received at least 90% of the time. Minimum sensitivity is a function of the modulation technique and FEC coding rate, and thus of the data rate of the packet. For the 10 MHz OFDM signal it varies from −85 dBm at 3 Mbps to −68 dBm at 27 Mbps.

Channel rejection is an indication of a receiver's ability to filter out energy that is outside the 10 MHz channel of interest. There are different specifications depending on whether the interfering transmitter is in an adjacent channel or not. For example, referring to the US DSRC band plan (see Section 10.3.1), assume that a receiver is tuned to Channel 172 and that the receive signal strength of a reference 1000 byte packet in Channel 172 is 3 dB above the minimum sensitivity (i.e. 3 dB above −85 dBm for 3 Mbps data rate). Channel rejection is concerned with the probability of successful packet reception in the desired channel as a function of the signal strength of an interferer in another channel. When the interferer is in Channel 174,

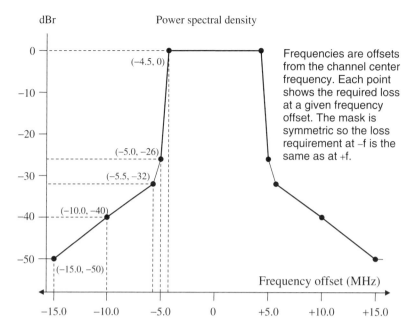

Figure 10.3 Transmit spectral mask C, for transmit powers up to 33 dBm.

the Channel 172 receiver must receive the reference packet with at least 90% probability for interference signal strengths (measured in Channel 174) up to -66 dBm. The difference between -66 and -82 dBm (16 dB) is called the *adjacent channel rejection*. If the interferer is in a channel not adjacent to Channel 172, e.g. Channel 176 or higher, the packet reception success rate of the Channel 172 receiver must be at least 90% for interference signal strengths (measured in the interfering channel) up to -50 dBm. The difference between -50 and -82 dBm (32 dB) is called the *nonadjacent channel rejection* (or sometimes the alternate adjacent channel rejection).

In the WAVE 802.11p standard, channel rejection performance is specified both in terms of a *required* level and a more stringent optional *enhanced receiver performance* level. The enhanced level was introduced to compensate for the more challenging communication environment associated with rapidly moving vehicles. For each of the data rates in Table 10.5 the recommended adjacent channel rejection levels are 12 dB higher than the required numbers. For example, the required adjacent channel rejection for rate 3 Mbps is 16 dB, and the recommended level is 28 dB. Similarly, the recommended nonadjacent channel rejection levels are 10 dB higher than each of the required numbers, e.g. 42 dB rather than 32 dB for 3 Mbps. As the standardization effort for IEEE 802.11p continues at the time of going to press, there is some debate about whether to require, rather than simply recommend, the higher channel rejection levels for a STA engaged in WAVE communication. The specific degree of enhancement, e.g. 12 dB for adjacent channel rejection, is also the subject of ongoing study.

There is one other aspect of the PMD for which the draft IEEE 802.11p amendment proposes constraints that differ from baseline IEEE 802.11: temperature range. The baseline IEEE 802.11 standard specifies three operating temperature ranges, the widest of which is −30°C to 70°C. The WAVE amendment proposes a fourth range for automotive and outdoor environments: −40°C to 85°C.

10.4.2 OFDM physical layer convergence procedure (PLCP) function

In a transmitter, the PLCP sublayer function is to process the bytes in a MAC frame so that they can be transformed into OFDM symbols for transmission over the air by the PMD sublayer. Technically the PLCP takes a physical layer service data unit (PSDU), i.e. the MAC frame, and creates a physical layer protocol data unit, PPDU. The MAC sublayer passes three parameters to the PLCP, in addition to the PSDU: a) the length of the PSDU, which 802.11 limits to a maximum of 4095 bytes (though practically it is rarely greater than 1500 bytes); b) the data rate at which the PSDU will be transmitted (see Table 10.5); and c) the power level to be used. The power level is selected from one of eight configured values.

In a receiver the PLCP performs essentially the inverse function to extract a PSDU from a PPDU. In addition to passing the received PSDU up to the MAC sublayer, the PLCP also provides the RSSI.

The sequence of steps to create a PPDU in the transmitter is as follows:

1. Create a PLCP preamble.

2. Add a PLCP header, consisting of a *SIGNAL* symbol and an additional 16-bit *SERVICE* field.

3. Continue with the MAC frame, i.e. physical layer service data unit (PSDU).

4. Add a trailer consisting of six *TAIL* bits and a variable number of *PAD* bits.

5. Perform bit scrambling on the *DATA* field (defined as the combination of the *SERVICE* bits, the PSDU, the *TAIL* and the *PAD* bits).

6. Perform forward error correction (FEC) coding on the *SIGNAL* and *DATA* fields.

The PPDU format is shown Figure 10.4. Here are some additional details about the PPDU.

Preamble:
The PPDU preamble is divided into two portions. The first consists of ten copies of a short *training sequence*, which is aimed at assisting the AGC, antenna diversity, clock acquisition, and coarse frequency acquisition functions in the receiver. Each short training sequence only utilizes one fourth of the subcarriers, and thus has a period of 1.6 µsec in a 10 MHz channel (compared with 6.4 µsec for the normal inverse FFT interval when all subcarriers are used). The length of the first portion of the preamble is 16 µsec.

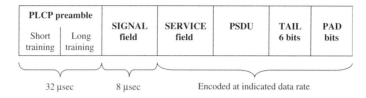

Figure 10.4 Physical protocol data unit format.

| Rate
Bits 0–3 | Reserved
Bit 4 | Length
Bits 5–16 | Parity
Bit 17 | Tail
Bits 18–23 |

Figure 10.5 *SIGNAL* field format.

The second portion begins with a guard interval (3.2 μsec) followed by two copies of a long training sequence, which is used for estimating the channel impulse response in each subcarrier and for fine frequency acquisition. The duration of the second portion is $3.2 + 2 \cdot 6.4 = 16$ μsec, and thus the entire preamble lasts 32 μsec.

One of the functions of the PHY layer receiver is to determine when the channel is busy and when it is idle. This is called the *clear channel assessment (CCA)* function in 802.11, and it is useful for the MAC medium access function described in Section 10.5.1. The 802.11 standard requires that an OFDM receiver in a 10 MHz channel be capable of recognizing the start of a PPDU and declaring the channel busy within 8 μsec from the start of the preamble, i.e. within the first five short training sequences.

SIGNAL:
Following the preamble is a set of bits that make up one OFDM symbol. These bits carry two pieces of information about the PPDU: the data rate (see Table 10.5) and the PSDU length. The data rate of the PPDU is encoded in four bits, and applies to the rest of the PPDU after the *SIGNAL* field, which is itself always sent at the lowest data rate (3 Mbps in a 10 MHz channel). The PSDU length is in units of bytes and is encoded in 12 bits. The data rate and length are provided to the PHY from the MAC, along with the PSDU. There is also one parity bit (even parity) and the remaining seven bits of the *SIGNAL* field are either reserved (bit 4) or defined to be zero (bits 18–23). The *SIGNAL* field is shown in Figure 10.5. It is not subject to bit scrambling.

SERVICE and TAIL:
The 16-bit *SERVICE* field and the 6-bit *TAIL* field facilitate the operation of the bit scrambling function. All 22 bits are set to 0 prior to scrambling and coding.

PAD:
The length of the entire PPDU, following FEC coding, must be an integer multiple of the number of bits per OFDM symbol, which depends on the modulation technique. For BPSK, QPSK, 16-QAM, and 64-QAM the OFDM symbol lengths, respectively, are 48 bits, 96 bits, 192 bits, and 288 coded bits. The *PAD* bits, each of which has

value zero, are added prior to scrambling and FEC coding so that this constraint is met. The minimum number of *PAD* bits is two, since the *TAIL* has six bits and all other fields are multiples of eight bits long. At the lowest data rate, 3 Mbps, there are 24 data (uncoded) bits per OFDM symbol, so the *PAD* will add 2, 10, or 18 bits, i.e. 0, 1, or 2 bytes beyond the 2-bit minimum. At the highest data rate, 27 Mbps, there are 216 data bits (27 bytes) per OFDM symbol, so the *PAD* adds between 0 and 26 additional bytes beyond the 2-bit minimum.

The overhead introduced by the PLCP can be calculated either in units of time or data. The latter is defined in terms of the Table 10.5 data rate, and thus represents the equivalent number of bits or bytes before coding. The coding bits are usually not counted as part of the overhead since their effect is already reflected in the data rate, e.g. the 3 Mbps (BPSK, 1/2 coding rate) option includes three million MAC and overhead bits per second and another three million coded bits per second. The overhead calculation is intended to determine how much of the 'data rate' is available for sending MAC data.

When operating at 10 MHz, the preamble consumes 32 μsec and the *SIGNAL* adds 8 μsec. The combination of the *SERVICE*, *TAIL*, and *PAD* add between one and two more OFDM symbol times, i.e. at least 8 and less than 16 μsec. At the minimum data rate, 3 Mbps, the longest *PAD* (18 bits) makes this portion of overhead $1\frac{2}{3}$ OFDM symbols, or 13.3 μsec. At the maximum data rate, 27 Mbps, this portion can be as long as $1\frac{26}{27}$ OFDM symbols, or 15.7 μsec. The total PLCP overhead is therefore in the range 48–55.7 μsec. Translated to bytes, this corresponds to a range of 18–20 bytes at 3 Mbps, and a range of 162–188 bytes at 27 Mbps.

10.5 DSRC Data Link Layer Standard (MAC and LLC)

Like the PHY layer, the data link layer is commonly divided into sublayers. The lower of these is the *medium access control (MAC)* sublayer, which defines the rules by which STAs compete to share a wireless medium. The upper sublayer is the simpler *logical link control (LLC)*. These two sublayers are described in the sections below.

10.5.1 Medium access control (MAC) sublayer

The purpose of the MAC sublayer is to establish rules for accessing the common medium so that it can be shared efficiently and fairly among a set of STAs. The 802.11 rules fall into two categories: the session-based rules that define steps that an STA must take before it is allowed to communicate information on behalf of Layer 3, and the frame-by-frame rules governing an individual transmission. The draft 802.11p amendment (IEEE P802.11p 2009) makes significant changes to the session-based rules, while using the frame-by-frame rules as defined in the baseline 802.11 standard (IEEE 802.11 2007).

Session-based rules

The 802.11 standard defines a concept called the *basic service set (BSS)*. A BSS is a set of STAs that agree to exchange data plane information. There are two types of BSS.

The more common type, the *infrastructure* BSS, has a special access point (AP) STA that announces the BSS, and establishes some parameters and constraints for using the BSS. The AP serves as a gateway to a *distribution system* (DS) that provides access to additional networks beyond the wireless LAN, for example to the public Internet.

Before a STA can transmit user plane data to the AP it must first hear the BSS announcement, called a beacon frame, and then go through a series of 'setup' steps: joining (which includes synchronizing with the AP STA's clock), authenticating, and associating.

The second type of BSS is the *independent* BSS, sometimes informally called an *ad hoc* BSS. An independent BSS has no AP STA, and thus has no DS to provide backhaul connectivity, so its utility is limited to direct communication between the STAs in the BSS. These STAs collectively shoulder the responsibility of announcing the existence of the BSS along with its parameters. Communicating within an independent BSS requires that the BSS first be announced, via a beacon frame, and that other STAs synchronize with the announcing STA.

Nodes in a VANET are free to use the infrastructure and independent BSS concepts in 802.11. However, there are concerns about the delays attendant to following the setup steps outlined above, especially in the case of communicating through an AP to a DS. In a highly mobile vehicular environment, the opportunity to communicate may be fleeting, lasting only a few seconds, so there is a desire to define alternative, 'lightweight' rules for accessing the medium.

That desire is in fact the primary motivation for the 802.11p WAVE amendment. The result of this effort is the definition of a new type of 802.11 communication 'outside the context of a BSS' (abbreviated here 'OCB'). In traditional 802.11, all data frames are sent between STAs that belong to the same BSS. By contrast, communication of data frames OCB is limited to STAs that do not belong to a BSS. There is no MAC sublayer setup required before STAs exchange data frames OCB.

The 802.11 frame header includes a six-byte BSS identifier (BSSID) field. Each BSS is assigned an identifier by the STA that sends the beacon: for example, in an infrastructure BSS the BSSID is the MAC address of the AP. Each frame sent within the context of a BSS includes the BSS's identifier in its header. The purpose of the BSSID is to allow a receiver to easily distinguish frames belonging to *its* BSS, which it will pass up the protocol stack for further processing, from frames belonging to *other* BSSs, which it will ignore. The BSSID field of a frame sent OCB is set to all 1 s, i.e. 0xFFFFFF in hex notation, which is called the *wildcard* value. A BSSID field set to the wildcard value plays the same role described above, i.e. it helps a receiver determine whether an incoming frame merits additional processing resources or should be ignored. It may be helpful to think of the BSSID field more generally as a frame filter field. A data frame sent OCB may be sent to a unicast destination address, to a multicast group, or to the broadcast address. Like the wildcard BSSID, the broadcast destination address is all 1 s. The basic safety message (see Section 10.7.2) will likely be encapsulated in a WAVE short message (see Section 10.6.2) and then sent OCB to the broadcast destination address.

The IEEE 802.11p amendment introduces a new management frame, the *timing advertisement (TA)* frame. While this can be used to announce information about the sender's time source, it has another potential use of more immediate interest to this discussion. If the TA frame includes one or more *vendor-specific information elements (VSIEs)*, these can be used to convey a WAVE service advertisement (WSA) (see Section 10.6.2) on behalf of the management plane. The contents of a VSIE are, as the name implies, determined outside of the 802.11 standard by a 'vendor,' i.e. a third party ('organization' would have been a better name than vendor, but the VSIE has been in the 802.11 standard for a long time). In this case the vendor is the IEEE 1609 standards working group. An identifier in the VSIE indicates the vendor, and thus the format of the vendor-defined information field within the information element. The length of the VSIE is constrained, so a given WSA may need to be segmented across multiple VSIEs, within a single TA frame, and then reassembled at the receiver. The TA frame could thus be used, for example, by a roadside unit (RSU) that has a traffic service or a commercial service to offer to passing vehicles. A WSA can also be sent in a *vendor-specific action frame*. The TA frame and vendor-specific action frame are discussed more below.

There are few rules governing communication OCB, other than that a STA cannot engage in OCB communication while it belongs to a BSS. The definition of communication OCB is notable more for what it leaves out than for what it adds, and this has proved somewhat controversial among long-time 802.11 standards participants. In particular, communication OCB does not:

- use a beacon to announce a BSS

- require one STA to synchronize with another before they can communicate

- use authentication at the MAC sublayer

- include any notion of the STAs 'associating' before they communicate.

Assuming that all those things are included in the main 802.11 standard for a good reason, it is fair to ask why it is acceptable to skip them for communication OCB. Here are some of the answers, for the specific points mentioned above:

Lack of beacon:
The 802.11 beacon periodically announces the existence of a BSS, and conveys parameters important to its correct operation, including the BSSID to be used in subsequent data frames sent 'within the context' of the BSS. The OCB type of communication does not define a BSS, and so does not use a beacon. Much of the contents of the beacon, defined in Table 10.7–10.8 of (IEEE 802.11 2007) are not important for communication OCB. Some of the beacon contents, however, are relevant to any type of communication, including OCB communication. For example, a beacon can convey a set of allowed data rates and a set of *quality of service (QoS)* parameters (this QoS function is called *enhanced distributed channel access (EDCA)* and is discussed in Section 10.5.1). Communication OCB avoids the need for a beacon by providing alternative means for determining those types of generic communication parameters.

One alternative is to rely on default values. As noted in Section 10.4, a 10 MHz 802.11 STA is required to support a default set of data rates: 3, 6, and 12 Mbps. A default set of EDCA parameters is defined as well; indeed, the 802.11p amendment defines a different default set of EDCA parameters for communication OCB than is defined for communication within a BSS. STAs willing to use default parameters for communication OCB have no need to exchange explicit parameters over the air.

Another alternative to the beacon is the TA frame mentioned above. This new, optional frame is capable not only of carrying the VSIE but also of conveying additional communication parameters. There are some overlaps between the contents of a beacon and a TA frame, but differences as well. One important common field is a local timestamp that can be used for synchronization as discussed below.

A final alternative to parameter exchange in a beacon is for STAs to agree externally (i.e. outside of the 802.11 standard) on OCB communication parameters. The agreement could be via a separate standard, or could be proprietary. The agreement may or may not require exchange of information over the air. A WSA is an example of such external information, in this case information that is specified in a higher layer standard, IEEE 1609.3. A VSIE could just as easily convey manufacturer-specific information between vehicles made by the same company. As noted above, it can be conveyed transparently within the VSIE of a TA frame.

The conclusion is that several alternatives to the beacon exist.

Synchronization:

MAC sublayer synchronization between STAs is required for operation within a BSS, but not for communication OCB. The primary reason for MAC sublayer synchronization is to facilitate 'power management' whereby a STA may alternate between 'awake' and 'doze' states, the advantage of the latter being low power consumption. In a vehicular environment, many instances of communication OCB will be among STAs that do not rely on limited battery power and furthermore do not wish to risk missing messages while dozing, so it was determined that MAC sublayer synchronization is not required. If the STAs involved in an instance of communication OCB do wish to be synchronized, that need will be determined at higher layers (above the MAC), and it is up to those higher layers to achieve that synchronization. The TA frame includes a timestamp field and a timing information element that can be used for that purpose.

Authentication:

Like synchronization, the need for authentication in communication OCB is determined at higher layers. In the DSRC model, a means of authenticating messages is provided in the 1609 part of the protocol stack, specifically via the IEEE 1609.2 standard (IEEE 1609.2 2006). This method, which is further discussed in Section 10.6.4, is preferable to that defined in 802.11 for efficiency and privacy reasons. OCB communication that uses IPv6 rather than the 1609 upper layers can utilize a variety of well-established techniques for authentication, if desired.

Association:

The association of STAs in an infrastructure BSS has a specific purpose, to help the AP bridge link layer frames between a non-AP STA within the BSS and a node on the other side of the DS, i.e. a node not part of the wireless LAN. Some types of communication OCB have no need for a bridging function, for example vehicle-to-vehicle safety messages. Other types of communication OCB do involve extending connectivity beyond the wireless LAN, e.g. to a remote server. A STA that communicates with other STAs OCB is not prohibited from performing a similar

STANDARDS AND REGULATIONS

bridging function to a backhaul network, but it does not do so as an AP (since the communication is not within an infrastructure BSS), and the backhaul does not satisfy the definition of a DS, so any such bridging is managed by means that are outside the scope of 802.11 or the 802.11p amendment. For example, the forwarding STA could be supplied by higher layers with the MAC address of a STA to which it may need to bridge a frame arriving from the backhaul. Alternatively, the forwarding STA may be a router that uses network layer information beyond the scope of 802.11. In any case, the association function defined for AP bridging is not needed for communication OCB.

In summary, the traditional 802.11 functions of beaconing, synchronization, authentication, and association are not needed at the MAC sublayer for communication OCB. The TA frame offers an optional, lightweight alternative to the beacon. The other functions are optionally implemented at higher layers, either as part of a separate standard or via proprietary means.

Medium access rules

The 802.11 standard defines a complex set of rules that allow STAs to efficiently share the wireless medium. The most important points are summarized here. The 802.11p amendment does not alter these rules, i.e. they apply identically to frames sent within and outside of the context of a BSS.

The basic medium access paradigm of 802.11 is carrier sense multiple access/ collision avoidance, or CSMA/CA. The simplest communication scenario under CSMA/CA is as follows:

1. A STA that has a frame to send first senses the wireless medium (i.e. the 'carrier') to see if another STA is already using it.

2a. If the STA senses that the medium is idle for a certain interval of time then the STA begins transmission of its frame.

2b. If the STA senses that the medium is busy, it performs a random 'backoff.' The STA chooses a random integer from a range. Each number in the range represents an interval of time called a slot time. A STA choosing the random number n waits for n slot times to elapse and then begins transmitting its frame. The countdown of the n slot times begins when the medium becomes idle, is interrupted during any non-idle interval, and resumes when the medium returns to an idle state.

3. If the frame is sent to a single STA, i.e. is a *unicast* frame, and is correctly received, the recipient sends an acknowledgement frame (ACK) after a short interval. If a unicast frame is not ACKed within a given timeout interval, these steps are repeated until either an ACK is received or a maximum number of retransmissions is attempted. If the frame is sent to a *group address*, i.e. it is multicast or broadcast, it is not acknowledged by any of the recipients, and it is never retransmitted.

Carrier sensing:
There are two criteria that must cause a STA to declare a busy medium on a 10 MHz

DSRC OFDM channel: a) if it detects the start of the PHY preamble (the first five short training sequences described in Section 10.4.2) with received signal strength at least −85 dBm, or b) if it detects signal energy of at least −65 dBm. The STA is allowed to declare the medium busy if it detects a PHY preamble even if the received power is below −85 dBm. For a frame sent to a group address the medium is considered busy until the end of the frame, which can be sensed, but can also be computed based on the rate and length values in the *SIGNAL* field of the header. For a unicast frame the medium is considered busy until the time when the ACK should return.

The 802.11 MAC protocol optionally allows the transmission of a unicast frame to be preceded by an exchange of short request to send (RTS) and clear to send (CTS) frames between the sender and receiver. A potential transmitter sensing the channel and detecting a valid RTS and/or CTS will compute the time until the ACK should return and declare the medium busy for that entire interval.

Interframe spacing:
The standard defines several interframe space (IFS) intervals. The length depends on the channel bandwidth, and all values cited here are for a 10 MHz channel (values scale inversely with channel bandwidth). The time over which an idle carrier is sensed is called the distributed IFS, or DIFS, and is equal to 58 μsec. The time between the end of a unicast frame and the start of the returned ACK is called a short IFS, or SIFS, and is equal to 32 μsec. The slot time is equal to 13 μsec.

Random backoff:
The 'collision avoidance' part of CSMA/CA stems from the random backoff. Other MAC protocols that use carrier sensing, such as Ethernet, allow a device to start transmission as soon as it detects a transition from busy to idle. This will result in a collision if two devices are waiting for that same transition. Ethernet has collision detection, allowing it to recover efficiently when there is a collision. 802.11 does not have collision detection, and so the protocol places more importance on avoiding collisions in the first place. Therefore, when an STA with a frame to send senses the medium transition from busy to idle it does not deterministically start its own transmission. Instead it chooses randomly from the next several slot times, and thus even if two or more STAs are waiting to send, there is a non-zero probability that one will gain access to the medium without experiencing a collision. There is still a chance that two STAs will choose the same random backoff number, in which case they will collide when the countdown reaches zero. The probability of a collision depends on the number of competing STAs and the range over which the random backoff is chosen. The latter range is called the *contention window (CW)*, and has default size of 16 slots, i.e. the random number is in the range [0,15].

Retransmission:
After sending a unicast frame a STA waits for an ACK. If the ACK does not arrive within a timeout interval, the STA goes through another backoff interval and then retransmits. The choice of the random backoff duration is always uniformly distributed across the contention window. A frame transmission may be unsuccessful for a variety of reasons, for example due to a collision or to unrecoverable bit errors induced by fading, interference, etc. Since the probability of collision is a function

STANDARDS AND REGULATIONS

of the CW size, the STA adapts CW in response to successful and unsuccessful transmissions. In particular, when a STA executes a backoff before retransmission it first doubles CW, to 32 for a first retransmission, to 64 for a second, and so on until CW reaches a maximum size of 1024. When a STA experiences a successful transmission, i.e. it receives an ACK, it resets CW to the minimum size 16. An STA will retransmit a frame until it is successful or until a configurable number of unsuccessful attempts have been made.

Enhanced distributed channel access (EDCA):
The 802.11 standard defines a mechanism by which one class of frames can be given priority over another in their competition to access the medium. This mechanism, called EDCA, defines up to four access categories (AC) of frames, each of which has its own queue. The head-of-line frames in the AC queues compete with each other for access. The competition is probabilistic in the sense that a frame from a higher priority AC has a higher probability of gaining access than a frame from a lower priority AC, but the advantage is not deterministic. The priority advantage is the result of modifying two parameters of the protocol. The first is the contention window size; in fact both the minimum and maximum CW values can be configured per AC. A smaller CW reduces the average backoff duration, and reducing the CW for AC X compared to AC Y is a way of increasing the probability that the head-of-line frame in the AC X queue will gain access to the medium. The second parameter is the delay after the medium goes idle before an STA either begins a transmission or initiates a backoff. Without EDCA that delay is fixed at DIFS. With EDCA the delay depends on the AC, with higher priority ACs given a lower delay. As noted earlier, the EDCA parameters for each of the ACs are either set to a default defined in the standard, or are established via an over-the-air communication. One set of default EDCA parameters is defined for frames sent within a BSS. Another set of default parameters is defined in IEEE 802.11p for frames sent OCB.

This section summarizes the 802.11 CSMA/CA protocol. The protocol itself is quite complex, and many details have been omitted from this presentation.

802.11 MAC frame format

Every 802.11 MAC frame (more formally the *MAC Protocol Data Unit or MPDU*), consists of a *header, frame body*, and a *frame check sequence (FCS)*.

The frame body consists of the *MAC service data unit (MSDU)*, possibly augmented with a few bytes of encryption overhead. The MSDU is passed into the MAC sublayer from a higher layer or from the management plane. The MSDU is limited in size to 2304 bytes. The encryption overhead, frame header and FCS together add only a few tens of bytes, so given the MSDU limit the largest MPDU is much smaller than the 4095 byte maximum PSDU allowed by the OFDM PHY that was discussed in Section 10.4.2. Furthermore, since IP packets are normally limited to 1500 bytes, and IEEE 1609 WSMs are even shorter, the typical 802.11 MAC frame used in DSRC will be less than 1600 bytes.

The FCS field is four bytes and carries a cyclic redundancy code (CRC) used for detecting bit errors in the MAC frame. The CRC is computed over the frame header and frame body, so that a bit error anywhere in the frame is detectable.

Bytes: 2	4	6	6	6	2	2	0–2304+	4
Frame control	Duration	Address 1	Address 2	Address 3	Sequence control	QoS control	Frame body	FCS

Figure 10.6 802.11 MAC header (most common form).

The frame header can have a variety of formats, depending on the type of the frame. Three frame types are defined:

1. control frames (includes RTS, CTS, ACK)

2. data frames

3. management frames (includes beacon, TA frame).

The most common frame header format is shown in Figure 10.6. The various header fields are:

Frame control:
This field includes a protocol version, a frame type and subtype, and several other bit fields. Two of importance are the *To DS* bit and the *From DS* bit. The meaning of these bits is only defined for frames sent within an infrastructure BSS. If the *To DS* bit is set, the AP will bridge the frame from the wireless medium to the DS. Conversely, if the *From DS* bit is set the AP will bridge the frame from the DS to the wireless medium. If both bits are zero the frame remains on the wireless medium. The state of those bits also determines the contents of the three address fields.

Duration:
This field has relatively complicated rules depending on the frame type and other QoS and address parameters. Loosely, it indicates the frame's time duration, possibly including some overhead beyond the physical transmission time.

Addresses 1, 2, and 3:
It is helpful to define some terms used by 802.11 to discuss these fields.

Transmitting STA address (TA) the MAC address of the STA that transmits the frame on the wireless medium. This is always an individual address.

Receiving STA address (RA) the MAC address of the STA that receives the frame from the wireless medium. If the MAC address is a group address it may identify multiple STAs.

Since an AP on an infrastructure BSS may bridge frames between the wireless medium and the DS, the standard also defines terms for the ultimate source or destination of the frame, i.e. a device that communicates with the AP across the DS.

Source address (SA): the MAC address of the device that created the MSDU in the frame body. This is the ultimate source of the frame that is sent with *From DS* $= 1$. For a frame that is sent with *From DS* $= 0$, this is the same as the TA.

Destination Address (DA): the MAC address of the device that is the ultimate destination of the MSDU in the frame body. If *To DS* = 1, this device is across the DS. If *To DS* = 0 this is the same as the RA. If the MAC address is a group address it may identify multiple devices.

Recall that for an infrastructure BSS the BSSID is equal to the MAC address of the AP. When *To DS* = 1, the AP bridges the frame onto the DS, so the RA identifies the AP, and thus the RA is equal to the BSSID. Conversely, when *From DS* = 1 the AP bridges the frame from the DS onto the wireless medium, so the TA is equal to the BSSID.

The general rule is that Address 1 always equals the RA, and Address 2 always equals the TA. Sometimes these correspond also to the DA and SA, respectively, and sometimes they do not.

Using these definitions, it is possible to further define the contents of the three address fields as a function of the *To DS* and *From DS* bits.

- *To DS* = 0, *From DS* = 0:
  This is a frame for which both the source and destination(s) are on the wireless medium. Address 1 (RA) carries the DA. Address 2 (TA) carries the SA. Address 3 carries the BSSID, which is discussed in Section 10.5.1.

- *To DS* = 0, *From DS* = 1:
  This is a frame that arrives at the AP from across the DS, where it was sent by the ultimate source. The AP bridges the frame onto the wireless medium. In this case, the TA and SA are not the same. The TA identifies the AP, since that is the STA that transmits the bridged frame onto the wireless medium. Recall that for an infrastructure BSS the BSSID is equal to the MAC address of the AP. Therefore, while Address 1 (RA) again carries the DA, Address 2 (TA) now carries the BSSID (i.e. AP MAC address). For this case, the SA is placed in Address 3.

- *To DS* = 1, *From DS* = 0:
  This is a frame that is sent by a STA onto the wireless medium, received by the AP, and bridged by the AP onto the DS toward the ultimate destination(s). In this case, the RA and DA are not the same. The RA identifies the AP, since that is the STA that transmits the bridged frame onto the DS. So, for this case, Address 1 (RA) carries the BSSID (i.e. AP MAC address), Address 2 (TA) carries the SA, and the DA is placed in Address 3.

- *To DS* = 1, *From DS* = 1:
  This case is covered for completeness, though it is not expected to be used in VANETs. The 802.11 standard states that it 'does not define procedures for using this combination of field values' (Clause 7.1.3.1.3 of (IEEE 802.11 2007)). However, the standard does define a frame format associated with this combination, and unlike the other frames this one has a fourth address field, because to use this combination it would be necessary to convey the RA, TA, SA, and DA separately.

Placing the RA consistently in Address 1 allows a STA to easily determine whether it ought to dedicate processing resources to receiving the frame or not. The address

Table 10.7 802.11 MAC frame address field contents.

To DS bit	From DS bit	Address 1 (RA)	Address 2 (TA)	Address 3
0	0	DA	SA	BSSID
0	1	DA	BSSID	SA
1	0	BSSID	SA	DA

field contents are summarized in Table 10.7 for the first three *To DS/From DS* combinations above.

Sequence control:
This field is used if a frame has to be fragmented.

QoS Control:
The contents of this field depend on the frame type, and involve details of QoS behavior beyond the scope of this chapter.

The overhead imposed by the MAC sublayer is 32 bytes when it uses the format shown in Figure 10.6.

10.5.2 Logical link control (MAC) sublayer

The logical link control (LLC) sublayer of the DSRC protocol stack uses the standard IEEE 802.2 (IEEE 802.2 1998) protocol with minimal changes. The frame format is identical. The LLC provides several types of service, but DSRC uses only the most common and the simplest of these, the *Unacknowledged Connectionless (a.k.a. Type 1) Service*. Data frames using Type 1 service are referred to as *Unnumbered Information (UI)* frames. The LLC frame format has a three field header: a one byte *Destination Service Access Point (DSAP)*, a one byte *Source Service Access Point (SSAP)*, and a one or two byte *Control* field. The UI format is indicated if the Control field is equal to 0x03.

The main service provided by Type 1 LLC is to identify the protocol associated with the payload of the UI frame. For this protocol identification function the LLC protocol is frequently supplemented by another protocol called the *Subnetwork Access Protocol (SNAP)* (IEEE 802 2002). If the LLC DSAP and SSAP fields are both set to 0xAA, the LLC header is followed by a five-byte SNAP header. The first three bytes of the SNAP header indicate whether the protocol identifier is part of a private range or is taken from a public registry. If the first three bytes are all zero (0x00 00 00), then the final two bytes represent a public protocol identifier called an Ethernet Type, or EtherType. The IEEE assigns EtherType values and manages the EtherType registry. In DSRC there are two protocols above the LLC in the stack (see Figure 10.2): IPv6 and WSMP. The EtherType assigned to IPv6 is 0x86DD. The EtherType assigned to WSMP is 0x88DC. The IEEE 1609.3 standard (see Section 4.3.2) requires that all Layer 3 data packets be encapsulated in LLC UI frames with a SNAP header indicating the protocol type. Only IPv6 and WSMP are recognized as valid Layer 3 data plane protocols in IEEE 1609.3, though of course many other protocols have assigned EtherTypes. In the case of IPv6 packets, IETF RFC 1042 defines additional

STANDARDS AND REGULATIONS

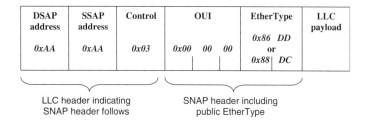

Figure 10.7 IEEE 802.2 LLC frame format used in DSRC. EtherType 0x86DD indicates IPv6. EtherType 0x88DC indicates WSMP.

rules for LLC-SNAP encapsulation, and IEEE 1609.3 requires that these be followed as well. MAC management frames, for example those carrying WSAs, do not use the LLC sublayer.

Figure 10.7 shows the LLC PDU format for DSRC, including the LLC and SNAP headers. This becomes the MAC SDU when passed to the MAC sublayer.

10.6 DSRC Middle Layers

As shown in Figure 10.2, the DSRC protocol stack above Layer 2 splits into two branches. The first uses protocols defined specifically for VANETs by the IEEE 1609 standards working group. IEEE 1609 standards are sponsored by the Intelligent Transportation Systems Committee of the IEEE Vehicular Technology Society. This part of the stack supports safety messages efficiently, as well as service advertisements for more general DSRC services. While those services could also be delivered using these new protocols, they will instead frequently run over traditional Internet protocols in the other branch of the stack. Since those protocols (principally IPv6, UDP, and TCP) are mature and well documented, this chapter will not describe them further.

Figure 10.8 shows the IEEE 1609 view of the protocol stack. The WAVE MAC is a set of one or more instances of the IEEE 802.11p MAC defined above, plus a 'channel switching' protocol that defines how a given device can operate on multiple DSRC channels, one channel at a time. This multi-channel operation concept is defined in the IEEE 1609.4 standard (IEEE 1609.4 2006). A separate logical instance of the MAC protocol, including queues and state variables, runs for each channel on which the radio device operates. Each of these logical MACs is conformant to the IEEE 802.11p standard. The channel switching protocol is explained below. This is best thought of as a management extension of the MAC layer. However, since it is standardized as part of the IEEE 1609 suite it is appropriate to describe it in this section. The physical layer shown in Figure 10.8 similarly consists of one instance of the IEEE 802.11p PHY protocol for each DSRC channel on which the radio device operates. The 802.11p MAC and PHY instances are unaware of being operated as part of a multi-channel solution.

Figure 10.8 also shows the *WAVE short message protocol (WSMP)* defined by the IEEE 1609.3 standard (IEEE 1609.3 2007). This is a simple, efficient alternative to

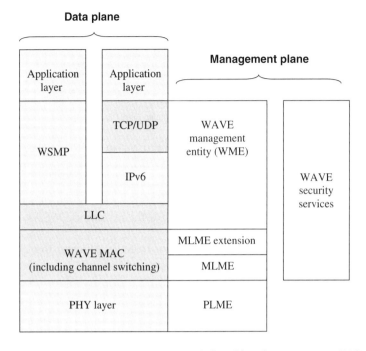

Figure 10.8 WAVE protocol stack defined by the IEEE 1609 WG.

Layer 3 and 4 Internet protocols in the vehicular environment. The third major function defined within the IEEE 1609 suite is Security, which is specified in the 1609.2 standard. This defines protocols for optional message authentication and encryption. The IEEE 1609 WG considers security services to be part of the management plane. Also in the management plane are several management entities associated with the various layers. The physical layer and MAC sublayer management entities (PLME and MLME, respectively) are defined in the IEEE 802.11 standard (IEEE 802.11 2007). The MLME extension is defined in the IEEE 1609.3 (IEEE 1609.3 2007) and 1609.4 (IEEE 1609.4 2006) standards, as is the WAVE management entity (WME).

In addition to the IEEE 1609.2, 1609.3, and 1609.4 standards, which are discussed in some detail in this chapter, the IEEE 1609 WG is also developing the following standards as this book goes to press:

1. IEEE 1609.0 WAVE architecture – this standard defines the overall IEEE 1609 framework

2. IEEE 1609.1 – low functionality WAVE devices (e.g. a 'smart' traffic cone)

3. IEEE 1609.5 – communications manager (not clear if this will progress to a standard)

4. IEEE 1609.11 – electronic toll/fee collection.

These are not covered in detail in this chapter.

10.6.1 MAC extension for multi-channel operation: IEEE 1609.4

IEEE 1609.4 is applicable when DSRC is operating in a multi-channel environment, as it is expected to be in most regulatory domains. Section 10.3.1 described the DSRC regulatory framework defined by the FCC in the US. The band plan consists of seven 10 MHz channels in the 5.9 GHz band. The European DSRC regulatory plans call for at least two channels, and possibly more. IEEE 1609.4 defines a mechanism by which a system with one or more radios can effectively switch among those channels. It is best described as an extension to the MAC, with a separate logical instance of the IEEE 802.11p MAC running in each channel.

The goal of IEEE 1609.4 is to define a mechanism by which devices that are switching among multiple channels will find each other, i.e. tune to the same channel at the same time so that they can communicate. The problem is especially challenging for devices that have a single radio; the situation when devices have multiple radios is discussed below. The solution involves two concepts: the rendezvous channel and time division. The rendezvous channel concept defines one channel in the band plan as a special channel that the devices will tune to on a regular basis. In IEEE 1609.4 this is called the *control channel (CCH)*. Other channels in the band plan are designated *service channels (SCHs)*. The time division concept assumes that all devices have access to a common time source, *universal coordinated time*, which is abbreviated UTC. UTC is available in a *global positioning system (GPS)* signal. A device without a GPS receiver could potentially still synchronize with UTC by receiving timing signals from another device. With all devices synchronized, IEEE 1609.4 imposes a division of time into control channel intervals and service channel intervals. During a CCH interval most devices rendezvous on the CCH. During an SCH interval devices may switch to one of the SCHs. The CCH is used primarily for two types of messages:

- Safety messages from one vehicle to another, i.e. basic safety messages as defined in (Section 10.7.2). Safety messages use the WAVE short message (WSM) format defined by the WSMP (Section 10.6.2).

- WAVE service advertisements (WSAs): A WSA (Section 10.6.2) is used to announce the availability of one or more WAVE services on the SCHs during the next SCH interval. Most DSRC services are expected to be offered by a roadside unit (RSU), but a vehicle on-board unit (OBU) can also advertise and offer a service.

Only WSMs and WSAs are allowed on the CCH; IP packets are not allowed. Some low-end devices that do not wish to send or receive either safety messages or WSAs might never tune to the CCH, for example an RSU providing a service that is advertised by an upstream RSU. In general, such devices are of little interest to this discussion on middle layers, and they are not referred to explicitly again.

In the US, Channel 178 is designated as the CCH and the other six DSRC channels are designated as SCHs (see Figure 10.9). IPv6 packets and WSMs are allowed on the SCHs. Figure 10.9 also shows the optional alternative assignments of the 20 MHz on either side of Channel 178 (designated Channels 175 and 181, respectively) and the 5 MHz reserved guard band at the lower end of the DSRC spectrum.

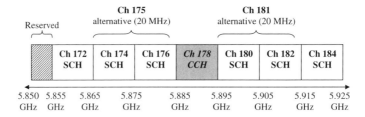

Figure 10.9 US DSRC band plan channel designations.

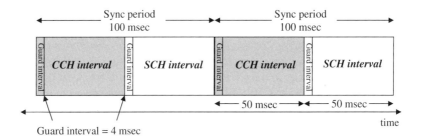

Figure 10.10 Division of time into CCH intervals and SCH intervals.

Figure 10.10 illustrates the basic time division concept defined in IEEE 1609.4. Time is segmented into 'sync periods,' which by default are 100 msec each. Each sync period consists of one CCH interval followed by one SCH interval. The default division is 50 msec for each.

Each CCH and SCH interval begins with a 4 msec guard interval, which is used by the device to switch control from one virtual MAC to another. The device may begin receiving frames as soon as it is ready within the guard interval, but it normally will not transmit until the guard interval is complete because it will assume that its neighbors are still performing their own transitions. The guard interval also accounts for small errors in a device's representation of UTC.

Switching to an SCH: If a device receives one or more WSAs during a CCH interval and determines that it is interested in accessing an advertised service (more on this process in the next section), it will switch to the relevant SCH at the end of the CCH interval. There are two primary reasons why a single-radio device might remain tuned to the CCH during the SCH interval:

- During the CCH interval the device does not hear an advertisement for a service that it is interested in accessing.

- The device temporarily loses synchronization with UTC, and thus its ability to discern the boundaries between CCH and SCH intervals; in this case the device must remain tuned to the CCH until it reacquires synchronization.

The first is the more common and interesting case. A device that remains tuned to the CCH during an SCH interval is allowed to transmit and receive frames, but such data should be limited to nonessential information since one or all neighboring devices may be tuned to an SCH. In general, essential information should be conveyed between the end of the CCH guard interval and the end of the CCH interval itself. This is the 'rendezvous' time when all devices are expected to be listening. Switching to an SCH is an entirely local decision; one device can never force another device to leave the CCH.

Multi-radio devices: IEEE 1609.4 defines the rules of operation for a 'WAVE device,' i.e. an OBU or an RSU. The device could have a single radio or multiple radios. The IEEE 1609 protocols do not currently provide a way for one device to determine how many radios a neighboring device has, so a multi-radio device must assume that at least some of its neighbors have only one radio. This is important for determining how to use a second radio, for example.

If a device has multiple radios, then the requirement to rendezvous on the CCH during the CCH interval can be satisfied by one of the radios. During the CCH interval a second radio could either tune to the CCH as well, presumably to give the device two chances to hear each broadcast message, or it could tune to one of the SCHs. With the 'early departure' exception noted below, a multi-radio device that tunes to an SCH during the CCH interval will only communicate with other multi-radio devices. As with the earlier case of a radio tuned to the CCH during the SCH interval, communications between multi-radio devices on an SCH during the CCH interval are limited to information that need not be heard by all neighboring devices. In particular, a second radio does not significantly improve the performance of safety communication, which largely consists of broadcasts of the basic safety message (see Section 10.7.2) on the CCH during the CCH interval. Under the current IEEE 1609.4 channel switching protocol, the main use of a second radio would appear to be to allow a device to access two services simultaneously, on two different SCHs, during one SCH interval. Researchers are actively studying ways to improve the utility of a second, optional radio, without requiring multiple radios.

Early departure from the CCH during the CCH interval: As this book goes to press, the IEEE 1609 group is planning to relax the requirement for a device to remain tuned to the CCH during the CCH interval. This feature would be of primary interest to a device (or a single radio within a multi-radio device) whose purpose is accessing general DSRC services, not engaging in safety communication, since it would hinder the ability to hear all of the basic safety messages of its neighbors. There are two early departure mechanisms. The management entity in a device can initiate a one-time *immediate* departure of the CCH for an SCH at any time during the CCH interval, and then resume normal channel switching between the CCH and SCH on the next sync period (the device is not precluded from undertaking another immediate departure in the next period). Alternatively, the management entity can execute an *extended access* departure for multiple sync periods, during which the device does not return to the CCH. While an immediate departure naturally terminates at the start of the next CCH interval, an extended access departure only terminates when the management entity actively causes a switch back to the CCH. The main purpose of these early departure mechanisms is to access services more quickly. The immediate departure mechanism simply avoids the latency of the residual CCH interval. The extended access departure is useful for a service that normally takes several sync periods to deliver; it roughly doubles the capacity of the SCH by allowing it to be used all the time until the service is delivered.

Frame collision issues: There are two frame-collision-related concerns with 1609.4 channel switching. The first is a 'synchronized collision' phenomenon that is relatively easy to avoid. The second is a 'congestion collision' problem that is likely to become a significant problem at high penetration rates.

Synchronized frame collisions: The nominal safety communication model is that each vehicle broadcasts one safety message on the CCH during each CCH interval, i.e. during the rendezvous time when all devices are listening. If the higher layers passing the message down the stack are unaware of the underlying time division, there is about a 46% probability that the message will be enqueued in the MAC between the CCH guard time and the end of the CCH interval, and about a 54% probability it will be enqueued in the MAC at a time when it cannot immediately be sent. The 1609.4 standard states that any frame enqueued for transmission on the CCH during either the SCH interval or the CCH guard interval will treat the channel as initially 'busy.' The frame will thus experience a normal CSMA/CA backoff when the CCH guard interval ends. The default backoff is a random number of time slots between 0 and 15. Any two packets that choose the same number of time slots are destined to collide when the backoff timer counts down (note: countdowns are suspended when the channel is busy). If a large number of packets enter backoff at the start of a CCH interval, the probability that any one of the 16 backoff time slots is chosen by exactly one frame is low, so most of the frames will be lost to collisions.

This problem is relatively easy to avoid if the message generation function in the higher layers is provided with a signal indicating the start of a sync period. Then it can choose to enqueue its message at the MAC layer during a random time within the 46 msec interval. It may still find the channel busy and enter backoff, but at reasonable channel loads it is far less likely to suffer a collision. The 1609.4 standard recommends, but does not require a device to take steps to avoid this phenomenon.

Congestion-related frame collisions: The 1609.4 channel switching scheme has (at least) one significant positive attribute. It allows a single-radio device to support both safety applications and general DSRC services. This is expected to be helpful for DSRC market penetration. DSRC technology can only facilitate collision-avoidance between two vehicles that are DSRC-equipped, so a DSRC device that cannot support applications other than safety delivers minimal value during an initial deployment phase. Market experts hope that access to other DSRC services will provide enough value to encourage early deployments, which will in turn lead to effective safety support as penetration rates increase. A multi-radio device can also support both safety and general DSRC services, but some manufacturers are concerned about the extra cost of the additional radio(s). So, 1609.4 channel switching was motivated by a desire to offer support for both types of applications with a single radio.

On the other hand, the 1609.4 channel switching scheme has (at least) one significant negative attribute as well. It wastes a lot of bandwidth; in particular, more than half of the CCH bandwidth cannot be used to carry essential safety information. As noted above, the safety communication model assumes that each vehicle sends one basic safety message per 100 msec sync period, and these are limited to the 46 msec phase from the end of the CCH guard interval to the end of the CCH interval. During the early deployment phase, this limitation is not likely to be significant since only a fraction of vehicles will be DSRC-equipped. However, for higher penetration rates the channel capacity needed to carry all of the safety messages becomes a concern, so limiting that capacity via channel switching can lead to safety performance problems.

The problem can be seen with a rough 'back-of-the-envelope' computation. Assume each safety message is 3000 bits, and that the DSRC PHY layer uses the 6 Mbps data rate. Therefore, an 'always-on' safety channel can support no more than 2000 messages per second, or no more than 200 vehicles sending one message in every 100 msec period. With channel switching constraining the messages to 46 msec intervals, the upper limit slips to 92 vehicles. The CSMA/CA MAC protocol suffers an increasing number of collisions as the offered load approaches 100% of capacity, so the reality is that the number of vehicles within a communication range is limited even below 92. There are steps that vehicles can take to control congestion, at the expense of reduced communication performance, but the main point is that channel switching imposes a significant capacity constraint for safety communication.

10.6.2 Network services for DSRC: network and transport layers, IEEE 1609.3

The Internet Protocol (IP) has become the default Layer 3 protocol in many networks today, especially those that are interconnected with other networks as part of the public Internet. The primary service that IP offers to higher layers is connectivity, i.e. the ability to find a path to a node anywhere, based only on its public IP address. The IP connectivity service is achieved via a set of highly successful IP routing protocols.

In the VANET environment, however, many packets are sent directly over the air from the source to the destination, so routing is less of an issue. In order to avoid the packet overhead associated with Internet protocols (a minimum of 52 bytes for a UDP/IPv6 packet), the IEEE 1609 WG defined a new Layer 3 protocol that is efficient for these one-hop transmissions: the *WAVE short message protocol (WSMP)*. Packets sent using WSMP are referred to as *WAVE short messages (WSMs)*. The minimum WSM overhead is eight bytes, and even with options and extensions it will rarely exceed 20 bytes. Given the congestion concerns noted in the previous section, the efficiency of WSMP is quite valuable.

The CCH carries data packets in WSMP format, and WAVE service advertisements (WSAs). The WSM and WSA packet formats are defined in the IEEE 1609.3 standard. IP packets are not allowed on the CCH. Packets on an SCH can use either WSMP or IPv6. Services that are provided directly between two DSRC-equipped nodes, e.g. from an RSU directly to an OBU, can use WSMP for efficiency. Services that rely on internetworking, e.g. with nodes in the Internet, will likely require IPv6.

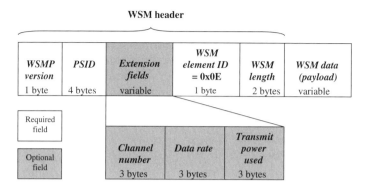

Figure 10.11 WAVE short message format.

WAVE short message format

The transmission of a WSM is initiated by a higher layer via a simple data-plane primitive that provides the content of the WSM plus additional protocol control information, including: data rate, transmit power, user priority (which can be translated to channel access priority in the MAC), and for unicast packets the destination MAC address. The WSM then becomes the payload of a subsequent primitive passed down to the LLC and MAC layers for transmission in a MAC frame. The process for a received WSM is complementary. The received WSM is passed up from the MAC and LLC, along with information about the frame in which it was extracted, and it is passed along to the higher layers in a data-plane primitive.

The WSM format consists of a variable-length header followed by a variable-length payload, as shown in Figure 10.11. The message format includes both mandatory and optional fields. The fields are defined below.

> **WSMP version:** This mandatory one-byte field contains a WSMP version number. The version number associated with the current standard is 1 and other values are reserved. The version concept is intended to distinguish incompatible variations on the WSM protocol. Minor additions and updates to the standard will not lead to a new version. A device that receives a WSM with a version number higher than it was designed to support will discard the packet.
>
> **Provider service identifier (PSID):** The mandatory four-byte PSID identifies the service that the WSM payload is associated with. This might be the safety service, or a service that is advertised in a WSA. A device creates a list of PSIDs that have active receive processes at higher layers. When a WSM arrives, if the PSID matches one of those on the list, the WSM payload is forwarded to that process. In this way, the PSID serves a purpose that is similar to a TCP or UDP port number. PSIDs are administered through the IEEE Registration Authority, which charges a fee to assign a PSID.

WAVE element ID	Length	Contents
1 byte	1 byte	Variable length

Figure 10.12 Extension field data structure.

WSM extension fields: In IEEE 1609.3 there is a facility for including an 'extension field' in a WSM or WSA header. An extension field is by definition optional. It consists of three pieces of information in a single data structure, as shown in Figure 10.12: a one-byte *identifier*, a one-byte *length*, and a variable length *contents* field. The identifier distinguishes one extension from another; identifiers are unique and are defined in the 1609.3 standard. The length indicates the number of bytes in the contents field. The extension field concept provides flexibility and extensibility over various releases of the 1609.3 standard. New extension fields can be defined as needed. If a receiver detects an extension field with an identifier that it does not recognize, it uses the length field to skip that extension field and continue processing the received message. In the current version of IEEE 1609.3, three extension fields are defined for the WSM. The contents of each uses one byte, so with the *identifier* and *length* each extension is three bytes long:

Channel number: This optional one-byte field holds an integer representing the channel on which the WSM is sent. The channel number is interpreted in the context of a particular regulatory domain. For example, the seven 10 MHz channels in the US DSRC band have channel numbers 172, 174, 176, 178, 180, 182, and 184.

Data rate: This optional one-byte field holds the data rate at which the frame containing the WSM is sent, using a format defined in IEEE 802.11. The data rate is 500 Kbps times the integer in the lower seven bits of this field, e.g. if the field value = 0x06 then the data rate is 3 Mbps.

Transmit power used: This optional one-byte field indicates the transmit power level at the output of the antenna connector. The power is represented as a signed integer in the range -128 dBm to $+127$ dBm.

WSM element ID for WSM data: This mandatory one-byte field indicates that there are no more extension fields.

WSM length: This mandatory two-byte field is the end of the WSM header. Its value is equal to the number of bytes in the WSM payload, which follows immediately. The minimum WSM payload length is 1 byte. A maximum value is defined in the MIB.

WSM data: This is the payload of the WSM. It is provided by a higher layer.

WAVE service advertisement format

The WAVE service advertisement (WSA) includes information about one or more DSRC services that are offered in an area. A service can be almost any information exchange that provides value to a vehicle's occupants. Example services include traffic alerts, tolling, navigation, restaurant and shopping information, entertainment, and Internet access. Most services are provided by a roadside unit (RSU) device, but in theory a service could be provided by another vehicle. The information exchange can be unidirectional or bidirectional. WSAs are sent on the CCH. The services they advertise are offered on one or more of the SCHs. One type DSRC communication that is not considered a service is the broadcast of basic safety messages from a vehicle to its neighbors. Those broadcasts are not advertised via a WSA.

A given RSU may support more than one service offering. For efficiency, it provides information about all the services that it supports in a single WSA. One RSU can also advertise services offered by another, nearby RSU. This can be useful in a high speed vehicular environment where the time available for the service data exchange is quite limited, especially for example if the provider RSU is on a curve in the road that obstructs its line-of-sight with oncoming vehicles until they are very close.

Services can be provided by either the IPv6 or WSMP part of the protocol stack. The information passed in the WSA will vary depending on the service protocol, as shown below. The data exchange associated with many services can be completed in one SCH interval; others take more than one SCH interval, in separate sync periods, to complete. The *immediate* and *extended access* variations on channel switching, discussed in Section 10.6.1, provide flexibility to the timing of providing a service. Typically a WSA is sent on the CCH during each CCH interval; indeed it may be sent more than once in an interval as explained below.

Transmission of a WSA is initiated by a primitive request in the management plane from the WAVE management entity (WME) to the MAC layer management entity (MLME). The resulting WSA will typically be transmitted in each sync period (perhaps more than once, according to the *Repeats* parameter) until service advertisements are cancelled by the WME. The MLME then initiates transmission of each individual WSA. This can be accomplished in a variety of ways. For example, the WSA can be encapsulated in

- a vendor-specific information element (VSIE) within a timing advertisement management frame, or

- the payload of a vendor-specific action frame (a type of MAC management frame).

At the receiving device, the process is reversed. A WSA can be extracted from a variety of MAC data structures and passed to the MLME, which then passes it up within the management plane. The WAVE management entity can then examine the WSA and decide whether to access any of the advertised services. Typically the receiver keeps a list of PSIDs for services that it is interested in, and performs an initial match with the PSID of each service advertised in a WSA. Even if there is a PSID match, the receiver may opt not to access a given service. For example, a

STANDARDS AND REGULATIONS 403

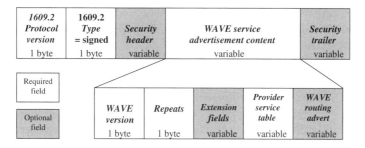

Figure 10.13 WAVE service advertisement format.

service providing information about restaurants and hotels may be repeated in each sync period, and a receiving vehicle might access that once and then ignore it. The process of deciding which services to access is subject to local policy in the receiver.

The WSA format is quite complex. The WSA is always sent within the *Secured-Message* format defined in the IEEE 1609.2 security standard (IEEE 1609.2 2006). The *SecuredMessage* always includes a *1609.2 version* field (currently set to 1) and a *type* field that supports three types of *SecuredMessage*:

- unsecured (type = 0)
- signed (type = 1)
- encrypted (type = 2).

The *SecuredMessage* containing a WSA will either have *type = unsecured* or *type = signed* (a signed message can subsequently be encrypted, see Section 10.6.4, but that is unlikely for a WSA and the *type* would still be *signed* in this part of the message). When *type = unsecured*, the WSA is an 'unsecured WSA.' When *type = signed*, the WSA is a 'secured WSA,' but note that both an unsecured WSA and a secured WSA are types of a 1609.2 *SecuredMessage*.

An unsecured WSA consists of only the *1609.2 version, type,* and *WAVE service advertisement content*. A secured WSA includes those fields, plus a 1609.2 security header and security trailer. The *WAVE service advertisement content* is identical in an unsecured WSA and a secured WSA.

The WSA format is shown in Figure 10.13, including the optional security header and trailer that appear in a secured WSA. The content from WAVE version through WAVE routing advertisement is independent of whether security is applied or not, and is explained here. The security information is presented in Section 10.6.4.

WAVE version: This mandatory one-byte field conveys the format version number of this WSA. The version number associated with the current standard is 1 and other values are reserved. This serves the same purpose for a WSA as the WSMP version serves in a WSM.

Repeats: This mandatory one-byte field indicates how many times the advertisement is repeated over the air in each sync period. The total number of transmissions is one more than the *repeats* value. Note: In future versions of the 1609.3 standard the *repeats* field will likely be redefined to permit fewer than one transmission per sync period.

Repeating an advertisement multiple times within a single CCH interval can serve two purposes. It increases the probability that the advertisement is successfully received at least once. It also provides a crude link quality measure at the receiver, which can compare the total number of transmissions with the number actually received. That measure, combined with other indications of link quality (e.g. RCPI), can assist a receiver in determining whether to attempt to access an advertised service.

There are problems with this approach to measuring link quality. One is that repeated transmissions of redundant information contribute to congestion during the CCH interval, which, as noted above, is a serious concern. Another problem is that the over-the-air success of advertisements on the CCH may not be a good indication of link quality on the SCH offering a service.

WSA header extension fields: As with the WSM, optional extension header fields are allowed in the WSA header. In the current standard, four extensions are defined:

- *Transmit power used:* the power with which the WSA's frame was transmitted, from -128 dBm to $+127$ dBm, as measured at the output of the antenna connector.
- *2D location:* the location of the transmitting antenna, encoded as a 32-bit latitude and 32-bit longitude.
- *3D location and confidence:* the 3D location of the transmitting antenna (32-bit latitude, 32-bit longitude, and 20-bit elevation) and a 4-bit location confidence value.
- *Advertiser identifier:* a text string of 1–32 bytes that is somehow associated with the device that sends the advertisement.
- *Country String:* a 3-byte field that identifies the regulatory domain of the sender, using a format defined in IEEE 802.11.

Provider service table: This is a complex field that consists of one or more *Service Info* segments and one or more *Channel Info* segments. There is one *Service Info* segment for each service advertised in the WSA. There is one *Channel Info* segment for each channel on which an advertised service is offered. Since more than one service can be supplied on a single channel, the number of *Channel Info* segments will be no greater than the number of *Service Info* segments.[4] The provider service table is the most important part of the WSA; this is where the services are actually advertised.

Each *Service Info* or *Channel Info* segment starts with a WAVE element ID indicating the segment type. This is necessary so that the receiver can determine

---

[4]If a provider supports services with the same PSID on multiple channels in the same sync period, these are considered distinct services and are advertised with separate *Service Info* segments.

what type of information follows. *Service Info* segments are identified with a WAVE element ID = 0x01. *Channel Info* segments are identified with a WAVE element ID = 0x02. Each *Service Info* or *Channel Info* segment includes a fixed length mandatory part, and a variable length optional extension part. The length of each extension field is indicated explicitly, so it is not necessary to also explicitly indicate the length of the overall *Service Info* or *Channel Info* segment.

The *Service Info* segment is composed as follows:

- One-byte *WAVE element ID* set to 0x01.
- Four-byte *provider service identifier* (PSID), which is described in the WSM header section above (Section 10.6.2).
- One-byte *service priority*, with values restricted to the range 0–63 (0 is lowest priority). This priority is associated with the higher layer process initiating the advertisement. It is used to help arbitrate access to competing advertised services.
- One-byte *channel index*. Indicates the *Channel Info* segment associated with this *Service* info segment.
- Variable length *service info extension fields*. These extensions are specific to a *Service Info* segment, and they utilize the same encoding format as other extensions. The service info extensions defined in the current standard are:
    - provider service context (PSC): a string of up to 31 bytes that provides additional information about the service; each PSID has a unique PSC format, which is defined by the organization to which the PSID is assigned
    - IPv6 address: a 16-byte address of the entity hosting the service if the service is provided using IPv6 rather than WSMP
    - service port: a two-byte port number for the transport layer protocol (UDP or TCP) if the service is provided using IPv6
    - provider MAC address: a six-byte IEEE MAC address of the device hosting the advertised service if different from the MAC address of the device sending the WSA.
    - RCPI threshold: recommended minimum WSA received power, in dBm from 0 to 110. If the RCPI is below this threshold, the recipient should not attempt to access the service. The format is defined in IEEE 802.11k.
    - WSA count threshold: recommended minimum number of WSAs for this service that should be received before attempting to access the service. The range is 0 to 255.

As noted, there is one *Channel Info* segment for each channel that supports a service advertised in the WSA. The format of the *Channel Info* segment is formatted as follows:

- One-byte *WAVE element ID* set to 0x02.

- One-byte *Regulatory Class*, which provides context for the channel number that follows. The format of *Regulatory Class* is defined in IEEE 802.11.

- One-byte *channel number*. For US DSRC, this will be an element of the set 172, 174, 176, 180, 182, 184. Notice that Channel 178 is not included because that channel is designated as the control channel (CCH) in the US, and this segment describes a service channel (SCH).

- One-byte *adaptable* field, of which only one bit is used, as a bit flag. If the flag is 0, then the data rate and transmit power level (see directly below) are fixed. If the flag is 1, then the data rate is a lower bound and the transmit power level is an upper bound. In that case, higher rates and/or lower power levels are allowed when accessing the service.

- One-byte *data rate*, the format of which is defined in IEEE 802.11. In most cases it is in units of 500 Kbps, so for example a value of 12 decimal (0x0C) indicates a data rate of 6.0 Mbps.

- One-byte *transmit power level*, with range -128 dBm to $+127$ dBm. This power is measured at the output of the antenna connector.

- *Channel info extension fields:* in the current standard two extension fields are defined for the *Channel Info* segment. The first extension field allows for the advertisement of an EDCA parameter set, which can be used to define channel access priority for up to four different priority classes. The format of the EDCA parameter set, as well as default EDCA parameter sets are defined in the IEEE 802.11 standard. This extension field provides an opportunity to advertise a non-default set. The second extension field indicates whether the service is provided on the SCH continuously or only during the SCH interval.

WAVE routing advertisement (WRA): This is an optional field within the WSA. It is only used when an advertising device offers a service that utilizes the IPv6 part of the protocol stack. The WRA provides information about how to connect to the Internet, which a receiver (e.g. a vehicle) can incorporate in its network configuration. If a WRA is included in a WSA, it begins with a *WAVE Element ID* set to 0x03, which tells the receiver that this is a WRA and not another *Channel Info* segment. As with the *Service Info* and *Channel Info* segments, the WRA has a fixed part and a variable-length optional extension part. Each extension includes an explicit length indicator, so it is not necessary to also explicitly indicate the length of the entire WRA. The WRA format is as follows (a 'mandatory' field is required whenever a WRA is included in the WSA):

- **WAVE element ID:** a mandatory one-byte field set to 0x03.
- **Router lifetime:** a mandatory two-byte field indicating how long, in seconds, the default gateway information that follows is valid.
- **IpPrefix:** a 16-byte IPv6 subnet prefix.

- **Prefix length:** a one-byte value indicating how many of the 128 bits in the preceding IpPrefix are significant.
- **Default gateway:** a 16-byte IPv6 address of the default router used to achieve Internet connectivity.
- **Gateway MAC address:** a six-byte MAC address associated with the default router, if different from the MAC address of the sender.
- **Primary DNS:** a 16-byte IPv6 address of a device that can serve as a domain name server.
- **WRA extension fields:** optional extension fields using the normal format; the current standard defines one extension field, which can be used to provide the address of a secondary DNS.

10.6.3 WSA length summary

The WSA format described above has a lot of flexibility, and the length is highly variable depending on what is included. Here is a summary of the lengths of various mandatory and optional components (the length of the security information is described in Section 10.6.4 below):

- Protocol version and type, one byte each (mandatory).
- WAVE version – one byte (mandatory).
- Repeats – one byte (mandatory).
- WSA extension fields – up to 55 bytes if transmit power (3 bytes), 3D location and confidence (13 bytes), a maximum size advertiser identifier (34 bytes), and a country string (5 bytes) are included, each in their *identifier-length-contents* format.
- Provider service table:
    - From 1 to 32 instances of a *Service Info* segment. Each *Service Info* segment includes seven mandatory bytes and up to 69 bytes of extension fields, for a maximum total length per segment of 76 bytes.
    - From 1 to 32 instances of a *Channel Info* segment. Each *Channel Info* segment includes six mandatory bytes and up to 22 bytes in an EDCA parameter set extension field 25 bytes of extension fields (including 20 bytes for an EDCA parameter set), for a maximum total length per segment of 31 bytes.
    - While the standard permits up to 32 *Channel Info* segments, there can be only one per SCH. In the US there are six SCHs, and in other geographic regions there are likely to be even fewer SCHs, so a more practical upper bound is six channel information segments. While it is also unlikely that a single WSA will advertise the maximum of 32 services, it is possible. With 32 *Service Info* segments and six *Channel Info* segments, each of maximum

size, the *provider service table (PST)* length could grow so large that a WSA could not fit within the maximum 802.11 MAC payload of 2304 bytes. A minimum size PST would be just 13 bytes for one minimum-size *Service Info* segment and one minimum-size *Channel Info* segment.

- WAVE router advertisement – This optional field includes 58 mandatory bytes and 18 optional bytes, for a maximum length of 76 bytes.

So, the *maximum* valid WSA exceeds the MAC frame limit of 2304 payload bytes. The *minimum* valid WSA is just 17 bytes.

Example WSA length

For illustration purposes consider a more '*typical*' WSA, advertising three services (two of which use IPv6) on two channels. Imagine that the optional fields are populated as follows (recall that each extension requires an extra two bytes for the type and length encoding):

- WSA extensions include: transmit power, 2D location, and 16 bytes of advertiser identifier, for a total of 31 bytes including encoding overhead, i.e. $3 + 10 + 18$.

- *Service Info* extensions include a 16-byte PSC, and for the IPv6 services also a 16-byte IPv6 address and 2-byte service port. The total length of *Service Info* extensions is 98 bytes, i.e. $18 + 2 \cdot (18 + 18 + 4)$. (This does not include the mandatory parts of the three *Service Info* segments.)

- *Channel Info* extensions: assume the default EDCA parameter set is used on each channel, so no channel info extensions are necessary.

- WRA extensions: assume a secondary DNS is not provided, so no WRA extensions are necessary.

The length of this hypothetical 'typical' WSA is:

- 2 bytes for protocol version and type
- 2 bytes for WAVE version and repeats
- 31 bytes of WSA extensions
- 21 bytes for mandatory parts of three *Service Info* segments
- 98 bytes of *Service Info* extensions
- 12 bytes for mandatory parts of two *Channel Info* segments
- 58 bytes for the WAVE router advertisement
- Total length (without security) is 222 bytes.

So, while a WSA can theoretically be anywhere from 16 to about 2000 bytes long, a more typical length will be a few hundred bytes, probably most often between 100 and 400 bytes if services are offered over IPv6 (and thus the WRA is included).

10.6.4 Middle layer security: IEEE 1609.2

The general topic of security in VANETs is the subject of another chapter in this book. This section explains how those principles are applied in the specific case of the IEEE 1609.2 standard: Security Services for Applications and Management Messages.

IEEE 1609.2 defines standard mechanisms for authenticating and encrypting messages. These are covered, in turn, in the following subsections.

IEEE 1609.2 Authentication

An authenticated message carries a *digital signature* that can be used to verify the identity of the sender. IEEE 1609.2 authentication uses the elliptic curve digital signature algorithm (ECDSA), which is an asymmetric cryptographic algorithm. Two different key lengths are specified, 224 bits and 256 bits. ECDSA is a relatively processor-intensive algorithm. This is a concern in a VANET environment, especially at a receiver that needs to authenticate hundreds of safety messages per second. Research is ongoing into authentication mechanisms that require less processing for use in safety applications.

To sign a message a sending device must have a signing key and a certificate associated with that key. There may also be a chain of certificates associated with the key, but that possibility is not explored in detail in this section. The certificate may carry various scoping restrictions, for example with regard to time or application type (or, in the case of an RSU, restrictions with regard to location).

Certificate format:
An RSU and an OBU use the same basic certificate formats, but have different substructures, notably with regard to scope. They will also tend to exercise different options. A typical OBU certificate is described below (see IEEE 1609.2 for more information on the RSU certificate):

- Certificate version: one byte, set to 1 in the current standard.

- Subject type: one-byte code to indicate the type of entity that owns the certificate, in this case an OBU.

- Signer_ID: an eight-byte hash of the certificate authority's own certificate. This is a way to identify the issuing authority.

- Application scope: a variable length list of applications for which this certificate is valid.

- Expiration: a time after which the certificate is invalid. This time is encoded in 32 bits in units of seconds, relative to UTC time. If the expiry time is 0 the certificate never expires.

- CRL series: This is a 32-bit integer that identifies a subset of certificates to which this certificate belongs. The *certificate authority (CA)* that generates a certificate may revoke it by placing it on a *certificate revocation list (CRL)*. To improve the scalability of CRL distribution, a CA may segment the CRLs it maintains into different series, each of which represents a subset of the CA's certificates. This field in the certificate identifies the specific CRL series that would be used to revoke it. If set to 0, this certificate will never appear on a CRL. The expiry time and the CRL series cannot both be zero, i.e. a certificate must either have a finite lifetime or be revocable, or both.

- One or two public keys being certified. If there are two, one key is for authentication and one is for encryption. The public key encoding has a one-byte length followed by the actual key. For a 224-bit (28-byte) key used for authentication the encoded length is 29 bytes. For a 256-bit (32-byte) key, used either for authentication or encryption, the encoded length is 33 bytes.

- Certificate signature – the signature of the certificate authority. A certificate signature always uses a 256-bit key and occupies 64 bytes.

According to IEEE 1609.2, a typical certificate that covers a single application at an OBU has a length of 125 bytes.

Signing an outgoing message:
To generate the authenticated message, the device first creates a `ToBeSigned-Message` data structure consisting of the following elements:

- Application information (PSID) for the application to which the message is destined. The indicated application must be within the scope of the certificates associated with the signing key. The PSID is four bytes long.

- A set of four message flags in a two-byte field. The first flag indicates whether the message has been fragmented. The remaining three flags indicate whether the `ToBeSignedMessage` includes, respectively, a generation time, an expiry time, and/or a use location. As with the application information, any timing or location information must be within the scope of the signing key certificates.

- The higher layer message content, i.e. payload. This is known as the 'application data.' A two-byte application data length field precedes the higher layer data.

- The optional message generation time, if the corresponding message flag is set to 1. This field is useful in defeating replay attacks as described in the receiver authentication section below. A sender cannot generate two messages with the same generation time. The field is set to the current time within the generation process, relative to UTC time. It is represented in 64-bits in units of microseconds.

- The optional expiry time, if the corresponding message flag is set to 1. The message is not valid after its expiry time. The expiry time uses the same 64-bit format as the generation time.

- The optional location and location confidence field, if the corresponding message flag is set to 1. This indicates the location of the sender, and it uses the same format as the 3D location and confidence field in the WSA, described in Section 10.6.2.

The generic signed message is then composed of the following elements:

- *Protocol version*, a one-byte mandatory field that is set to 1 for the current standard.

- *Type*, a one-byte mandatory field that indicates what type of security is applied to this message. The code point indicating a 'signed message' is applied for authentication.

- *Signer information:* This variable-length field starts with a one-byte *signer identifier type*, which usually indicates that either the message carries a full certificate or that the message carries a certificate digest. As shown above, a typical OBU certificate is 125 bytes long. A certificate digest is an eight-byte hash of the certificate, so it is more bandwidth efficient. The choice of using a certificate or a digest is discussed further below.

- The `ToBeSignedMessage` data structure discussed above.

- The *digital signature* computed over the entire `ToBeSignedMessage` data structure. The signature computation uses either the 224-bit or the 256-bit ECDSA algorithm. The signature length is twice the key length, i.e. 56 bytes for a 224-bit key or 64 bytes for a 256-bit key.

The generic signed message can be sent as noted above. However, if the message that is being signed is a WSM, there is a further level of efficiency that can be gained due to some replication between elements in the signed message and elements in the WSM header (see Section 10.6.2). In particular, the PSID and the length of the application data are redundant with WSM header information, so a *signed WSM* removes these six bytes from the generic signed message, after the message is signed. The signature, then, covers these values as well.

Full certificate or certificate digest:
A certificate digest is a short reference to a certificate. It is useless at a receiver that has not previously seen the full certificate to which it refers. When a series of messages are sent as part of a service, it is possible to include the full certificate in the first and then only a certificate digest in the remaining messages. The assumption is that if the receiver somehow misses the first message of the series, the service will not be successfully delivered even apart from security considerations.

The policy for vehicle-to-vehicle broadcast safety messages (i.e. the basic safety message) sent once per sync period is more complicated. Since receiving vehicles are moving in and out of range of a given sending vehicle quickly and without coordination, each BSM is potentially the first that is heard by some receiver. There could be value, then, in including a full certificate with each WSM carrying a BSM. However, there is a steep bandwidth penalty attached to a full certificate (e.g. 125 bytes) compared to a digest (e.g. 8 bytes). An alternate policy is to interleave BSMs with full certificates and BSMs with digests according to some schedule that provides a reasonably small interval between the BSMs that carry full certificates. For example, if a vehicle sends BSMs at 100 msec intervals, and if every third BSM carries a full certificate and the other two carry digests, then the average amount of bandwidth consumed by the certificate portion of the message is $(125 + 8 + 8)/3 = 47$ bytes. If a receiver enters the range of a sender, it may have to wait for up to three BSMs before receiving one that it can authenticate. And, of course, since no individual over-the-air packet is guaranteed to be received correctly, the receiver may need to wait for even more BSMs before successfully receiving one with a full certificate. Once it receives that full certificate, it can authenticate succeeding BSMs from that same sender, so long as the sender continues to sign with the public key authenticated in that certificate. There is a clear engineering trade-off between bandwidth consumed and latency before the first authenticated message from a sender is received. At present there is no consensus on how to implement that trade-off, and this is a topic of active research.

Authenticating an incoming message:
A device receiving a signed message executes the following steps to authenticate the signature:

- Verify that the message has the correct format.

- Verify that the message was issued recently but is not identical to a previously received message. This is usually done by comparing a timestamp in the message with the current time, and comparing the actual message content with a cache of recently received messages. The purpose of this verification is to defeat a temporal replay attack.

- If the location of the transmitter is included in the message, compare it with the receiving device's location to ensure that the sender is within a reasonable range. The purpose of this verification is to defeat a spatial replay attack.

- If the message contains a certificate digest rather than a full certificate, verify that the full certificate associated with the digest is in the device's local received certificate cache.

- If the certificate contains a geographic scope, verify that the sender's location is within that scope.

- Verify that the application associated with the message is consistent with the application scope in the certificate.

- Verify that the certificate has not appeared on a certificate revocation list (CRL).

- Verify the actual signature on the message using the ECDSA 224-bit or 256-bit algorithm, as appropriate.

IEEE 1609.2 encryption

The encryption process in IEEE 1609.2 uses a combination of symmetric and asymmetric cryptography. The sender of an encrypted message must know a public key associated with the desired recipient. It learns this from a certificate issued by the recipient (for example, when the recipient device signed an earlier message).

The encrypted message can be sent to more than one recipient, but a public key is needed for each so it cannot be sent to an unknown set of receivers.

Symmetric cryptography is computationally more efficient than asymmetric cryptography. So, the approach to encryption in IEEE 1609.2 uses two steps:

1. The sender selects a new symmetric key, and uses this key to encrypt the message according to a symmetric algorithm.

2. The sender then performs an asymmetric encryption of the symmetric key, using the receiver's public key.

With this approach, the bulk of the cryptography is done with an efficient symmetric algorithm.

In the current standard, one symmetric algorithm and one asymmetric algorithm are specified. The symmetric algorithm is the advanced encryption standard with 128-bit keys in CCM mode, i.e. AES-CCM. The asymmetric algorithm is the elliptic curve integrated encryption scheme (ECIES).

Encrypting an outgoing message:
The specific steps at the sender are as follows:

- retrieve the recipient's private key from a certificate cache

- check to make sure the public key has not been revoked

- generate a random symmetric key

- encrypt the message using the random symmetric key and the AES-CCM algorithm, generating ciphertext from plaintext

- encrypt the random symmetric key using the ECIES algorithm and the recipient's public key

- pack the ciphertext and the encrypted symmetric key into a message that can be sent over the air.

Decrypting an incoming message:
The specific steps at the receiver are as follows:

- the receiver uses its own private key to decrypt the encrypted symmetric key using the ECIES algorithm

- the receiver then uses the decrypted symmetric key to decrypt the ciphertext using the AES-CCM algorithm.

Signing and encrypting messages:
It is possible to both sign and encrypt a message. If a message is to be both signed and encrypted, the unsigned message is first processed according to the authentication mechanism described in the previous section. The product of the first step is a signed message. Then, the signed message is encrypted using the mechanism described

in this section. Thus, the message is first signed and then encrypted. Note that the encryption phase does not need to know that it is acting on a signed message.

At a receiver the process is reversed. The receiver detects that the message it has received is encrypted. The receiver decrypts the message using the steps outlined above, to produce an unencrypted signed message. The IEEE 1609.2 encrypted message format includes a one-byte *content type*. The two content types defined in the current standard are *application data* and *signed*. If the type is *signed* the embedded unencrypted message is signed. In that case the receiver authenticates the unencrypted message using the mechanism described in the previous section. Thus, the message is first unencrypted and then authenticated.

10.7 DSRC Message Sublayer

At the top of the protocol stack in Figure 10.2, above the middle layers, is the application layer. This is where the application protocols themselves are defined, and in some cases additional protocols that provide direct support to applications. The very notion of what constitutes an application is somewhat ambiguous because there is often a hierarchy of applications, with one sublayer providing a 'service' to a higher sublayer.

In the case of a VANET, the application layer is where one sees the network's full range of diversity. VANET applications range from those that help vehicles to avoid colliding with each other, to those that provide travel information to drivers, and those, like basic Internet access, that have no vehicle-specific function. The latter are well described in many other places, and so receive less description in this book. Some of the applications more directly related to the vehicular environment are described elsewhere in the book. This chapter concentrates on one set of protocols that specifically enable some important VANET applications: the DSRC messages defined in the SAE J2735 *DSRC message set dictionary* standard.

The mission of the J2735 standard is to define the over-the-air structure of messages that are known to be of interest for certain applications expected to utilize the DSRC communication system. This section describes the set of messages in the J2735 standard, and discusses three of them in some detail.

10.7.1 SAE J2735 DSRC message sets

The SAE J2735 (SAE J2735 2009) standard is described as a *DSRC message set dictionary*. In the terminology of the standard, a *message set* is a set of message types, a *message type* is a generic structure and a message is a specific instantiation of a message type. In this section the names of data structures defined in the J2735 standard are represented in this font: `Sample`.

Table 10.8 lists the 15 message types that are defined in the current version of the SAE J2735 standard (SAE J2735 2009). Messages marked 'informative' are not considered normative; all others are 'standard' and carry an expectation of backward compatibility in future versions of J2735.

The J2735 message set dictionary defines the format of each of the message types listed in Table 10.8. Each message type is defined as a collection of constituent data

Table 10.8 SAE J2735 DSRC standard message sets.

Message Set	Purpose
`A La Carte Message`	Generic message with flexible content.
`Basic Safety Message`	Conveys vehicle state information necessary to support vehicle-to-vehicle safety applications.
`Common Safety Request` (informative)	A vehicle uses this to request specific state information from another vehicle.
`Emergency Vehicle Alert Message`	Alerts drivers that an emergency vehicle is active in an area.
`Intersection Collision Avoidance`	Provides vehicle location information relative to a specific intersection.
`Map Data`	A roadside unit uses this to convey the geographic description of an intersection.
`NMEA Corrections`	Encapsulates one style of GPS corrections – NMEA (National Marine Electronics Association) style 183.
`Probe Data Management`	A roadside unit uses this to manage the collection of probe data from vehicles.
`Probe Vehicle Data`	Vehicles report their status over a given section of road to allow a roadside unit to derive road and traffic conditions.
`Roadside Alert`	A roadside unit uses this to alert passing vehicles of hazardous conditions.
`RTCM Corrections`	Encapsulates a second style of GPS corrections – RTCM (radio technical commission for maritime services).
`Signal Phase and Timing Message`	A roadside unit at a signalized intersection uses this to convey the signal's phase and timing state.
`Signal Request Message`	A vehicle uses this to request either a priority signal or a signal preemption.
`Signal Status Message`	A roadside unit uses this to convey the status of signal requests.
`Traveler Information`	A roadside unit uses this to convey advisory and road sign types of information to passing vehicles.

Table 10.9 Content of `ApproachesObject` data frame.

Name of constituent	Frame or element	Purpose
`ReferencePoint`	Frame	Anchor for other points
`LaneWidth`	Element	Width of lanes
`Approach`	Frame	Describes approach lanes
`Approach`	Frame	Describes egress lanes

structures called *data elements* and *data frames*. A data element is the most basic data structure in the J2735 standard. A data frame is a more complex data structure, composed of one or more data elements or other data frames. The J2735 standard defines the syntax (length, format) and semantics of each data element and data frame.

An example of the relation between data elements and data frames can be seen in the `ApproachesObject` data frame, which is used as part of the description of an intersection. As shown in Table 10.9, this data frame is made up of four constituent parts. One of these is an atomic data element, while the other three are other complex data frames, including two instances of a data frame called `Approach`. The `Approach` data frame is itself composed of a collection of data elements and data frames. The `ApproachesObject` data frame shown in Table 10.9 is a constituent of a larger frame called `Intersection`. Ultimately, message types are composed of collections of data elements and data frames. The hierarchical structuring of data elements, data frames, and message types encourages reuse of data structures. The J2735 message set dictionary defines approximately 150 standard data elements and 70 standard data frames. It also includes several additional elements and frames that are defined in other SAE standards, so-called 'external' data structures that can also be included in J2735 message sets. The data elements, data frames, and message sets in J2735 are defined in terms of a formal language called *Abstract Syntax Notation One (ASN.1)*. ASN.1 is defined in the ITU-T X.680 series of standards. The J2735 dictionary standard also calls for use of the *distinguished encoding rules (DER)*, a subset of the *basic encoding rules (BER)*, to translate the ASN.1 into over-the-air bits and bytes. BER is defined in the ITU-T X.690 standard (ITU-T X.690 2008). DER encodes each data item (element or frame) in a three-part structure consisting of an *identifier*, a *length*, and the *contents*. These are also sometimes called the *tag*, *length*, and *value*, which is abbreviated *TLV*.

The tag identifies the type of data in the value part of the structure, e.g. in Table 10.9 the three types of items: `ReferencePoint`, `LaneWidth`, and `Approach`, would have unique tags, but the two instances of the `Approach` frame would have the same tag since they identify the same type of data. The length simply indicates how many bytes are in the value field. The value field contains the contents of interest, e.g. the actual width of the lane conveyed by the `LaneWidth` element. Each data frame is DER-encoded, and the constituent parts are also DER-encoded. So, there is a natural recursive nature to the encoding. In the example of Table 10.9, the outer encoding would identify this as an `ApproachesObject` data frame, and within the value field of the frame would be found four TLVs, one for each constituent part. The TLV for the `LaneWidth` element would convey the lane width value. Since the other constituents are all data frames, each of their TLVs would have subordinate TLVs in the value field, recursing down in a tree-like structure with each branch ultimately ending with a data element.

Writing the content definitions in ASN.1 and calling for a specific encoding has three principal advantages:

- *Interoperability:* By using standard definitions of terms like integer and octet, the standard ensures that a conformant sender and a conformant receiver will have a common understanding of the value that the bits convey.

- *Efficient parsing:* The TLV encoding lends itself to efficient parsing, even for complex data structures. The parser logic follows directly from the data dictionary definitions.

- *Extensibility:* It is easy to maintain backward compatibility while adding new data frames and elements to the dictionary. Imagine that a receiver built to support one version of the dictionary standard encounters a TLV identifying a frame or element defined in a later version. The tag value indicates that this is a newer data type, which the receiver cannot understand. The length value allows the parser to efficiently skip the remainder of the TLV and resume parsing the message.

There are other advantages of DER encoding as well; for example the variable length capability allows for a more compact representation of a number that does not utilize its maximum dynamic range. The encoding can also impose a data size (and therefore bandwidth) penalty, however. Each tag and length represents overhead, and if the value fields are short the overhead can be significant. In some cases, notably the `Basic Safety Message`, some of the flexibility of DER encoding is sacrificed in the name of bandwidth efficiency. The next three subsections describe specific message sets in more detail.

10.7.2 Case study: The basic safety message

The `Basic Safety Message` (BSM) is one of the most important message types in the J2735 standard. It conveys core state information about the sending vehicle, namely its position, dynamics, system status, and size. It also has the flexibility to convey additional information at the discretion of the sender.

The BSM is a key part of the communication model employed by a vehicle-to-vehicle collision avoidance system. The concept is that all vehicles will broadcast their state information frequently in BSMs. A receiving vehicle tracks its neighbors, predicts a trajectory for each, and compares these to its prediction of its own trajectory. If its model predicts that it will collide with another vehicle then the system takes some corrective action, e.g. warning the driver or directly controlling the vehicle.

There has been extensive research into the content of safety messages for collision avoidance (DOT HS 810 591 2006). This research demonstrated that although there are many distinct collision avoidance applications, there is a significant overlap in the state information that each application in a receiving vehicle needs from its neighbors. This commonality led to the definition of the BSM for support of all vehicle-to-vehicle safety applications, rather than defining a group of application-specific messages.

The common requirements only go so far, however. The BSM has been designed with two parts. Part I includes core state information that is thought to be necessary with each update, and so it must be sent in every BSM. The data structure for Part I emphasizes compactness and efficiency. Part II is an optional area where additional data elements and frames can be included. Part II provides three forms of flexibility.

- It allows for the inclusion of some data types at a reduced frequency compared to the overall BSM transmission rate. This is especially important for large data frames.

- It provides for evolution in the definition of new safety applications, in the understanding of how best to accommodate existing applications, and in the types and precision of sensors whose output may be of use to other vehicles (e.g. camera data, advanced positioning data).

- It supports some degree of message customization, which could be useful for enabling company-specific features.

This section describes the content of the BSM Part I, and of a few data types that are thought to be necessary for inclusion in Part II, albeit at a reduced rate. The general topic of message flexibility and message composition is addressed in Chapter 8.

The nominal transmission rate of BSMs from a given vehicle is 10 Hz. However, if the channel is congested the vehicle may decrease the actual transmission rate below 10 Hz as a means of alleviating the congestion. The BSM rate is not part of the J2735 standard. The SAE DSRC Technical Committee is in the early stages of defining a new standard related to minimum communication requirements for vehicle-to-vehicle safety. This standard might specify constraints regarding the minimum BSM transmission rate, as well as the minimum rate at which certain data types are included in Part II of BSM transmissions. These Part II data types are also discussed below.

Table 10.10 lists the content of Part I of the BSM. This content is fixed and is present in every instance of the BSM. The first column of the table uses the official data element and data frame terminology from the standard. The `Acceleration Set4Way` and `VehicleSize` items are based on data frames, and the remaining items are based on data elements. The constituents of Part I are an exception to the recursive encoding rule mentioned above. There is a heightened sensitivity to the bandwidth consumed by BSMs in general and Part I of BSMs in particular. For that reason, the constituent pieces of the BSM Part I are not individually DER-encoded, since that would add at least two bytes for the tag and length of each. The content shown in Table 10.10 consumes 39 bytes. If the items were individually DER-encoded, the over-the-air bandwidth would include an approximately equal number of tag and length overhead bytes. So, instead, DER-encoding is applied to just two items. The `DSRCmsgID` must be separately DER-encoded because it is parsed independently of the rest of the content. The remainder of Part I is defined

Table 10.10 J2735 basic safety message, Part I.

Data element/frame and length (bytes)	Description
DSRCmsgID element, 1 byte	This is the first element in every message. It is used by the parser to determine to which message type this particular message belongs.
MsgCount element, 1 byte	This is a sequence number that is incremented with each successive transmission of a BSM by a given vehicle. It is used primarily to estimate packet error statistics.
TemporaryID element, 4 bytes	This is under control of the sending vehicle. Typically it is a random constant for a period of time, and is changed occasionally for privacy reasons. It makes it easier for a receiver to correlate a stream of BSMs from a given sender.
DSecond element, 2 bytes	This is a clock signal, modulo one minute, with resolution 1 millisecond.
Latitude element, 4 bytes	Geographic latitude, with resolution 1/10 microdegree.
Longitude element, 4 bytes	Geographic longitude, with resolution 1/10 microdegree.
Elevation element, 2 bytes	Location above or below sea level, with resolution 0.1 meter.
PositionalAccuracy element, 4 bytes	A complex element that records the one standard deviation error along both semi-major and semi-minor axes, and the compass heading of the semi-major axis. This is used to convey how precise the latitude and longitude values are.
TransmissionAndSpeed element, 2 bytes	Three bits encode the vehicle's transmission setting. The remaining 13 bits convey the unsigned vehicle speed, with resolution 1 cm/second.
Heading element, 2 bytes	Compass heading of vehicle, with resolution 1/80 degree.
SteeringWheelAngle element, 1 byte	Current position of the steering wheel. A clockwise rotation is expressed as a positive angle. The resolution is 1.5 degree, allowing a range of approximately ±180 degrees.

Table 10.10 Continued.

Data element/frame and length (bytes)	Description
`AccelerationSet4Way` frame, 7 bytes	This is a complex frame that provides longitudinal acceleration, lateral acceleration, vertical acceleration, and yaw rate.
`BrakeSystemStatus` element, 2 bytes	This complex element conveys whether or not braking is active on each of four wheels, and also conveys the status of the following control systems: traction control, anti-lock brakes, stability control, brake boost, and auxiliary brakes. For each system, the frame indicates whether the vehicle is equipped with the system, if it is turned on, and if it is active.
`VehicleSize` frame, 3 bytes	The vehicle's length and width, with resolution 1 cm.

as one complex element (called the *BSMblob*), to which one DER tag and length are applied. Adding the *type* and *length* of the DSRCmsgID and the BSMblob, the total size of the BSM Part I is 43 bytes.

While it is theoretically possible to include a wide variety of standard data elements or data frames in Part II of the BSM, the reality is that experts focus on just a few items. Given the sensitivity to creating congestion on the channel that carries BSMs, senders should include additional items only if there is good reason to believe that receivers can use the information to enhance safety.

There are four data items that are most often discussed for inclusion in Part II of the BSM. These are collected in a data frame called `VehicleSafetyExtension`, which is shown in Table 10.11. A given BSM may or may not include the `VehicleSafetyExtension` frame, and a given `VehicleSafetyExtension` frame may be composed of any combination of the four data items shown. The first item reports the occurrence of one or more 'events,' and is included in the frame only when there is such an event. The remaining three items, collectively, can be somewhat long (several dozens of bytes, up to a few hundred), so even though they are required for the operation of some safety applications they are not included in every BSM. The subrate necessary for each of these items is an open research question. Part II of a BSM may also include a complex data frame called `VehicleStatus`, which is composed of dozens of less important data elements and frames. As with `VehicleSafetyExtension`, the elements and frames in a given instance of `VehicleStatus` are all optional, so the BSM Part II provides a lot of flexibility in reporting information that may be of use in preventing collisions. Examples of data items in the `VehicleStatus` frame are: coefficient of friction, bumper height, and wiper rate.

Table 10.11 Vehicle safety extension data frame, required for some safety applications, sent in Part II of some BSMs.

Data element/frame and length (bytes)	Description
`EventFlags` element, 2 bytes	This is a set of bit flags, each of which can convey the occurrence of a given 'event.' Each event has a set of minimum activation criteria, which must be met before the sender can set the associated event bit flag. The sender is not required to set the event flag bit, or to include this element. Some event flags of note are: • hard brake (braking exceeds 0.3 g) • hazard lights on • sender is emergency response vehicle • stop line violation (sender expects to enter intersection against red light or stop sign). These flags can be used to elevate the priority of a received message, even if the information is encoded elsewhere in the BSM.
`PathHistory` frame, variable length (typically on the order of 10 bytes for a straight path and less than 100 bytes for a curved path)	This is used to convey where a vehicle has been, in the form of individual data structures sometimes called 'breadcrumbs.' Each breadcrumb includes position information, which is encoded as an offset to the full-precision position information in Part I of the BSM. In addition, each breadcrumb can include a time (offset) and an indication of position accuracy. The collection of breadcrumbs in a `PathHistory` frame can be used in the prediction of a possible collision. The number of breadcrumbs in a frame is variable, and should be chosen based on the deviation from a predictable trajectory (e.g. deviation from a straight line).
`PathPrediction` frame, 3 bytes	This frame is used to indicate the path that a sender expects to traverse. This is represented as a two-byte radius of curvature (an infinite radius code is used to indicate a straight line prediction) and a one-byte prediction confidence.

Table 10.11 Continued.

Data element/frame and length (bytes)	Description
RTCMPackage frame, variable	Like the RTCMcorrections message type, this frame allows a sender to convey GPS correction data in the RTCM style. The length is variable and depends on the number of GPS satellites in view of the sender's GPS unit.

10.7.3 Case study: The probe vehicle data message

The Probe Vehicle Data message is sent by a vehicle to an RSU that has indicated that it is interested in receiving probes. These RSUs will typically be deployed in areas of specific interest to, for example, a state's Department of Transportation. An individual Probe Vehicle Data message includes information about the vehicle's current and past state. By gathering this information an RSU can learn about real-time road, weather, and traffic conditions. With knowledge of local conditions, the roadside network of devices can advise approaching vehicles and suggest appropriate action (e.g. reduce speed or re-route due to congestion, avoid traffic hazard in left-most lane, etc.).

Table 10.12 shows the structure of the Probe Vehicle Data message in the J2735 standard. Some vehicles, for example state-owned or licensed vehicles, may be configured to provide probe data. These will usually provide a unique identifier that can be traced to the vehicle, using the VehicleIdent frame. Other vehicles that agree to act as probes may choose to include a non-traceable identifier using the ProbeSegmentNumber element. The probe vehicle then conveys both its current position and up to 32 snapshots of its past state. The latter are conveyed using the Snapshot frame.

10.7.4 Case study: The roadside alert message

The RoadSide Alert message is typically sent by a roadside unit to vehicles within a given range to alert them of nearby hazards. These may be of the form 'bridge icing ahead,' 'train approaching,' etc. The specific event types and descriptions come from a separate SAE standard, J2540-2 (SAE J2540-2 2006), which lists International Traveler Information Systems (ITIS) 'phrases.' Table 10.13 shows the structure of the J2735 roadside alert message.

10.8 Summary

This chapter explains the role of regulations and cooperative standards in promoting interoperability between nodes in a VANET. The layered architecture concept is shown to be a useful way of dividing the functional requirements of a VANET into

Table 10.12 J2735 probe vehicle data message.

Data element/frame and length (bytes)	Description
`DSRCmsgID` element, 1 byte	Identifies the message type.
`ProbeSegmentNumber` element, 2 bytes	An optional random identifier that changes on the order of a few seconds or a few hundred meters.
`VehicleIdent` frame, variable length	An optional frame carrying information that identifies a specific vehicle. This will not be used by probe vehicles desiring anonymity. The information can be conveyed in a variety of ways, including: a human-readable name, a vehicle identification number (VIN), a vehicle owner code, a Temporary ID (same ID used in BSM). The data frame can also convey the vehicle class.
`FullPositionVector` frame, variable length, up to 29 bytes	This frame conveys the vehicle's 2D position, and optionally includes time, elevation, speed, heading, and various confidence and accuracy measures.
`VehicleType` element, 1 byte	This element conveys the 'type' of probe vehicle, which can include motorcycle, car, and various combinations of axle count, tire count, and trailer type.
`Snapshot #1` frame, variable length ...	First snapshot. Each snapshot is a data frame that conveys the full position vector and a set of vehicle status data associated with a previous point in the vehicle's path. The sender has the flexibility to select any of a large number of optional elements related to vehicle status.
`Snapshot #n` frame, variable length	Final snapshot. There can be up to 32 snapshots in a single `Probe Vehicle Data` message.

several manageable pieces, and a specific layered protocol stack is introduced for the DSRC use case.

The regulatory framework for DSRC in the US is codified by the FCC in CFR 47 Part 90 (for RSUs) and Part 95 (for OBUs), both of which require adherence to the ASTM E2213 standard. These regulations focus primarily on the allocation of spectrum for DSRC in the 5.9 GHz band, the functional designation of channels within that spectrum, and the power emission rules that govern specific equipment types in specific channels. The regulatory structure in Europe is not as mature.

Table 10.13 J2735 roadside alert message.

Data element/frame and length (bytes)	Description
`DSRCmsgID` element, 1 byte	Identifies the message type.
`MsgCount` element, 1 byte	Sequence number of roadside alert messages sent from this roadside unit.
`ITIS.ITIScode` externally defined structure, 2 bytes each, between 1 and 9 codes are in the message	This data comes from the SAE J2540-2 standard. The first ITIScode is mandatory. It lists a category of alert and a specific item from that category. Up to eight additional ITIScodes are allowed. These can be used to further describe the event that triggered this message, including perhaps providing guidance to a driver on how to avoid a hazard.
`Priority` element, 1 byte	This is an optional element to convey the 'transit signal priority' of the indicated event, relative to other ITIS events. This does not define local priority among messages from a given device.
`HeadingSlice` element, 1 byte	This optional element conveys a coarse indication of the heading(s) of interest for the indicated event. The resolution is 22.5 degrees of arc.
`Extent` element, 1 byte	This optional element conveys the spatial distance over which the message applies, using an encoding with increasing granularity. The finest extent is '3 meters' and the largest is '10 kilometers'. It is also possible to convey an immediate or an unlimited extent.
`FullPositionVector` frame, variable length, up to 29 bytes	This frame is optional. If included, it conveys the vehicle's 2-D position, and may also include time, elevation, speed, heading, and various confidence and accuracy measures.
`FurtherInfoID` element, 2 bytes	This element is a two-byte string that can be used to provide a link to any other incident information that may be available.

The most significant step so far is the allocation by the EU of 30 MHz of bandwidth in the 5.9 GHz band for ITS operations.

The discussion of standards proceeds from lower to higher layers in the stack, and DSRC is again the focus. The physical layer uses the OFDM protocol defined in the IEEE 802.11p amendment. Since the wireless channel is a shared medium, the data link layer includes a medium access control sublayer. This protocol uses the CSMA/CA concept, and is also defined in IEEE 802.11p. A MAC extension protocol that defines how a device can switch among the various channels in the DSRC band is standardized in IEEE 1609.4. At the network and transport layers DSRC communication can either rely on standard Internet protocols (i.e. IPv6, TCP, and UDP), or on the efficient WAVE short message protocol defined in the IEEE 1609.3 standard. IEEE 1609.3 also defines a management plane message called the WAVE service advertisement that is used to advertise the availability of specific DSRC services in a given area. Protocols that allow for the authentication and encryption of DSRC messages are defined in the IEEE 1609.2 standard. At the application layer, the SAE J2735 standard promotes interoperability by defining several message types and a large number of small data structures that can be combined flexibly into these messages. Most of the data structures convey some aspect of a vehicle's state. The details of three specific message types are presented as case studies: the Basic Safety Message the Probe Vehicle Data Message, and the RoadSide Alert Message.

10.9 Abbreviations and Acronyms

2D	two-dimensional
3D	three-dimensional
AC	Access category
ACK	Acknowledgment
AES-CCM	Advanced encryption standard – counter with CBC MIC mode
AGC	Automatic gain control
AP	Access point
ASN.1	Abstract syntax notation one
BER	Basic encoding rules
BPSK	Binary phase shift keying
BSM	Basic safety message
BSS	Basic service set
BSSID	BSS identifier
CA	Certificate authority
CALM	Continuous air interface for long and medium distance
CCA	Clear channel assessment
CCH	Control channel
CRC	Cyclic redundancy code
CRL	Certificate revocation list
CSMA/CA	Carrier sense multiple access/collision avoidance
CTS	Clear to send
CW	Contention window
DA	Destination address
dB	decibels

dBm	dB relative to 1 milliwatt
dBr	dB relative to a stated reference
DER	Distinguished encoding rules
DIFS	Distributed IFS
DNS	Domain name server
DS	Distribution system
DSAP	Destination service access point
°C	degrees Celsius
DSRC	Dedicated short-range communications
ECDSA	Elliptic curve digital signature algorithm
ECIES	Elliptic curve integrated encryption scheme
EDCA	Enhanced distributed channel access
EIRP	Equivalent isotropically radiated power
ETSI	European Telecommunications Standards Institute
EU	European Union
FCC	Federal Communications Commission
FCS	Frame check sequence
FEC	Forward error correction
FFT	Fast Fourier transform
GHz	Gigahertz (10^9 Hertz)
GI	Guard interval
GPS	Global positioning system
IETF	Internet Engineering Task Force
IFS	Interframe space
IP	Internet Protocol
IPv6	IP version 6
ISO	International Standards Organization
ITIS	International traveler information systems
ITS	Intelligent [transport or transportation] systems
ITU-T	International Telecommunication Union – Telecommunication standardization sector
Kbps	Kilobits per second (10^3 bits/second)
KHz	Kilohertz (10^3 Hertz)
LAN	Local area network
LLC	Logical link control
MAC	Medium access control
Mbps	Megabits per second (10^6 bits/second)
MHz	Megahertz (10^6 Hertz)
MIB	Management information base
MLME	MAC layer management entity
MPDU	MAC protocol data unit
MSDU	MAC service data unit
μsec	microsecond (10^{-6} second)
msec	millisecond (10^{-3} second)
NMEA	National Marine Electronics Association
OBE	On-board equipment

OBU	On-board unit	
OCB	Outside the context of a BSS	
OFDM	Orthogonal frequency division multiplexing	
OSI	Open systems interconnection	
PHY	Physical layer	
PLCP	Physical layer convergence procedure	
PMD	Physical medium dependent	
PPDU	Physical layer protocol data unit	
PSC	Provider service context	
PSD	Power spectral density	
PSDU	Physical layer service data unit	
PSID	Provider service identifier	
QAM	Quadrature amplitude modulation	
QoS	Quality of service	
QPSK	Quadrature phase shift keying	
RA	Receiving STA address	
R&TTE	Radio & telecommunications terminal equipment	
RCPI	Received channel power indicator	
RF	Radio frequency	
RFC	Request for comment (a type of IETF document)	
RSE	Roadside equipment	
RSSI	Received signal strength indication	
RSU	Roadside unit	
RTCM	Radio Technical Commission for Maritime Services	
RTS	Request to send	
SA	Source address	
SAE	Society of Automotive Engineers	
SCH	Service channel	
SIFS	Short IFS	
SNAP	Subnetwork access protocol	
SSAP	Source service access point	
STA	Station (an IEEE 802.11 device)	
TA	Timing advertisement or transmitting STA address	
TCP	Transmission control protocol	
TLV	Tag (or type), length, value	
TSF	Timing synchronization function	
UDP	User datagram protocol	
UI	Unnumbered information	
UTC	Universal coordinated time	
VANET	Vehicular ad-hoc network	
VIN	Vehicle identification number	
VSIE	Vendor-specific information element	
WAVE	Wireless access in vehicular environments	
WG	Working group	
WRA	WAVE routing advertisement	
WSA	WAVE service advertisement	

WSM   WAVE short message
WSMP   WSM protocol

References

ASTM E2213-03, 2003. *Standard Specification for Telecommunications and Information Exchange Between Roadside and Vehicle Systems – 5 GHz Band Dedicated Short Range Communications (DSRC) Medium Access Control (MAC) and Physical Layer (PHY) Specifications.* ASTM Committee E17 on Vehicle-Pavement Systems.

CFR 47 Part 90. *Code of Federal Regulations, Title 47, Part 90, Private Land Mobile Radio Services.* http://www.access.gpo.gov/nara/cfr/waisidx_08/47cfr90_08.html.

CFR 47 Part 95. *Code of Federal Regulations, Title 47, Part 95, Personal Radio Services.* http://www.access.gpo.gov/nara/cfr/waisidx_08/47cfr95_08.html.

DOT HS 810 591, 2006. *Vehicle Safety Communications Project – Final Report.* US Department of Transportation, National Highway Traffic Safety Administration.

ETSI EN 302 571 v1.1.1, 2008. *Intelligent Transport Systems (ITS); Radiocommunications equipment operating in the 5 855 MHz to 5 925 MHz frequency band; Harmonized EN covering the essential requirements of article 3.2 of the R&TTE Directive.*

EU Commission Decision of 5 August, 2008. *On the harmonized use of radio spectrum in the 5 875-5 905 MHz frequency band for safety-related applications of intelligent transport systems (ITS).* Official Journal of the European Union.

FCC 06-110, 2006. *Amendment of the Commission's Rules Regarding Dedicated Short-Range Communication Services in the 5.850-5.925 GHz Band (5.9 GHz band).* FCC Memorandum Opinion and Order, adopted July 20, released July 26. WG Docket no. 01-90.

IEEE Std 802-2001, 2002. *IEEE Standard for Local and Metropolitan Area Networks: Overview and Architecture.* IEEE Computer Society.

IEEE Std 802.2, 1998 (R2003). *IEEE Standard for Information Technology – Telecommunications and information exchange between systems – Local and metropolitan area networks – Specific Requirements; Part 2: Logical Link Control.* IEEE Computer Society. Also adopted by the ISO/IEC and redesignated as ISO/IEC 8802-2:1998.

IEEE Std 802.11, 2007. *IEEE Standard for Information Technology – Telecommunications and information exchange between systems – Local and metropolitan area networks – Specific requirements; Part 11: Wireless LAN Medium Access Control (MAC) and Physical Layer (PHY) Specifications.* IEEE Computer Society.

IEEE Std 1609.2, 2006. *IEEE Trial-Use Standard for Wireless Access in Vehicular Environments – Security Services for Applications and Management Messages.* IEEE Vehicular Technology Society.

IEEE Std 1609.3, 2007. *IEEE Trial-Use Standard for Wireless Access in Vehicular Environments – Networking Services.* IEEE Vehicular Technology Society. Note that some information in this chapter is taken from a draft of a newer version of IEEE 1609.3, which should be available in 2010.

IEEE Std 1609.4, 2006. *IEEE Trial-Use Standard for Wireless Access in Vehicular Environments – Multi-channel Operation.* IEEE Vehicular Technology Society. Note that some information in this chapter is taken from a draft of a newer version of IEEE 1609.4, which should be available in 2010.

IEEE P802.11p/D8.0, 2009. *Draft Standard for Information Technology – Telecommunications and information exchange between systems – Local and metropolitan area networks – Specific requirements; Part 11: Wireless LAN Medium Access Control (MAC) and Physical Layer (PHY) specifications; Amendment 7: Wireless Access in Vehicular Environments.* Note: This is a work in progress. This document is available to members of the IEEE 802.11 WG and is not available to the general public. A public version should be available in 2010.

IETF RFC 768, 1980. *User Datagram Protocol.* J Postel.

IETF RFC 793, 1981. *Transmission Control Protocol.* J Postel.

IETF RFC 1042, 1988. *A Standard for the Transmission of IP Datagrams over IEEE 802 Networks.* J Postel and J Reynolds.

IETF RFC 2460, 1998. *Internet Protocol, Version 6 (IPv6) Specification.* S Deering and R Hinden.

ITU-T Recommendation X.200, 1994. *Information Technology – Open Systems Interconnection – Basic Reference Model: The Basic Mode.*

ITU-T Recommendation X.690, 2008. *Information Technology – ASN.1 Encoding Rules: Specification of Basic Encoding Rules (BER), Canonical Encoding Rules (CER) and Distinguished Encoding Rules (DER).*

SAE J2540-2, 2006. *ITIS Phrase Lists (International Traveler Information Systems).* SAE Advanced Traveler Information Systems Committee.

SAE J2735, 2009. *Draft SAE J2735 Dedicated Short Range Communications (DSRC) Message Set Dictionary, Revision 32.* Society of Automotive Engineers, DSRC Committee. Note: This is a work in progress. This document is available to members of the SAE DSRC Committee and is not available to the general public. A public version should be available in 2010.

Index

3G systems, 87

Aggregation
    Geographical data, 66
    Hierarchical, 69, 73
    Landmark-based, 99, 101
    Probabilistic, 93
    Travel times, 99
    Using sketches, 94
Applications
    Advertisements, 88
    Blind spot warning, 13
    Control loss warning, 13
    Curve speed warning, 275
    Do not pass warning, 13
    Electronic payments, 10
    Emergency electronic brake lights, 12, 275, 281
    Entertainment, 88
    Forward collision warning, 13, 149, 275
    Green light optimal speed advisory, 150
    Hazardous location notification, 149
    Heartbeat, 11
    In-vehicle signage, 10
    Internet access, 87, 150
    Intersection movement assist, 13
    Intersection violation warning, 281
    Lane change warning, 13, 275
    Lane merging assistance, 150
    Left turn assist, 275
    Payment services, 88
    Pre-crash sensing, 149, 275
    Probe data collection, 10
    Stop sign assist, 275
    Traffic information systems, 88, 92
    Traffic signal indication, 11
    Traffic signal violation, 275
    Traveller information, 11
    Vehicular safety, 274
        Communication requirements, 23, 274
        Principle benefits, 22
        Software architecture, 26
ARIB, 9
ASTM, 7, 158, 205, 229, 372
Attacker, 309
Attacks, 312
    Breach of privacy, 312
    Denial of service (DoS), 312
    Eavesdropping, 312
    Extraction of secret keys, 313
    Masquerading and Sybil attacks, 313, 336
    Message manipulation, 312
    Replay and tunnel attacks, 312
Authentication, 314, **329**
    Digital signature, 329
    Identification, 332
    Message authentication code (MAC), 330
    TESLA, 331

Beaconing, 54, 59, 237, 274, 277
Behavioral model, 135
    Agenda-based, 138
    Artificial intelligence, 135
    Multi-agent, 135
    Non-normative motion patterns, 137
    Social network, 135
Blind averaging, 63
Broadcasting, 54

CALM, 9, 376
Capacity constraints, 49, 82
Capturing effect, 234, 244
Car2Car Communication Consortium, 37, 149, 301

CDMA, 224
Certificate, 318, **318**
    Distribution, 320
    Issuance, 320
    Renewal, 323
    Revocation, 321
Certificate authority (CA), 317
Channel
    Bandwidth, 184
    Channel access time, 221
    Channel busy time, 264
    Congestion, 255
    Control channel, 230
    Equalization, 158, 184, 206
    Estimation, 183, 206
    Linear channel, 159
    Measurement, 184
    Sounding, 157, 184
    Spacing, 159
    Tap, 171, 172
    Time-varying channel, 158
Classic bicycle model, 31
Clear channel assessment, 227
Client server systems, 89
Clustering, 58, 67, 68
    Hierarchical, 71
Coherence
    Coherence bandwidth, 175
    Coherence time, 170, **181**
Communication density, 249
Confidentiality, 314
Connectivity constraints, 49, 83
Controller-area network (CAN), 51
CSMA/CA, 227, 387
Cumulative noise, 234
Cyclic prefix, 164

D-FPAV, 259
Data elements, 276, 279
    Dictionary, 279
DCF, 227
Dedicated short-range communications, *see* DSRC
Delay
    Delay spread, 163
    Maximum excess delay, 175
    Mean excess delay, 175
    RMS delay spread, 175
Delay tolerant network, 58
Design framework, 147

External influence, 147, **149**
Motion constraints, 147, **147**
Time, 147, **149**
Traffic generator, 147, **148**
Differential GPS (DGPS), 24
Dissemination-based system, 50
Doppler
    Doppler shift, 170
    Doppler spread, 157, 170, 171, **179**
DSRC, 7, 367, **369**, 375
    Data link layer, 383
        MAC, 383
    Physical layer, 376
        PLCP, 381
        PMD, 377

EDCA, 229, 385, **389**
Embedded models, 141
    MoVes, 141
    NCTUns, 142
    VCOM, 142
EMDV, 263
ETSI, 375
Exponential backoff, 227

Fading
    Fast fading, 182
    Flat fading, 176, 183
    Frequency-selective fading, 159, 175, 176, 183
    Slow fading, 182
    Time-selective fading, 159
Fatalities, vehicle related, 22
Federated mobility models, 143
    CARISMA, 144
    Federated simulations development kit (FDK), 143
    MSIE, 143
    TraCI, 143
    TraNS, 143
    VGrid, 143
    VSimRTI, 144
Flajolet Martin sketches, 94
Flooding, 54
Flow model, 115
    Cellular automaton (CA), 120
        Nagel–Schreckenberg, 120
    Impact on vehicular mobility, 129
    Intersection management, 128
    Lane changing, 124

Gap Acceptance, 125
Macroscopic, 127
Mesoscopic, 127
Microscopic, 125
MOBIL, 126
Macroscopic, 116, **121**
Conservation of vehicle, 121
Lighthill–Whitham–Richard (LWR), 121
Mesoscopic, 116, **123**
Gas-kinetic, 123
Phase-space-density (PSD), 123
Queue, 123
Microscopic, 115, **116**
Car following model, 116
Collision avoidance, 117
Gazis–Hermann–Rothery (GHR), 118
General Motor (GM), 118
Gipps, 117
Intelligent driver model (IDM), 118
Krauss, 119
Pipe, 117
Psycho-physical, 119
Safety distance, 117
Stimulus-response, 117
Wiedemann, 119
Fourier transform
Discrete Fourier transform (DFT), 160
Fast Fourier transform (FFT), 159
Inverse fast Fourier transform (IFFT), 162
Futurama exhibit, 4

Geocast, 54
Global navigation satellite system (GNSS), 23
GPS/IMU integration, 24
Group key agreement, 328
Guard interval, 163

Headway
Distance, 117
Time, 117
Hidden terminal, 206, 220, 222, 238, 243

Identification, 314

IEEE 1609, 14, 142, 158, 229, 231, 301, 305, 323, 325, 337, 356, 369, 386, 392–395, 399, 409
IEEE 1609.2 authentication, 409
IEEE 1609.2 encryption, 412
WSMP, 393, **400**
IEEE 802.11p, 7, 14, 142, 158–160, 203, 225, 228, 232, 237, 368, 373, 376, 380, 383, 387
IFS, 227
Distributed IFS (DIFS), 227
Short IFS (SIFS), 227
Information dissemination, 49, 85
Epidemic, 55
Gossiping, 55
Rumor spreading, 55
Information kiosks, 60
Information sharing, 58
Information transport, 54
Integrity, 314
Intelligent transportation systems, 4, 375
Intercarrier-interference (ICI), 204
Intersymbol-interference (ISI), 163, 175
Isolated mobility models, 141

Kalman prediction, 35
Key assignment, 344

Maximum beaconing load (MBL), 261
Measurements
Aggregation, 63
Summarization, 63
Medium access, 220, 221
Message composition, 284
Message construction, 280
Message dispatcher, 273, 278, 294
Message handler, *see* Message dispatcher
Message set, 276
Misbehavior detection system (MDS), 336

Navigation system, 52
Network connectivity, 59
Network simulator, 139
GTNetS, 140
NS-2/NS-3, 140
EDCA, 140
IEEE 802.11p, 140
OMNeT++, 140
OPNET, 140
Qualnet, 140
SWANS, 140

SWANS++, 140
Non-repudiation, 314

OFDM, 228, 377
    Differential OFDM, 207
    Orthogonal frequency division multiplex, 159
    Subcarrier, 159

PCF, 227
Peer-to-peer systems, 90
PKI, 319
Power
    Power delay profile (PDP), 175
    Power spectral density (PSD), 173
Prediction of parameters, 53
Predictive coding, 287, 290, 291
    Linear, 288
Privacy, 303, **337**
Projects
    ASV, 7, 9, 38
    CACS, 5
    CarTalk, 7
    CICAS, 10, 276, 280
    Coopers, 9
    CVIS, 9
    DSSS, 9
    ERGS, 5
    FleetNet, 7
    IntelliDrive, 10, 12
    IVBSS, 10
    NOW, 9, 301
    PRE-DRIVE-C2X, 9
    PREVENT, 9
    PROMETHEUS, 6
    PROMOTE CHAUFFEUR, 7
    Safe Trip-21, 12
    SAFESPOT, 9
    SeVeCom, 9, 301
    VII, 10, 301
    VSC, 7, 12, 38, 257, 274, 301
    WILLWARN, 9
Proof of concept, 10
Public-key cryptography, 315
    Diffie–Hellman, 315, **327**
    Elliptic curve cryptography (ECC), 316

Radio propagation
    Multipath
        Deterministic multipath models, 166
        Statistical multipath models, 171
    Nakagami-m, 236
    Path loss, 173
    Rayleigh, 172
    Shadowing, 174
    Two-ray ground, 235
Random mobility model
    BonnMotion, 115
    Freeway, 114
    Manhattan, 114
    Random walk, 113
    Random waypoint, 113
    Reference point group mobility model, 113
Rate control, 265
Real time kinematik (RTK), 24
Relevance function, 62
Reliability, 304
Request/reply protocols, 56
Risk, 302
Road-segment averaging, 65

SAE J2735, 14, 276, 279, 280, 285, 291, 369, 425
SDMA, 223
Sensors, 51
    Computer vision, 26
    Noise of, 26
    Ranging sensors, 26
    Sensor fusion, 51
Simulation, 108
    Application-centric, 145, 146
    Computational costs, 241
    Congestion control, 262
    High level architecture, 144
    IEEE 802.11p, 232, 236, 237, 248
    Network-centric, 145
    Traffic, 122
Single-hop broadcast, *see* Beaconing
SINR, 234
Spectrum allocation, 207, 231, 368, 370, 375, 395
Store and forward protocols, 59
Symmetric-key cryptography, 314

TDMA, 222
TIGER, 115
Timestamp-based averaging, 64

INDEX

Timestamp-based comparison, 64
Traffic information system, 52
Traffic model, 131
    Dynamic traffic assignment (DTA), 131
    Impact on vehicular mobility, 134
    Macroscopic, 131, 133
    Microscopic, 133
    Path, 131, 133
        Agent-centric, 133
        Dijkstra, 133
        Flow-centric, 133
    Time influence, 134
    Trip, 131, 132
        OD matrix, 132
        Point-of-Interest (PoI), 132
        Stochastic turn, 132
Traffic Simulator
    AIMSUN, 134
    Carisma, 133
    CORSIM, 134, 143
    MATSim, 133, 136
    SUMO, 119, 134, 143
    TRANSIMS, 120
    VanetMobiSim, 119, 133
    VISSIM, 120, 134, 142, 143

Traffic theory
    Benchmark test, 129
    Flow-density, 129
Transmit power control, 258
Travel times, 73

Unicycle model, 30, 32
UTRA-TDD, 225

Vehicular mobility
    Networking shape, 134
Vehicular path history, 29
Vehicular path prediction
    Model choice, 34
    Non-parametric path prediction, 30
    Parametric path prediction, 31
    Stochastic path prediction, 35
Vehicular Traces
    Cabspot, 138
    CRAWDAD, 138
    Dieselnet, 138
    GATech Vehicular Traces, 138
    MIT Reality Mining, 138
    MMTS, 139
    USC Mobilib, 138

WAVE, 7, 377